C0-AWC-818

Advances in
MARINE BIOLOGY

VOLUME 51

Advances in MARINE BIOLOGY

Edited by

ALAN J. SOUTHWARD

Marine Biological Association, The Laboratory, Citadel Hill, Plymouth, United Kingdom

DAVID W. SIMS

Marine Biological Association, The Laboratory, Citadel Hill, Plymouth, United Kingdom

ELSEVIER

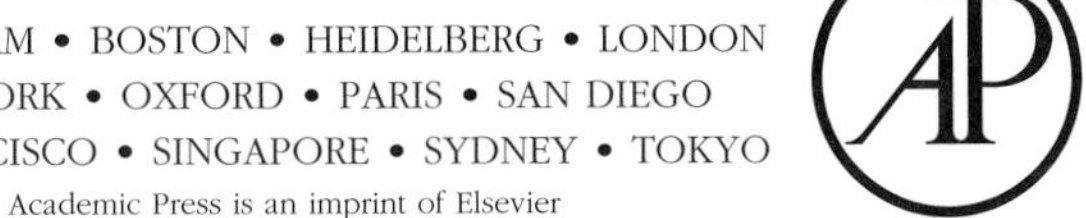

AMSTERDAM • BOSTON • HEIDELBERG • LONDON
NEW YORK • OXFORD • PARIS • SAN DIEGO
SAN FRANCISCO • SINGAPORE • SYDNEY • TOKYO
Academic Press is an imprint of Elsevier

Academic Press is an imprint of Elsevier
525 B Street, Suite 1900, San Diego, California 92101-4495, USA
84 Theobald's Road, London WC1X 8RR, UK

This book is printed on acid-free paper. ∞

For information on all Academic Press publications
visit our Web site at www.books.elsevier.com

ISBN-13: 978-0-12-026152-9
ISBN-10: 0-12-026152-9

PRINTED IN THE UNITED STATES OF AMERICA
06 07 08 09 9 8 7 6 5 4 3 2 1

CONTRIBUTORS TO VOLUME 51

CAROLIN E. ARNDT, *School of Zoology, University of Tasmania, Hobart, Tasmania, Australia; University Center on Svalbard, Longyearbyen, Norway; World Climate Research Programme, Geneva, Switzerland*

ELENA GUIJARRO GARCIA, *Marine Research Institute, Reykjavík, Iceland*

ANDREW JEFFS, *National Institute of Water and Atmospheric Research, Newmarket, Auckland, New Zealand*

JEFFREY M. LEIS, *Ichthyology, Australian Museum, Sydney, Australia*

MARK MEEKAN, *Australian Institute of Marine Science, Northern Territory, Australia*

JOHN C. MONTGOMERY, *Leigh Marine Laboratory and School of Biological Sciences, University of Auckland, Auckland, New Zealand*

STEPHEN D. SIMPSON, *School of Biological Sciences, University of Edinburgh, United Kingdom*

KERRIE M. SWADLING, *School of Zoology, University of Tasmania, Hobart, Tasmania, Australia; Tasmanian Aquaculture and Fisheries Institute, University of Tasmania, Hobart, Tasmania, Australia*

CHRIS TINDLE, *Department of Physics, University of Auckland, Auckland, New Zealand*

CONTENTS

The Fishery for Iceland Scallop (*Chlamys islandica*) in the Northeast Atlantic

Elena Guijarro Garcia

Are Larvae of Demersal Fishes Plankton or Nekton?

Jeffrey M. Leis

Sound as an Orientation Cue for the Pelagic Larvae of Reef Fishes and Decapod Crustaceans

John C. Montgomery, Andrew Jeffs, Stephen D. Simpson, Mark Meekan and Chris Tindle

Crustacea in Arctic and Antarctic Sea Ice: Distribution, Diet and Life History Strategies

Carolin E. Arndt and Kerrie M. Swadling

Series Contents for Last Ten Years*

*The full list of contents for volumes 1–37 can be found in volume 38.

The Fishery for Iceland Scallop (*Chlamys islandica*) in the Northeast Atlantic

Elena Guijarro Garcia

Marine Research Institute, Reykjavík, Iceland

This review focuses on the history and management of the Iceland scallop fishery in Iceland, Greenland and Norway (including the Svalbard archipelago and the islands of Bjørnøya and Jan Mayen), with information on research into each stock. The start of the fishery in all these regions followed the

ADVANCES IN MARINE BIOLOGY VOL 51

0065-2881/06 $35.00
DOI: 10.1016/S0065-2881(06)51001-6

discovery of virgin scallop beds made up of old, large specimens in very high densities. Despite the apparent similarity of original conditions, the fishery has followed very different trends in each region, with drastic declines in Iceland, Svalbard and Jan Mayen. The general biology of the Iceland scallop is summarised and compared with the biology of other North Atlantic species of pectinids.

The Icelandic fishery dates from 1969. There was a steady decrease in catch from 1985, when >16,000 tonnes were caught. By 2004 the stock had declined to 35% of its average size during the period 1993–2000 and a zero quota was advised. This decline is thought to have resulted from overexploitation, combined with a protozoan infestation and increasing sea bottom temperature.

Scallop dredging commenced in west Greenland in 1983. The stock is not very large, but fishing was driven by social factors. Catches ranged from 400 to 1900 tonnes during the period 1988–1992 and from 1200 to 2600 tonnes since 1995. There are indications that each scallop bed is extensively dredged before the fleet moves on to new areas, but nevertheless catches have been rather stable over the past decade.

The scallop stocks in Svalbard, Bjørnøya and Jan Mayen were depleted in three fishing seasons between 1985–1987, when up to 45,000 tonnes of scallops were dredged in a single season. Following a survey carried out in 1991, Bjørnøya was open to the fishery with a maximum quota of 2000 tonnes, but the stock off Svalbard was found to be not large enough to sustain a fishery.

1. SCALLOP BIOLOGY IN THE STUDY AREA

The Iceland scallop (Figure 1), *Chlamys islandica* (Müller, 1776), is a widely distributed sub-Arctic species. Its current distribution in the northeast Atlantic includes the southwestern Kara Sea, White Sea, Svalbard, Bjørnøya, Barents Sea along the western coast of Novaya Zemlya, Jan Mayen, most of Iceland except for the south coast and the western coast of Norway. It has been reported to exist as far south as Bergen and Stavanger Fjord, but these are relic populations 1000 km away from populations further north (Greve and Samuelsen, 1970). Wiborg (1963a) found only empty shells in many of the southern localities and reported that living scallops occurred only in fjords with shallow sills (10–15 m) at the entrance. Jensen (1912) found subfossil shells in the Faeroes, and Ockelmann (1958) believed that the species existed there, but *C. islandica* is currently absent from the Faeroes (K. Gunnarsson and U. Matras, personal communication). In the northwest Atlantic, *C. islandica* is found in west Greenland from Thule (77° N) to Cape Farewell (60° N). The Arctic Current that flows south along the east Greenland coast prevents the build up of scallop beds there

Figure 1 *Chlamys islandica*. Group of recruits to the fishery, at least 60 mm shell height. This size corresponds to an age of approximately 6.5–7 yr for the Breidafjörður area scallops.

because the water temperature is too low. Thus, scallop distribution in east Greenland is restricted to the King Frederick VI coast in the southeast, with the exception of a population found in the inner part of Franz Josef Fjord, regarded as a relic of a warmer interglacial period by Ockelmann (1958). Along the east coast of Canada, Iceland scallops have their northern distribution limit in Cumberland Peninsula, Hudson Bay and Foxe Channel, and they are found as far south as Cape Cod on the American East Coast (Simpson, 1910; Ockelmann, 1958; Wiborg, 1963a; Greve and Samuelsen, 1970; Lubinsky, 1980; Eiríksson, 1986; Pedersen, 1994). *C. islandica* has been often described as a circumpolar species, but Waller (1991) maintains that *C. islandica* is absent from the north Pacific, where it is replaced by very closely related species, *Chlamys behringiana, Chlamys albida, Chlamys rubida* and *Chlamys hastata*, all of which have been considered to be subspecies of *C. islandica* in the past (Waller, 1991). A description of these other *Chlamys* species and their distribution range can be found in Grau (1959) and Bernard (1983).

Fossil *C. rubida* and *C. hastata* dating from the Miocene have been found in California (Moore, 1984), whereas fossil records of *C. islandica* in the north Atlantic date back to the Pleistocene and Holocene, suggesting that the *Chlamys* complex originated in the Pacific and advanced eastwards until reaching the north Atlantic (Hopkins, 1967; MacNeil, 1967; Masuda, 1986 [all cited in Waller (1991)]). The Pleistocene and Holocene distribution of *C. islandica* was much wider than at present. It was found as far south as New York in the northwest Atlantic, and it was present in Sweden, the

North Sea, Scotland, Ireland, France, the Azores and the Mediterranean, where it penetrated during the Günz glaciation. Its distribution range off Norway and Greenland was also more extensive than it is nowadays (Dall, 1898 [cited in Waller (1991)]; Ockelmann, 1958; Richards, 1962; Pérès and Picard, 1964; Wagner, 1970; Lubinsky, 1980; Theroux and Wigley, 1983; Pérès, 1985; Pedersen, 1988a).

Along the Norwegian coast from the Lofoten Islands to Stavanger, *C. islandica* overlaps with two boreal scallop species of commercial interest, *Aequipecten opercularis* and *Pecten maximus*. There is, however, no large-scale fishery for *P. maximus* in Norway, where it is not a common species (Parsons *et al.*, 1991), but there is some fishing carried out by SCUBA divers (Wiborg and Bøhle, 1968). It has, however, the potential to be cultured, although production is still low (Bergh and Strand, 2001). *A. opercularis* is not harvested or cultured within the study area, but it has been fished on a small scale in Denmark and the Faeroe Islands (Parsons *et al.*, 1991).

The *Aequipecten* group is geologically younger than the *Chlamys* group, dating from the early Paleocene. *A. opercularis* has its northern distribution limit in the Lofoten Islands, Norway, and is found along the Norwegian coast, the Skagerrak, Faeroe Islands, North Sea, Irish Sea, Atlantic European coast and the Mediterranean. There are reports from the Azores, Canary Islands, Cape Verde Islands and the Atlantic African coast to roughly 30° N, but they seem to be based on juvenile, subfossil or misidentified specimens (Pérès and Picard, 1964; Brand, 1991; Waller, 1991). *A. opercularis* is abundant in Plio-Pleistocene and Holocene deposits in the southern North Sea, where it is now rare (Hickson *et al.*, 1999; Johnson *et al.*, 2000). Fishable beds of *A. opercularis* around the British Isles are few and widely spread (Sinclair *et al.*, 1985), but it is abundant in the north Irish Sea, where it is found with *P. maximus* and has been fished since 1969 (Ansell *et al.*, 1991; Brand, 2000). It has been suggested that the reduction in the distribution area of the species in the North Sea could be caused by increased pollution (Hickson *et al.*, 2000; Johnson *et al.*, 2000).

The *Pecten* group is the youngest of the three, dating from the Miocene. The distribution area of *P. maximus* ranges from Myken, Norway (roughly 67° N), to Cap Blanc in Mauritania (roughly 21° N) and the Canary Islands, and it is present in the North Sea, Irish Sea and Atlantic European coast. *P. maximus* penetrated the Mediterranean during a glacial period in the Pleistocene (Pérès and Picard, 1964), and currently it is found mostly off the Málaga province. *P. maximus* has occasionally been fished off the Costa del Sol following the discovery of several fishable banks in 1978, but catches have always been <1000 tonnes (Cano and Garcia, 1985; Cano *et al.*, 1999).

The largest Iceland scallop densities have been found between 20 and 110 m depth, although some beds extend down to nearly 600 m (Eiríksson, 1986; Hansen and Nedreaas, 1986; Rubach and Sundet, 1987; Pedersen, 1994).

Growth rate of *C. islandica* varies seasonally but also between the areas under study, probably in relation to feeding conditions and temperature (Vahl, 1980; MacDonald and Thompson, 1985). Most growth takes place during the phytoplankton bloom, from April to June (Vahl, 1978; Thorarinsdottir, 1993). There are also age-related changes in growth rate. In Iceland, growth rates decrease from 8–10 mm yr^{-1} during the first year to 0–3 mm yr^{-1} for the older shells, older than 10–15 yr (Eiríksson, 1986).

As a result of variation in growth rate between regions, age and size at maturity and maximum age are also different in the Iceland scallop populations of Norway, Greenland and Iceland. Iceland scallops reach maturity at 40–50 mm shell height (SH) at an age of 5–7 yr in Norway and Iceland, but a lower size for age is found in Greenland, where 30–55 mm SH, corresponds to an age of 4–9 yr (Wiborg, 1963a; Eiríksson, 1986; Pedersen, 1989, 1994). Iceland scallops grow in most areas to a maximum size of 80–110 mm, although individuals measuring 140 mm have been found in northwest Iceland (H. Eiríksson, personal communication). Regarding maximum age, there are records of Iceland scallops up to age 20 yr in Iceland (Eiríksson, 1986) and 40% of the scallops in the Nuuk area, in Greenland, were estimated to be older than 21 yr (Pedersen, 1989). The age estimation in the Icelandic stock was based on shell growth rings validated with mark-recapture methods (H. Eiríksson, personal communications), whereas winter checks in the ligament at the umbo were used in Greenland (Pedersen, 1989). Data availability on the biology of *A. opercularis* and *P. maximus* within the study area are scarce, but it is known that *P. maximus* and *A. opercularis* also show seasonal variability in growth rates linked to water temperature and food availability (Ursin, 1956; Gibson, 1956). *P. maximus* reaches an SH of 150 mm and some individuals have been recorded to be older than 20 yr (Tang, 1941), but average lifespan is probably much shorter even in unexploited populations (Dupuoy *et al.*, 1983). Growth rates vary among populations in different geographical locations (Ansell *et al.*, 1991).

A. opercularis is smaller and has a shorter lifespan than *P. maximus*, reaching 90 mm and 8–10 yr (Taylor and Venn, 1978; Mason, 1983; Ansell *et al.*, 1991). In Denmark and Faeroe Islands, this species grows rapidly until age 2 yr (40 mm), but afterwards growth rates decrease and cease completely at age 6 yr and an SH of 70 mm (Parsons *et al.*, 1991). Growth rates seem to be correlated with seawater temperature (Ansell *et al.*, 1991).

1.1. Reproduction

Sexes are separate in *C. islandica* (Wiborg, 1963a), with a sex ratio approaching 1:1 in Icelandic and offshore Norwegian populations (Eiríksson, 1986; Rubach and Sundet, 1987). Spawning takes place between the end of June and end of July, the precise timing being apparently influenced mostly by rising sea temperatures (Skreslet and Brun, 1969; Skreslet, 1973; Sundet and Lee, 1984; Eiríksson, 1986; Pedersen, 1988a; Oganesyan, 1994) and feeding conditions (Thorarinsdottir, 1993; Arsenault and Himmelman, 1998). Survival and growth rates of larvae also depend greatly on these factors (Dickie, 1955; Claereboudt and Himmelman, 1996; Tammi *et al.*, 1997; Robinson *et al.*, 1999). The spat settles 6 wk after spawning (Gruffydd, 1976; Thorarinsdottir, 1991), although this period increases to 10 wk in Greenland (Pedersen, 1988a). Prior to metamorphosis, the larvae attach themselves by byssal threads, preferably to hydroids and filamentous benthic algae (Harvey *et al.*, 1993; Harvey and Bourget, 1995). Recruitment is negatively correlated with abundance of adults (Vahl, 1982; Smith and Rago, 2004).

P. maximus is hermaphroditic. The spawning cycle is very variable among populations from different areas. In the Irish Sea and off west Ireland, there are two distinct spawning events, in spring and during July/August, although juveniles spawn only once in autumn (Wilson, 1987). There are also two spawning peaks in Danish waters (Ursin, 1959). In west Scotland and Norwegian waters, spawning occurs in summer only, but synchrony is very low, so the timing of spawning events varies among years (Ursin, 1959; Mackie, 1986 [cited in Ansell *et al.*, 1991]). Populations of *P. maximus* in the English Channel show further variation in the number and timing of spawning events and the degree of synchrony. Thus, maturity of scallops in the Bay of St. Brieuc is highly synchronised and the main spawning event occurs in spring, followed by secondary spawning cycles until August (Bergeron and Buestel, 1979; Ansell *et al.*, 1988; Paulet *et al.*, 1988), whereas in the Bay of Brest, synchrony is very low and there are repeated spawning cycles throughout the year (Cochard, 1985). *P. maximus* populations also show variability in egg size, larval lifespan and size at metamorphosis. Although these differences are partly related to the variability of environmental conditions throughout the distribution range of the species, it is thought that they may also reflect certain genetic isolation of the different local self-recruiting stocks (Mason, 1958, 1983; Ansell *et al.*, 1991).

A. opercularis is also hermaphroditic, and the timing of spawning events varies among populations. In the north Irish Sea and Clyde Sea, there are two distinct spawning peaks, in summer and autumn, the first being the larger (Taylor and Venn, 1978; Brand *et al.*, 1980; Paul, 1981). Spawning takes place once a year between June and October in Scandinavia (Ursin, 1956), whereas in the Mediterranean the main spawning events take place from February to June and in October, with secondary settlements the rest of the year (Peña *et al.*, 1993, 1996).

The duration of the pelagic stage is very variable, depending on temperature, food availability and possibly genetic factors (Paulet *et al.*, 1988).

1.2. Environmental factors

Scallops are filter feeders and, therefore, favour areas with strong currents and low sediment load (Mason, 1983). Iceland scallops are found on sandy, gravelly or shell sand bottoms, and they thrive best in temperatures between −1.4 and 10 °C (Wiborg, 1963a; Nicolajsen, 1984; Eiríksson, 1986; Pedersen, 1994) but can cope with small increases above 10 °C (Jónasson *et al.*, 2004). The distribution of *P. maximus* and *A. opercularis* is also restricted mainly by temperature and salinity (Maack, 1987 [cited in Parsons *et al.* 1991]), and they do not tolerate low temperatures as well as *C. islandica*. *P. maximus* shows poor survival when sea temperature is <4 °C, and the salinity must be >27‰; therefore, this species is more abundant along the Norwegian coast than in the Skagerrak, where salinity is very variable (Wiborg and Bøhle, 1968; Parsons *et al.*, 1991). *P. maximus* is usually found on soft bottoms, but its distribution is very uneven. In some areas, 10–100 individuals could be found, with maximum densities of 2–3 shells m^{-2}, but such areas alternated with long stretches where scallops were absent (Wiborg and Bøhle, 1968; Ansell *et al.*, 1991). *A. opercularis* is found within a temperature range of 3–16 °C and salinity >25‰, preferably in fine sand and gravel areas between 20 and 130 m depth (Ursin, 1956).

Natural mortality of Iceland scallops is estimated by the occurrence of cluckers, which are empty shells still joined by the hinge. These indicate a recent death, since the hinge parts within roughly 200 days (Jónasson *et al.*, 2004). In the Gulf of St. Lawrence, natural mortality was estimated to be between 0.024 and 0.084 in virgin beds and between 0.14 and 0.20 on dredged beds (Naidu, 1988).

2. ICELAND

2.1. The stock and the fishery

Dredging of Iceland scallops took place for the first time in 1969, in Jökulfjörður and Ísafjarðardjúp, both in northwest Iceland (Eiríksson, 1970a, 1971a,b; Anonymous, 1973, 1974, 1994). During the 1970s and early 1980s, the fishery extended to other areas of the country as new scallop beds were found (Figure 2), namely Breiðafjörður, Hvalfjörður and Faxaflói in West Iceland (Eiríksson, 1970b, 1971a; Anonymous, 1973, 1974, 1994), Húnaflói and Skagaströnd in North Iceland (Eiríksson, 1970c, 1971a; Anonymous, 1973, 1981) and the bays and fjords from Þistilfjörður to Berufjörður, in northeast to east Iceland, of which Vopnafjörður had the

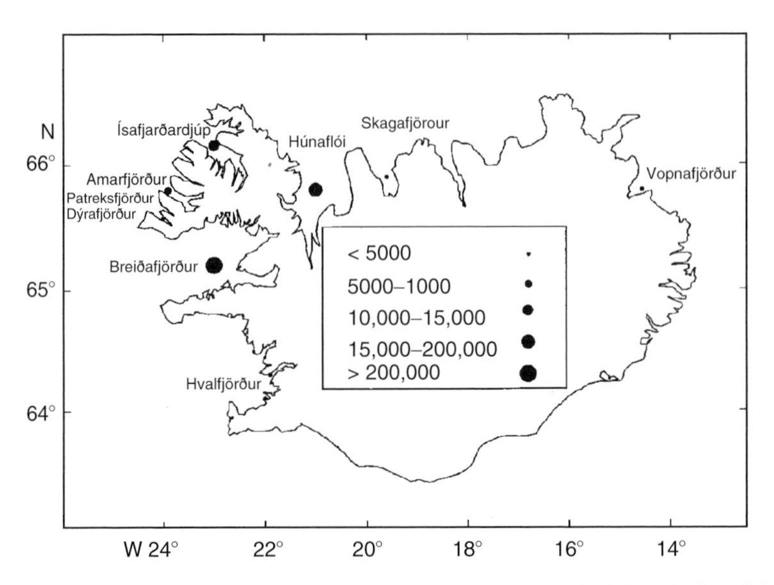

Figure 2 The scallop fishing areas around Iceland and their relative fishery importance, based on the total catch since the beginning of the fishery in 1969.

best scallop beds (Anonymous, 1973, 1982). Nevertheless, the scallop beds found in Breiðafjörður, West Iceland, were the largest by far and they have consisted of between 60 and 100% of the annual landings since the beginning of the fishery (Anonymous, 2003). During the first years of the fishery, the stock consisted of old scallops that had accumulated with time. There was a steady increase in the size of the fishing fleet operating in Breiðafjörður and catches from this area increased nearly every year up to a maximum in 1986 (Figure 3). The decrease between 1973 and 1975 was due to low market prices.

Technical improvements such as more efficient dredges, more powerful engines and better positioning equipment on board allowed the increase of catch per unit effort (CPUE) during the 1990s despite the continuous decline of the stock index (Figure 4) and average SH of the scallops, which decreased as the older individuals became rarer (Figure 5).

The stock index is an estimate of the density of scallops within the beds. It is estimated for each bed with the formula

$$([w\ A_{bed}]A_{swept}{}^{-1})q^{-1},$$

where w is the average weight of scallop catch from all the tows taken within a bed, A_{bed} is the area of the scallop bed in km^2, A_{swept} is the area swept within the bed in km^2 and q is the catchability index. Thus, the total stock index in Breiðafjörður for any given year is the sum of the indices of all beds.

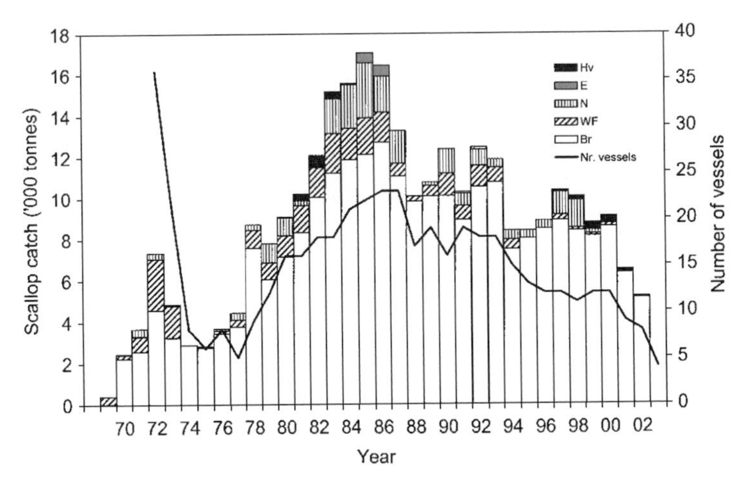

Figure 3 Scallop catches and fleet size since the beginning of the fishery. Br, Breidafjörður; WF (Western Fjords): Arnarfjörður, Ísafjarðardjúp, Patreksfjörður and Dýrafjörður; N (North): Húnaflói and Skagafjörður; E (East): Vopnafjörður; Hv, Hvalfjörður; Sh, number of vessels fishing in Breidafjörður. (Catch data from Anonymous, 2003.)

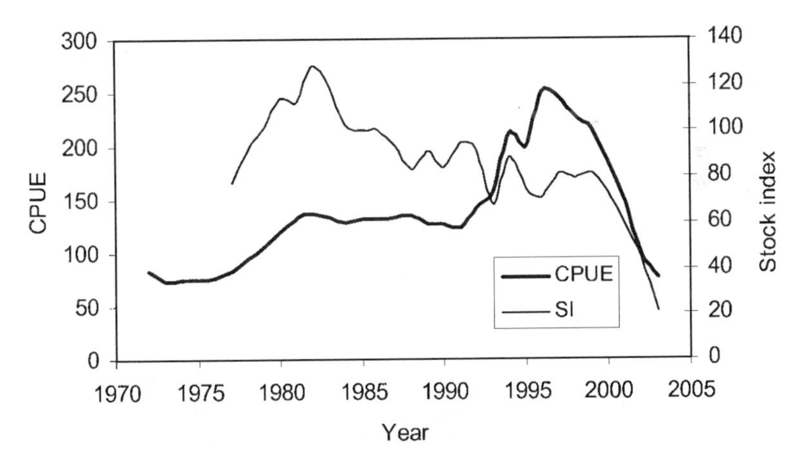

Figure 4 Catch per unit effort (kg h^{-1}) and stock index trends since the beginning of the fishery. (Stock index data from Anonymous, 2003.)

In 1990 it was estimated that the stock index had diminished by 30% compared with its estimated size in 1982 (Anonymous, 1990). The fleet kept decreasing simultaneously with catches, and during 2003, only three boats were dedicated exclusively to scallop dredging (Figure 6). The rest of the fleet

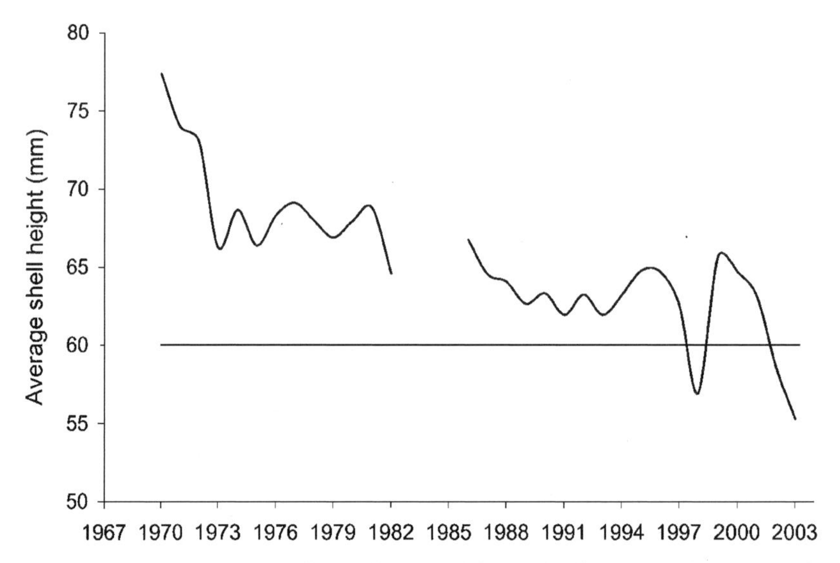

Figure 5 Average shell height of scallops in Breidafjörður during the period 1970–2003. The horizontal line indicates the minimal landing size, 60 mm shell height.

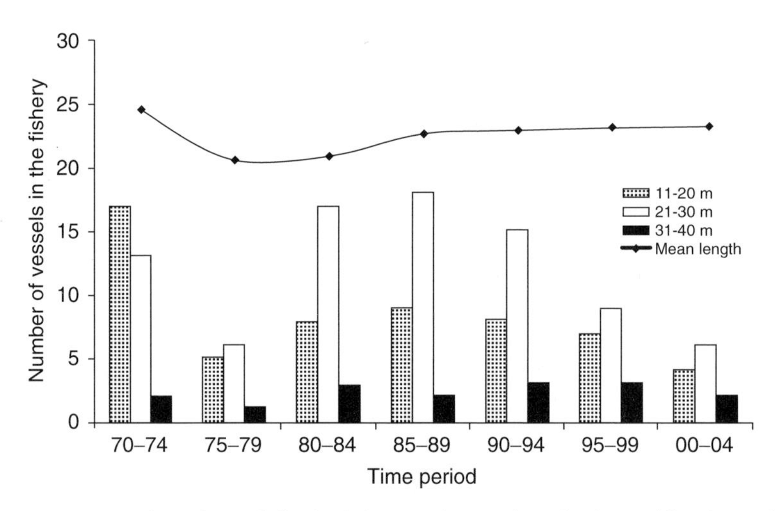

Figure 6 Number of vessels in the fishery and mean length of vessel for the scallop fleet since 1970, grouped in 5-yr intervals.

were multipurpose boats that fished cod with line or gillnets and/or inshore/offshore shrimp during winter (Eiríksson, 1997).

Catches have been unusually low since 2000 onwards, coinciding with unusually high natural mortality among the older scallops. Between spring 2002 and spring 2003, the stock biomass declined to one-third of its average size during the period 1993–2000. As a precautionary measure, the fishery has been closed since the fishing season 2003/2004.

The causes for this decline are being investigated but are still unclear. It is most likely that several factors have contributed to the current situation:

1. The fishing effort may have been too high considering the trend of the stock index (Figure 4) and annual variability of environmental factors.
2. Bottom-sea temperature in summer has increased by 2–3 °C during the past years (Eydal, 2003; Jónasson *et al.*, 2004).
3. Cyst-forming protozoan parasites (*Coccidia*) have been detected in adult scallops (Kristmundsson *et al.*, 2004).

However, the increase of natural mortality has shown great variability among the different scallop fishing grounds within Breiðafjörður (Jónasson *et al.*, 2004), so there might be other reasons to explain the observed mortality.

Scallops were sorted from the by-catch onboard and processed on land with mechanical shuckers for production of meats (adductor muscle). Cleaning machines separated the viscera from the meats, which were then fine-trimmed manually, mechanically size-graded and individually quick-frozen. Most of the production was exported to the United States until 1988, when sales to France started to increase. This new market led to the production of roe-on meats, which required manual handling following the mechanical shucking but increased the production by 15–18% (Eiríksson, 1997).

2.2. Fishing gear

During the first years of the fishery in Breiðafjörður, a rigid box-type dredge of Icelandic design was used until two Isle of Man models, the Blake and the Connolly dredges, were introduced in 1972. These scallop dredges consisted of a heavy metal frame to which a bag with a belly of steel rings joined with chain links was attached. The Blake was a sledge-type dredge, whereas the Connolly was a roller dredge. Roller dredges slide better over the seabed because of the wheels at each side of the frame. The width of the metal frame varied with boat size, and together with the diameter of the steel rings (subjected to management rules) determine the efficiency of the gear. The Blake dredge was widely used and adapted to the fishery in Breiðafjörður, becoming stronger and up to three times heavier (800–1000 kg), with sizes from 1.5 to 2.7 m width. The frame had two runners strengthened by

Figure 7 The roller dredge used in Iceland. The stone guards can spin around the axis, allowing the dredge to slide over boulders.

horizontal bars, and a stone guard to prevent big boulders from entering the bag. In front of the ring bag, a tickler chain disturbed the bottom to increase the catch, and the end of the bag is attached to a tail bar that kept the bag spread. However, an improved design of roller dredge was created in North Iceland in 1988 and has since become the most widely used dredge in Iceland (Eiríksson, 1997) (Figure 7). The Icelandic roller dredge is equipped to fish on both sides, and with its articulated bar, chain ground rope and wheels on both sides of the frame, it has proven to be far more efficient than other types. First trials took place in Húnaflói, North Iceland, and by 1991/1992 all scallop boats in Breiðafjörður had this new dredge. Since 1995, it has been used by all scallop boats (Eiríksson, personal communication).

2.3. Management

The fishery was open to all boats at the beginning of exploitation, and 48 licenses were issued for the fishing season in 1972, although not all of them were used. The fleet landed 3000 tonnes in only 2 mo, prompting managers to restrict access to the fishery from 1973 onwards. Only local boats could apply for licenses and the catch had to be processed in towns in Breiðafjörður. In addition, log books were made mandatory to monitor catches and the fishery was limited to a 5000-tonnes quota. Nevertheless, landings remained below this figure from 1973 to 1977 because of low market prices. The Ministry of Fisheries also restricted the number of processing plants after 1976, and from 1978 onwards each plant had a catch quota related to the total allowable catch

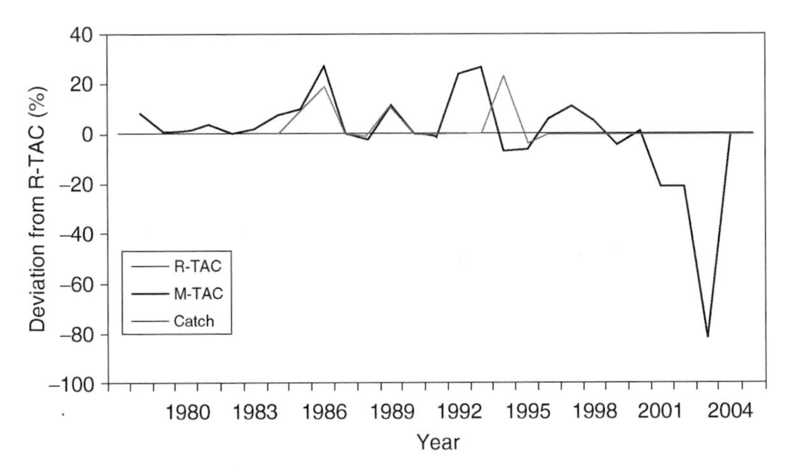

Figure 8 Comparison of scallop catches and total allowable catch for Breiðafjörður. R-TAC, total allowable catch recommended by the Marine Research Institute; M-TAC, total allowable catch limit issued by the Ministry of Fisheries.

(TAC) for the nearby scallop beds. The system changed again in 1984, and since then, catch quotas have been allocated to the boats, which can transfer quotas among themselves. The TAC is set for one fishing year, from September 1 to August 31, and the quota given to each boat is based on its average share of landings in years previous to the enforcement of the individual boat quota system. The Marine Research Institute (MRI) of Iceland recommends an annual TAC to the Ministry of Fisheries, which has accepted the recommendations of the MRI most years. Catches often exceeded the quota, but the extra catch has usually been <15% of the recommended TAC (Figure 8).

To estimate the TAC, detailed CPUE reports from the log books (daily catch, fishing hours, gear, crew, fishing ground, boat) are added to the information collected on annual surveys during which 120 standardised tows are taken. The minimum landing size is 60 mm SH. Given the stock size and CPUE reached during the 1990s, the only possibility to increase catches would have been developing more efficient gear that would reduce shell breakage and indirect fishing mortality (Eiríksson, 1997).

2.4. Research

The Danish Ingolf Expedition collected Iceland scallop samples in 1895 and 1896 in many locations all around Iceland except in the south coast, where it was absent (Jensen, 1912). Scallops were mentioned also in general studies on the Icelandic marine fauna (Madsen, 1949; Óskarsson, 1952; Hauksson and Gunnarsson, 1973). Surveys to investigate the possibility of commercial use of the stock commenced in 1969 in the western Fjords and continued in the early 1970s in Breiðafjörður, North and East Iceland (Eiríksson, 1970a, b,c, 1971a,b; Anonymous, 1973, 1974, 1981, 1982, 1990). Several surveys from 1971 to 1986 included searches for new scallop beds in different areas. From 1973 onwards, research efforts have focused mostly on age structure and stock size of the scallop population in Breiðafjörður and to a lesser extent in other areas. In addition, a project was run between 1989 and 1994 to investigate settlement, growth and age at maturity of scallops to assess the feasibility of suspended culture of this species. Growth rates were indeed higher than on the sea bottom, and in 3 yr cultured scallops reached the same size as 5-year-old scallops collected from the nearby beds. However, it was considered that the 4 yr that cultured scallops need to reach market size would make the enterprise very risky economically (Thorarinsdottir, 1991, 1993, 1994; Anonymous, 1994). Research effort has increased considerably because of the sudden decrease in stock size in Breiðafjörður since 2000 (Figure 4), coinciding with observed high natural mortality among older scallops. Thus, in addition to annual stock assessment surveys, samples are taken at roughly 2-mo intervals for monitoring of shell condition factors (weight of muscle, gonads, etc.) and clucker ratio (% of live vs recently dead individuals by size groups). Moreover, bottom temperatures are being registered daily at two to three major scallop fishing areas. In addition, the infection of adult scallops by a protozoan parasite is being thoroughly investigated (Kristmundsson *et al.*, 2004).

3. GREENLAND

3.1. The stock and the fishery

The commercial fishery began in 1983 as a private initiative in the Nuuk area (Eiríksson and Nicolajsen, 1984; Nicolajsen, 1984), although the occurrence of Iceland scallop in West Greenland as far north as 76° had been long known (Posselt, 1898; Jensen, 1912). The Greenland Home Rule carried out a series of surveys between 1984 and 1988 (Figure 9) to assess the possibility of developing a commercial scallop fishery in West Greenland (Pedersen, 1987a, 1988b). There were scallop beds in most of the communities investigated, some of them apparently with potential for a small-scale commercial fishery. The stock consisted mostly of old scallops, between 60 and 100 mm, which had accumulated at high densities in small areas. Growth rates and

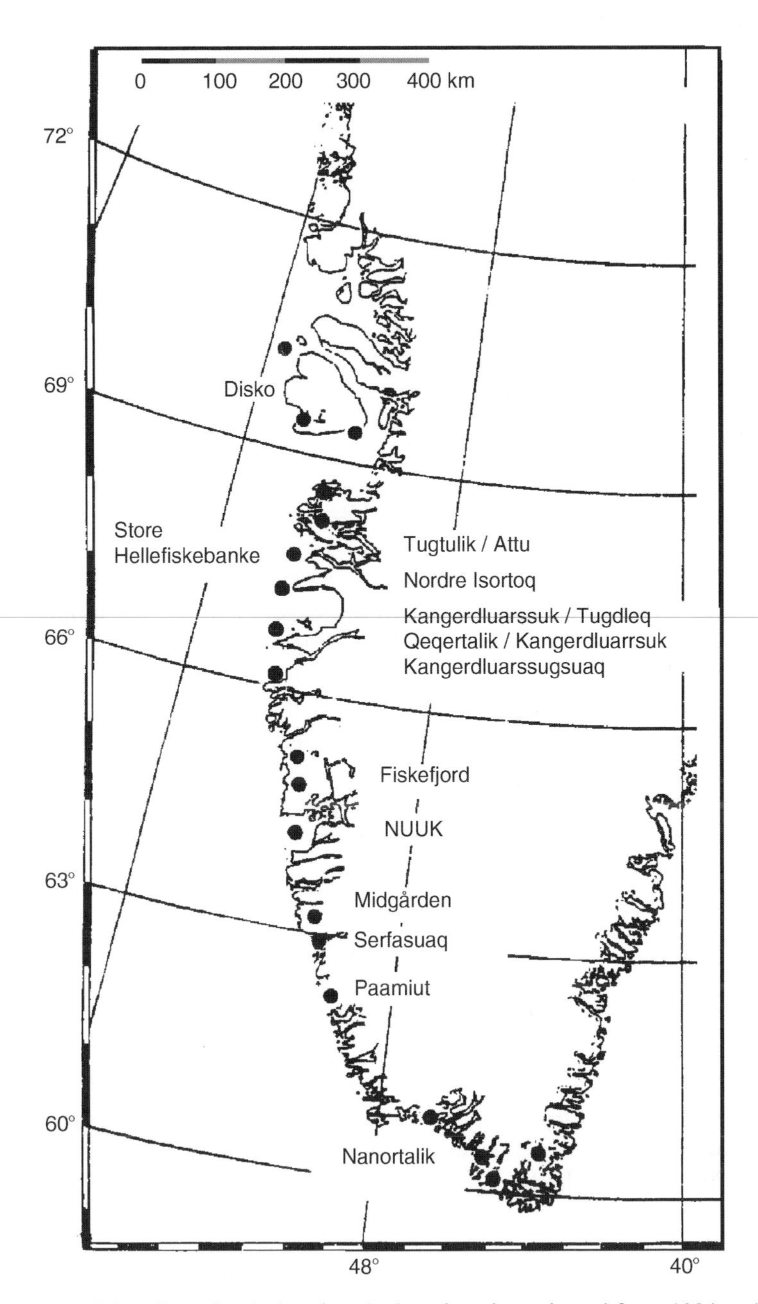

Figure 9 West Greenland, showing the locations investigated from 1984 to 1988. (Data from Pedersen, 1994.)

yield of scallops varied among areas, as both are affected by feeding conditions and sea temperature. These two factors differ among years, depending on inflow of polar water carried by the East Greenland current after it turns around Cape Farewell (Pedersen, 1988a; Buch, 2000). It was estimated that on average, scallops in West Greenland would take 15 yr to reach the minimum landing size, 65 mm SH (Pedersen, 1987b). The scientists applied a conservative approach when advising the TAC, limiting catches to 10% of the scallop stock >65 mm length, yet most scallop beds showed signs of overfishing 2 yr after the surveys.

Stock assessment was very difficult because of the lack of knowledge of the scallop populations and their recruitment processes. The length distributions indicated that recruitment was very low (Pedersen, 1987b). Juvenile scallops were sometimes found among the catch, but they could not be considered a representative sample, as the scallop dredge is not appropriate for sampling undersized scallops. In addition, it was suspected that juveniles might have a distribution different from that of adults. In any case, the scarce data on the younger year-classes collected during the surveys were not reliable enough to estimate year-class strength (Nicolajsen, 1984; Pedersen, 1986; Engelstoft, 1995). Yet for stock assessment, it had to be assumed that recruitment was constant. The collapse of several scallop fishing grounds shortly after their discovery suggests that the extraction of adult scallops did not lead to an increase of year-class strength influenced by density-dependent processes, as had been observed elsewhere (Vahl, 1982; Smith and Rago, 2004).

The scientists who analysed the survey data assumed the conditions listed below and admitted that the reliability of their estimations would be greatly jeopardised by any departure of the scallop stock from these assumptions, especially numbers 1 and 2 (Pedersen, 1988c,d):

1. No other species but scallops are present in scallop beds.
2. Scallops are not very mobile, so if an area is dredged over, the scallops will not be replenished from neighbouring areas.
3. The average catch per m^2 is known.
4. Dredge efficiency (% of available scallops actually caught) is known.
5. The total area of the scallop bed is known.

Uncertainty about dredge efficiency was another major drawback. There were differences in catchability not only among dredges but also among tows with the same dredge on different bottom types, currents and slopes (Pedersen, 1988a).

Stock assessment surveys and fishing were carried out in the localities in the following subsections. Since the beginning of the fishery, the catch has been processed manually or mechanically, depending on market demands. Manual processing takes place on board or in land-based processing facilities and generates three products: meats, meats with gonads and one of the shells containing the muscle and gonads. About 3000–4000 scallops, or 300 kg, can be processed per day per person (Pedersen, 1988a,e). Mechanical processing is carried out on land. First, scallops are killed in a sodium hydroxide solution, then they are steamed to open completely the shells and separate them from the muscle. A rotating separator discards the shells. The meats are then washed and go through a solution of citric acid to neutralise the previous sodium hydroxide treatment. The meats are again washed several times in clean water before trimming, which is manually done. The clean meats are frozen in blocks. The processing plant in Maniitsoq could process 25 tonnes a day and had space for 34 people to trim the meats (from Ott *et al.*, 1987 [cited in Pedersen, 1988f]).

3.1.1. Nuuk

Nuuk has always been the main fishing ground with the largest scallop beds (Figure 10). Unlike other areas, the fishery can operate here year-round because it is permanently free of sea ice.

The first survey was conducted in 1984 and used four dredge designs (Table 1) but without replicated tows. The results were published in two reports (Eiríksson and Nicolajsen, 1984; Nicolajsen, 1984) that are not easily compared because the 18 scallop beds found were grouped in 6 management areas in the first report but 11 in the second. The first report estimated the stock at 9000 tonnes and the second at 13,000 tonnes (Table 2). Both stressed the need to regulate the fishery and the uncertainty of the method used to calculate the stock size and recommended a maximum TAC equal to 10% of the initial stock size. Scallops were found in greater densities on hard bottoms with sand, gravel and boulders, in fjord areas and at 20–80 m depth, with great variability in growth rates from place to place (Eiríksson and Nicolajsen, 1984; Nicolajsen, 1984).

The second survey, in 1986, set out to investigate whether the data collected since 1984 in the surveys and the commercial fishery could detect changes in length distribution and density of the stock. Catches decreased considerably in 1986 compared with 1985 (Table 3) and a precautionary approach was advised again because the scarce data available, from both surveys and logbooks, were not reliable. Besides, for a stock constituted mostly of scallops 20 yr old (Tables 4 and 5), even a TAC of 10% could lead to overfishing, especially under the poor recruitment conditions observed (Pedersen, 1986).

As feared, the stock had decreased to 8500 tonnes by 1987 and to 4400 by 1994. The length distribution of scallops was very similar between years for most of the locations investigated, indicating that recruitment remained very low. At this point there were two possibilities: decrease the quota further or

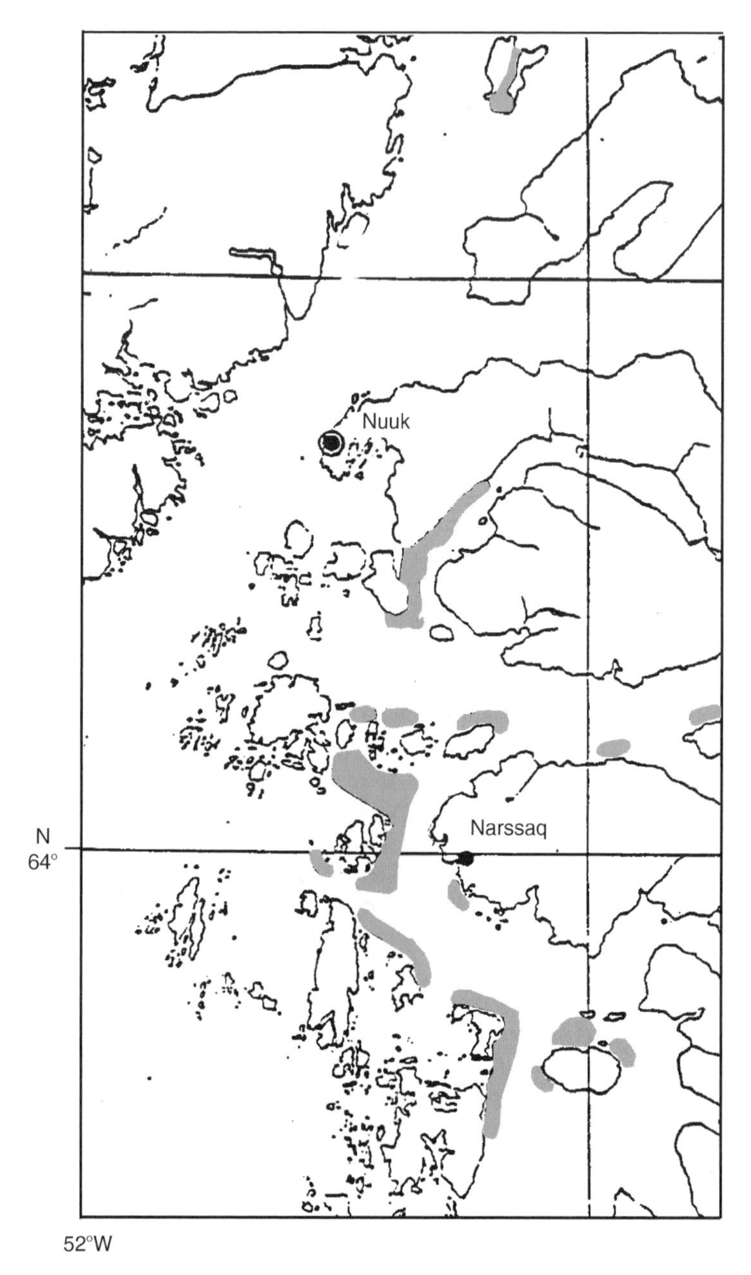

Figure 10 The scallop beds (shaded in grey) around Nuuk. The total stock size for these grounds was estimated as 13,000 tonnes. (Data from Eiríksson and Nicolajsen, 1984.)

Table 1 Dredges used in stock assessment surveys in Greenland

Survey	Dredge	Width (m)	Mesh size (mm)	Ring ø (mm)	Catchability	Type	References
Nuuk, 1984	Manx	1.9	125	75	0.1	Sledge	Nicolajsen (1984)
Nuuk, 1984	KIS	1.9	125	75	0.155	Sledge	Eiríksson and Nicolajsen (1984)
	Manx	2.4	120	65		Sledge	
	Saarullik	2.3				Sledge	
	Polarfisk	2.0				Sledge	
Paamiut, Qaqortoq, Nanortalik, 1984	Manx	1.9	120–140		0.155	Sledge	Nicolajsen (1985a)
Maniitsoq, Kangerdluarssugssuaq, 1985	KIS	2	140		0.1	Sledge	Nicolajsen (1985b)
Attu, 1986		1.9	120		0.2		Pedersen (1987b)
Aasiaat, Attu, Kangaatsiaq, 1988	KIS	0.9	120				Pedersen (1988c)
Sisimiut, 1988		1.9	120–140	65	0.2	Icel. roller	Pedersen (1988e)
	KIS	0.9			0.2		
			120			Sledge	
Fiskanæsset, 1988	KIS	0.9	120		0.2	Sledge	Pedersen (1988d)
West Greenland, 1988		2.4	120	75		Icelandic sledges	Pedersen (1988a)
		3.6	95	55			
Maniitsoq, 1987		1.9	120–140	65		Icel. roller	Pedersen (1988f)
		2	140			Sledge	
		1.2	140			Sledge	
	KIS	0.9	120			Sledge	

Note: Catchability was fixed in each survey.

Table 2 Stock parameters from the communities investigated[a]

Comm	Year	Man. unit	M catch	Density (kg m^{-2})	T catch (tonnes)	Stock size (tonnes)	Area (km^2)	References
Nuuk	1984	a		0.248		1984	8.00	Nicolajsen (1984)
		b		0.317		3804	12.00	
		c		0.879		879	1.00	
		d		0.160		320	2.00	
		e		0.289		1156	4.00	
		f		0.642		642	1.00	
		ALL	0.197	0.422		9044	28.00	
Nuuk	1984	1	0.118		20	129		Eiríksson and Nicolajsen (1984)
		2	0.203		347	2239		
		3	0.171		29	187		
		4	0.122		42	271		
		5	0.155		239	1542		
		6	0.192		263	1697		
		7	0.212		212	1368		
		8	0.206		423	2729		
		9	0.289		297	1916		
		10	0.158		79	510		
		11	0.195		65	419		
		ALL	0.183	1.27	2016	13,017		
Nuuk	1987	1	0.150			168		Pedersen (1987b)
		2	0.164			1813		
		3	0.191			213		
		4	0.027			58		
		5	0.092			916		
		6	0.183			1619		
		7	0.103			665		
		8	0.131			1735		
		9	0.152			1006		

		10	0.091			290		
		11	0.077			168		
		ALL	0.124			8651		
Paamiut	1984	1	0.004	0.024	2	13		Nicolajsen (1985a)
Nanortalik	1984	2	0.009	0.055	8	52		Nicolajsen (1985b)
		3	0.003	0.017	1	6.5		
		4	0.016	0.103	45	290		
		5	0.039	0.252	5	32		
		6	0.037	0.239	4	26		
		ALL	0.0207	0.133	63	404	4.88	
Kang.	1985	A	0.111	1.110	7.3	73		Nicolajsen (1985b)
		B	0.224	2.240	1.9	19		
		C	0.240	2.400	1.1	12		
		ALL	0.191	1.917	10.3	104	0.65	
Attu	1986		0.284	1.420	30	298	0.21	Pedersen (1987b)
	1988	1	0.084	0.429		90	0.21	Pedersen (1988c)
		2	0.225	1.124		236	0.21	
		3	0.135	0.680		34	0.05	
		4	0.169	0.850		85	0.10	
		5	0.245	1.230		123	0.10	
Maniitsoq	1985	Kang.				750		Nicolajsen (1985b)
Maniitsoq	1987	Kang.				150		Pedersen (1988e)
Fiskenæsset	1988	Mid				375	0.75	Pedersen (1988e)
Fiskenæsset		Ser				100	0.20	
Sisimiut	1988	Qeq. and Kgr.	0.190	0.950	0.783	3800	4.00	Pedersen (1988f)
Sisimiut		Kgr Td	0.164	0.821	0.292	370	0.45	
Sisimiut		N Is	0.262	1.308	0.35	785	0.60	

[a]Average data for some of the communities as given in the reports are shown in italics. Most of them do not give detailed information for each management unit, so average data cannot be estimated from the numbers in this table.

Note: Man. unit, management unit; M catch, mean catch; T catch, total catch; Kang, Kangerdluarssugssuaq; Qeq, Qeqertalik; Kgr, Kangerdluarssuk; Kgr Td, Kangerdluarssuk Tugdleq; N Is, Nordre Isortoq; Mid, Midgarden; Ser, Serfasuaq.

Table 3 Catches and total allowable catch (TAC) in Greenland

Year	Nuuk	Nan	Paam	Man	Tu/At	Qeq	SMud	Mud	Saq	Kan	NDB	SDB	SHe	Unk	Total	TAC	References
1984	300	100	10		?										410	1300	Pedersen (1988a)
1985	1333			340	?	20									1693		Pedersen (1989)
1986	510			254										100	864		Pedersen (1989)
1987	500					29								100	629	850	Pedersen (1989)
1988	700					20									720		Pedersen (1988a)
1989														641			Pedersen (1994)
1990														737			Pedersen (1994)
1991	502				190			439	19					816	1966		Pedersen (1994), *
1992	963				910				24					17	1913		Pedersen (1994), *
1993	600				346			175	30					256	1407		*
1994	341				88			55	94					873	1453		*
1995	534				132			48	28					1601	2342	1800	*
1996	184													992	1176	1800	*
1997	370							68	463					1044	1944	1800	*
1998														2200	2200	1800	*
1999	680				76	30	2	13	751	6	938	65	0	49	2610	1800	Siegstad (2000),*
2000	586				185	10	60	45	475	6	555	100	0	31	2053	1800	Siegstad (2001),*
2001	676				21	141	0	99	140	542	125	1	5	4	1754	2320	Siegstad (2001),*
2002	501				132	75	0	0.2	340	1135	3	16	29	9	2240	2320	*

Note: Nan, Nanortalik; Paam, Paamiut; Man, Maniitsoq (Kangerdluarssuqssuaq area); Tu/At, Tugtulik/Attu; Qeq, Qequertarsuatsiaat (Fiskenæsset); SMud, South Mudderbugten; Mud, Mudderbugten; Saq, Saqqaq/Sisimiut; Kan, Kanaarsuk/Sisimiut; NDB, North DiskoBay; SDB, South, Disko Bay; She, Store Hellefiskebanke; Unk, origin unknown; *, Siegstad personal communication; ?, some dredging took place in Tugtulik/Attu in 1984 and/or 1985, but figures are unknown.

Table 4 Population parameters estimated for the stock in Nuuk under assumption of $F = 0.15^a$

Mean height (mm)	85.6
Mean weight (g)	120
Age	15–25
Age at recruitment	8.8
Natural mortality	0.12
Biomass/recruit (g/m)	315
Production/recruit (%)	12
Yield/recruit (g/m)	15
Annual yield (tonnes)	1221
Muscle (% of weight)	10

[a]Data from Nicolajsen (1984).

fish the stock down, close the area to all fishing and wait about 10 yr for its recovery (Pedersen, 1987b, 1988a; Engelstoft, 1995).

The number of small boats (8–20 BRT) that joined the fishery in winter fell from 10–15 in 1985 to 3–4 in 1987 as catches declined. Eventually the processing plant where small boats took their catch had to close in 1998, when prices of fresh scallops grew higher than those for the processed meats (Pedersen, 1988a). The literature does not mention whether any of the three to four boats (100 BRT) still working in the area in 1985 and 1986 pursued other fisheries afterwards.

3.1.2. Nanortalik and Paamiut

The survey carried out in 1984 found only small scallop beds in Nanortalik and Paamiut. The stock in Nanortalik was estimated at 400 tonnes, whereas that in Paamiut was only 13 tonnes (Table 2). Around 10% of the stock was fished, and since then, there has been no dredging (Table 3). It was thought that scallops were very scarce here because of poor growth in the cold and nutrient-poor water from the Polar current, which surrounds the southern tip of Greenland and reaches these two areas. The bottom topography, with very sharp slopes and uneven surfaces, also made dredging very difficult (Nicolajsen, 1985a).

3.1.3. Maniitsoq and Kangerdluarssuqssuaq

Maniitsoq and Kangerdluarssuqssuaq were surveyed first in 1985. Close to Maniitsoq, there were extensive areas of seabed covered with empty shells, but a bed of live scallops was found in the Kangerdluarssuqssuaq area. The stock was estimated to be 750 tonnes (Table 2) and a TAC of 75 tonnes was

Table 5 Parameters for the von Bertalanffy growth equation estimated for scallops from different locations

Community	Year	Area	H8(mm)	K (yr^{-1})	T_0(yr)	Age 65 R(years)	References
Nuuk	1984	A	173.0	0.081	−0.235		Nicolajsen (1984)
		B	147.0	0.108	−0.285		
		C	131.0	0.121	−0.106		
		E	131.0	0.121	−0.106		
		F	131.0	0.121	−0.106		
		ALL	128.5	0.073	−0.875	9	
W Greenland	1986		94.4	0.094	1.100	12.5	Pedersen (1987c)
Attu	1986		99.3	0.160	−0.600	7	Pedersen (1987a)
Maniitsoq	1987	Kang	90.7	0.187	−1.940	9	Pedersen (1988d)
Maniitsoq		Fiskefjord	102.9	0.124	−1.560	10.5	
Fiskenæsset	1988	Mid	86.3	0.190	−2.090	13	Pedersen (1988e)
Fiskenæsset		Ser	93.9	0.140	−1.320	13	
Sisimiut	1988	Qeq and Kgr	111.6	0.070	0.080	13	Pedersen (1988f)
		Kgr Td	89.0	0.190	−2.130	9	
Sisimiut	1988	N Is	99.4	0.120	−1.490	11	

Note: $H\infty$, maximum height; K, growth coefficient; T_0, theoretical age at zero length; Age 65 R, age at 65 mm estimated from the age vs length plots shown in the reports; Kang, Kangerdluarssugssuaq; Qeq, Qeqertalik; Kgr, Kangerdluarssuk; Kgr Td, Kangerdluarssuk Tugdleq; N Is, Nordre Isortoq; Mid, Midgarden; Ser, Serfasuaq.

advised (Nicolajsen, 1985b). However, landings during that year amounted to 300 tonnes of scallops (Table 3). The community was optimistic regarding the possibilities of employment linked to the new fishery. Despite the scarcity of the resource, in 1986 the local processing plant in Maniitsoq obtained equipment to process 25 tonnes of scallops a day. The Maniitsoq commune paid for this and for the dredges. The reasons behind this investment were several. Scallops had to be processed to get higher returns, but it was difficult to establish a work routine because the number of workers in the processing plant varied. Some of them only worked there when they could not go to sea. Also, the fishing grounds were 4 hr away, the boats were small (8–14 m), and therefore, fishing was irregular, being dependent on the weather.

In 1986, 28 boats were given licenses to catch scallops, and catches declined by 25% soon afterwards. The processing method changed then to optimise the use of the raw scallops and the percentage of meats increased from 16 to 32% of the landing weight. However, market prices started to decline in autumn 1986 and eventually led to the closure of both the fishery and the processing plant in 1987 (Pedersen, 1988a,f).

A new survey was carried out in 1987 to assess the Kangerdluarssuqssuaq stock, which was found to be only 150 tonnes. There had been a significant decrease in the average size of scallops, and the percentage of small scallops had increased considerably. Thus, in 1985, >28% of the scallops were 90 mm in height, whereas in 1987, 25% of them were 12 mm (2–3 yr old). Yield also decreased. In 1985, for scallops >65 mm SH (10 yr old), gonad and muscle accounted for 30% of the weight, but in 1987, this was only 22%. New recruits would take around 10 yr to reach the minimal landing size and make the stock fishable again (Table 5) (Pedersen, 1988f).

Scallop beds in the neighbouring areas of Kangerdluarssuk, Alangua and Fiskefjord were also investigated, but none of them could be exploited commercially. The scallops were big enough, but the beds were small and widely distributed over the area.

In Kangerdluarssuk, it was not possible to estimate the stock size outside the sampled stations. Scallops of marketable size were found only in one station, but their yield was around 25%—low compared to other localities.

Scallops from Alangua were larger; indeed, few of them were <80 mm SH, but the stock size and biological parameters were not investigated. Besides, previous fishing in the area had shown poor results.

The stock in Fiskefjord had been previously fished, but it could not be estimated outside the sampling area. Most scallops were of marketable size, but recruitment was low. Yield in 1986 was 16%, but this was probably because the samples were collected after the scallops had spawned. Fishermen considered that a commercial fishery was not viable, as marketable scallops had a very limited distribution in this area (Pedersen, 1988f).

3.1.4. *Qequertarsuatsiaat*

Some minor scallop beds were found in 1985 and surveyed again in 1987 prior to the start of a very small scale winter fishery (<30 tonnes/yr) in May of that year (Figure 11). The stock estimation was just <500 tonnes, and a TAC of 48 tonnes was recommended (Table 2). The 1988 landings were hand-processed in the fish factory in Ervanga and sold as muscle and

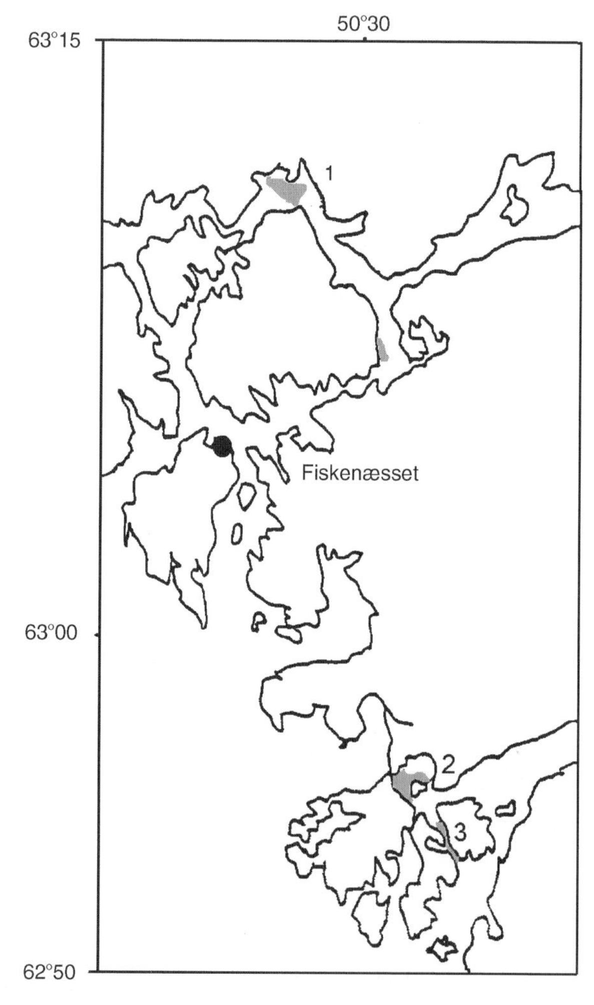

Figure 11 The scallop beds (shaded in grey) at Qequertarsuatsiaat (Fiskenæsset). The overall stock size was 475 tonnes. (Data from Pedersen, 1988d.)

as muscle with gonads. The very good quality and low price of these products enabled them to compete successfully in the U.S. market with those from Iceland and Norway (Pedersen, 1988d; R. Frandsen, personal communication).

Dredging started again in 1999 after a 10-yr pause with rather high catches considering the estimated size of the stock (Table 3). The scallops were on average 85 mm SH and had the lowest growth rates from all areas that were investigated between 1984 and 1988 (Table 5).

Several other locations in this area were surveyed in 1988, but nothing was found south of Qequertarsuatsiaat, probably because of the high suspended load in the water caused by the nearby glacier (Pedersen, 1988d).

3.1.5. Attu

Experienced skippers carried out the first survey to search for scallops in this area back in 1985, using private boats with facilities for hand-processing of the catch on board. Only one scallop bed suitable for commercial fishing was found (Figure 12), with an estimated stock size of 300 tonnes (Pedersen, 1987c).

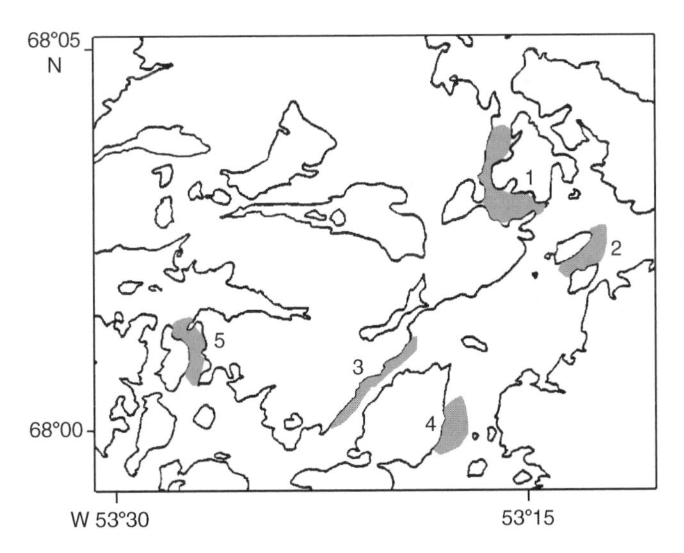

Figure 12 The scallop beds (shaded in grey) at Attu. The bed discovered in 1986 (no. 4) was estimated to hold nearly 300 tonnes, but the stock had decreased to 90 t by 1988. The other beds found in 1988 totalled about 500 tonnes. (Data from Pedersen, 1988c.)

A second survey took place in 1988 around Aasiaat and Attu in areas 30–60 m deep that were not marked on sea charts. Scallops were found in Aasiaat but not in sufficient numbers to start a fishery.

Five localities around Attu had scallop beds, three of them already mapped. One of the localities, which had its stock estimated at 300 tonnes in 1986, had decreased to 90 tonnes in 1988 (Table 2). It was unclear whether this difference in stock size had been caused by fishing, as there was no information on scallop dredging in the area during the previous 2 yr. It may have been a result of the uncertainty of the methods used to estimate the stock size. The remaining four scallop beds summed just <500 tonnes. The recommended annual catch was 20 tonnes because recruitment was very low. Dredging resumed in 1991 with rather variable catches (Table 3). Most scallops were around 80 mm SH except from one location where they were 100 mm because of very good environmental conditions. Growth rates in Attu (Table 5) were higher than in any of the other areas investigated (Pedersen, 1988c).

3.1.6. Sisimiut

Sisimiut was explored in 1985 and 1988. Private boats surveyed several locations in 1985, Strømfjord, Itiudleq, Kandgerdlussuaq and Ikertoq, but scallops were found only in Kandgerdlussuaq. A total of 6.6 tonnes were landed in 10 days. The remaining boats found only mussels.

The 1988 survey was carried out with a research vessel, and scallop beds with some potential for commercial fishing were found in three localities: Qeqertalik, Kangerdluarssuk Tugdleq and Nordre Isortoq. The stock for all three was estimated at 5000 tonnes, so about 500 tonnes could be fished, although it was pointed out that these figures were uncertain (Pedersen, 1988e).

The beds at Qeqertalik were at 15–60 m depth, and the scallops were relatively small, about 75 mm SH. Many 60-mm individuals were 10–20 yr old, so it was estimated that it might take 15 yr for the recruits to enter the fishery. This stock was estimated to be 3800 tonnes (Figure 13, Table 2).

Scallops in Kangerdluarssuk Tugdleq were larger, most of them being 87 mm SH, and faster growing, requiring 11 yr to recruit to the fishery. However, the stock was much smaller, 370 tonnes, and average catches of live scallops were low because of the abundance of empty shells (Figure 14, Table 2).

Nordre Isortoq had a stock of 785 tonnes, but the bottom was hard and uneven, making catches rather variable (Figure 15, Table 2). Scallops were on average 80 mm SH, there were small scallops, and the estimated time needed to reach marketable size was 13 yr (Table 5).

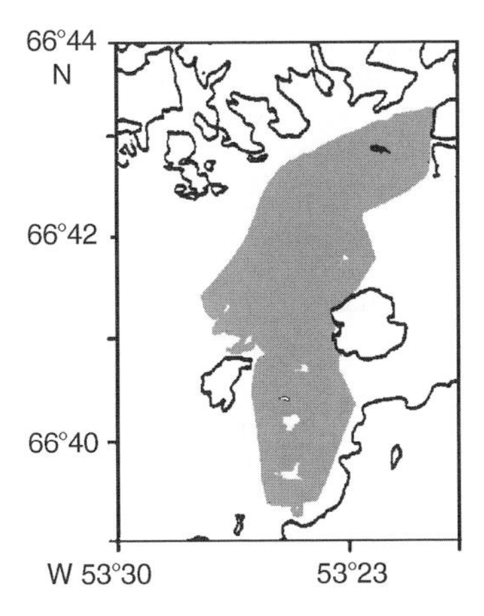

Figure 13 The scallop beds (shaded in grey) at Kangerdluarssuk and Qeqertalik, estimated to be 3800 tonnes in 1988. (Data from Pedersen, 1988e.)

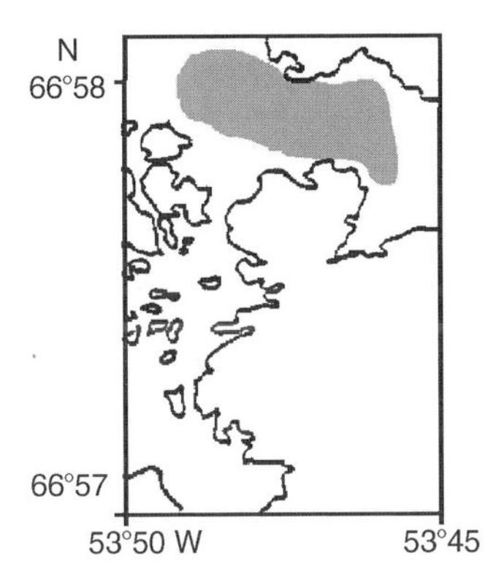

Figure 14 The scallop beds (shaded in grey) at Kangerdluarssuk Tugdleq. The stock size was estimated to be 380 tonnes. (Data from Pedersen, 1988e.)

There are no figures of catches from these locations in the early years, but new beds within the Sisimiut area, Saqqaq and Kanaarsuk, have been dredged since 1991 and 1999, respectively (Table 3). The scallop beds in most of these locations, except for Nuuk, Attu and Sisimiut, were exhausted in a very short time, probably a result of both overfishing and the lack of sound stock estimations. Pedersen (1986) pointed out that stock sizes had been calculated using the method described by Beverton and Holt, despite that this equation assumes constant recruitment, which seems to be nonexistent or very rare in Greenlandic scallop beds.

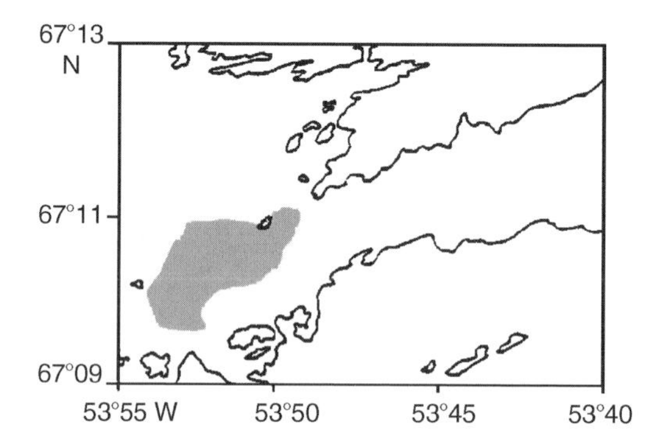

Figure 15 The scallop beds (shaded in grey) at Nordre Isortoq. The stock size was nearly 800 tonnes, but the sea bottom was hard and uneven and catches were very variable. (Data from Pedersen, 1988e.)

Since 1995, and following the discovery of new scallop beds by the fishermen, there has been a northward shift of the fishery. Fishing in Nuuk and Attu/Tugtulik (south of Disco Bay) continued, but fishing now also takes place in Mellemfjord, Mudderbugten and Søndre Upernavik, all of them north of Disko Bay (Engelstoft, personal communication) (Table 3).

3.2. Fishing gear

Several dredge types were used in the surveys (Table 1), depending on the size and horsepower of the vessels and on location, as it was found that dredges varied in performance under different conditions, such as bottom type, currents and slope (Pedersen, 1994).

Pedersen and Boje (1987) carried out a comparative study between one Manx dredge manufactured in Iceland and two Greenlandic KIS dredges, aided by underwater photography and divers and concluded that catchability was better with the Manx dredge, 37% versus 24% and 21% for the KIS dredges.

3.3. Management

Licenses have been issued since the beginning of the fishery to limit the number of vessels operating to four or five in the Nuuk area. The minimal landing size was set at 65 mm SH and annual catch quotas have been set every year since the beginning of the fishery. These quotas were to be 10% of the stock >65 mm SH (Pedersen, 1989, 1994; Anonymous, 1998).

A quota is still set every year for Nuuk, Attu, Sisimiut and Mudderbuggten, based on the stock assessment studies carried out during the 1980s and in 1994. There are no quotas for the fishing grounds discovered after 1995, but the Greenland Institute of Natural Resources advises a total catch of 2000 tonnes for all West Greenland. This figure is based on landings during the last 10–15 yr and the stability of catches (Engelstoft, personal communication), and it seems to be appropriate, as there have been no major fluctuations in overall annual landings (Table 3).

3.4. Research

There are old records of scallop beds found in West Greenland, dating from the end of the nineteenth century, as cited in Posselt (1898) and Jensen (1912). However, research on scallop beds with a view to commercial use of the stock did not start until 1984 with mapping of resources, stock size estimations (Eiríksson and Nicolajsen, 1984; Nicolajsen, 1984) and a study on catchability of different dredges (Pedersen and Boje, 1987). Pedersen (1994) published a study on population parameters based on survey data from 1984 to 1988.

The most recent report described a study to transfer scallops from areas with lower growth rates to other places where growing conditions were better in order to regenerate exhausted fishing grounds. However, the high mortality among the juveniles after transplantation and the small growth rates of adult scallops made the project unfeasible (Engelstoft, 2000).

The reports cited in this chapter provide nearly all the scientific knowledge accumulated during the last 20 yr concerning the scallop stocks in Greenland. Other aspects of the fishery such as by-catch have not been studied (Engelstoft and Siegstad, personal communication).

4. NORWAY: SVALBARD, BJØRNØYA AND JAN MAYEN

4.1. The stock and the fishery

Scallops have been traditionally used as bait in Norway, but this fishery decreased after World War II and stopped during the 1950s (Sundet, 1985). The possibility of starting a commercial fishery was not considered until the 1960s. The first surveys took place in 1961 and 1962 in Tromsø, Kvænangen fjord and between Vestfjord and Kirkenes, in North Norway (Wiborg, 1962, 1963a). The best scallop bed was found in Balsfjord, near Tromsø, covering

70,000–100,000 m^2. Most shells were >60 mm SH and all individuals reached maturity at 50 mm SH (Wiborg, 1963a). Density of scallops >65 mm SH was 33 individuals m^{-2} (Venvik and Vahl, 1979), and it was estimated that the Balsfjord scallop beds could support a fishery of 500–1000 tonnes yr^{-1}. In this location, CPUE reached 80 kg min^{-1} when using a 9-ft wide Icelandic dredge and 67% of the catch were scallops >65 mm SH (Venvik, 1982). However, later assessments recommended a maximum catch of 250 tonnes (Sundet, 1997, 2003). Another survey in North Norway found several beds with shells >75 mm SH and in densities of 35–40 shells m^{-2}, but no information is given about their distribution (Wiborg, 1963b). Sør-Varanger was also surveyed, but scallop density was very low and unsuitable for commercial fishing (Sundet, 1996).

The offshore scallop beds in Spitsbergen and Bjørnøya seemed to have better potential to develop a profitable fishery, but the catch had to be processed on board to prevent loss of quality (Venvik, 1979, 1982). Thus, most of the boats in the scallop fleet had onboard mechanised processing facilities. The catch was first heated at 30 °C, then at 80 °C to open the shells, which were then mechanically separated from the meats. The viscera were removed mechanically and manually fine-trimmed. The clean meats were then frozen and size-graded prior to packing (Oterhals, 1988).

4.1.1. Jan Mayen

In the mid 1980s, traditional fisheries were declining and the industry was looking for alternative resources. Boats turned to the offshore scallop grounds at Jan Mayen in 1985, starting the Norwegian large-scale scallop fishery. A total of 27 licenses were issued the first year and 20 boats between 40 and 70 m in length took part in the fishery. Seven of these boats were specially built for the scallop fishery, while the rest came from other fisheries or had been previously used as supply boats. The fishery was not as profitable to vessels without onboard processing because quality declined during the long time at sea, yet five such vessels took part during 1986. Catches approached 13,000 tonnes in this year (Anonymous, 1988) (Table 6).

The first survey in Jan Mayen, also in 1986, estimated the stock at 31,000 tonnes distributed over two scallop beds, one of them being 60 km^2 and the other 15 km^2 (Rubach and Sundet, 1987) (Figure 16). The stock was made up of large old scallops with growth rates as slow as those in West Greenland (Table 7). The second stock assessment, in May 1987, estimated that there were 8000 tonnes left and showed a decline in number of scallops <65 mm SH (Anonymous, 1988). The area was, therefore, closed

Table 6 Scallop catches (tonnes) in Bjørnøya, Svalbard (areas I–XII)[a] and Jan Mayen

Catch area Year	20 Bjørnøya	21 IX	22 X, XII	23 XI, XII	25 I–VII	35 Jan Mayen	36 Jan Mayen	Total
1987	12,227	3840	9392	621	16,397	1487	134	44,098
1988	5041	1166	2533		10,295			19,035
1989			843		3755			4598
1990	3269	656			2325		587	6837
1991	5510				479			5989
1992	5140			732	524			6396
1993	2277							2277
1994	982			2085				3067
1995	2135				1921			4056

[a]Data from the Norwegian Directorate of Fisheries.
Note: Arabic numbers indicate the spatial division of the Directorate of Fisheries and Roman numbers the division used in the reports (see Figures 16 and 17).

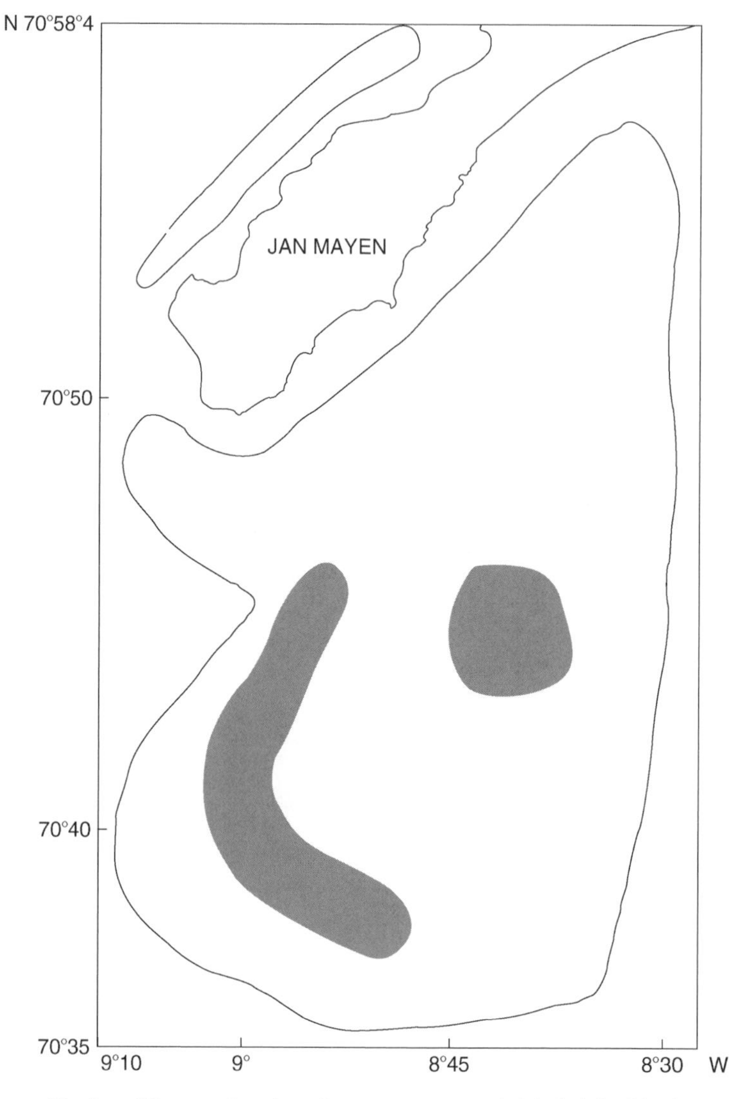

Figure 16 Jan Mayen, showing the areas surveyed (circled in black) and the scallop beds (shaded in grey). (Data from Rubach and Sundet, 1987.)

to dredging in August 1987 (Aschan, 1991), after a catch of 1500 tonnes had been taken (Table 6).

Subsequently, 13 of the 26 Norwegian boats were sold to Greenland and several of the remaining ships were sold in Norway to pursue other fisheries (Anonymous, 1988).

Table 7 Population parameters and additional information from the different scallop beds in Jan Mayen, Bjørnøya and Svalbard[a]

Bank area	Jan Mayen	Bjørnøya XII	Sørkapp IX	Sjubreflaket V	Nordgat IV	Klovningen III	Breibogen II	Moffen I
>65 mm (%)	64	91.5	50.6	64	60	93	82.5	55.8
Mean length	77.2	76.6	71.5	71.2	69.3	77.1	73.4	71.7
ML <65 mm	50.1	60.4	58.9	59.1	60.1	60.3	58.1	55.8
% Males	44.8	43.9	54.2				53.1	53.9
Age at 65	9	9	11.5				13.5	12
Muscle 65 (g)	5	7	5				6.5	7
Muscle 85 (g)	10	13	10.5				12.5	14
Bank area (km^2)	60	343	85	300*	55**	**	340*	680
Stock 1986 (tonnes)	31,000		11,400	45,000	11,100		146,000	205,000
Stock 1987 (tonnes)	8000	23,000	14,000				70,000^{+}	98,000
Depth MD	70–100	70–110	60–100	26–96	24–116		40–70	30–70
Dredging	Suitable	Very good	Good	Good	Favourable	Variable	Favourable	Boulders >20 cm

[a]Data from Rubach and Sundet (1987) and Anonymous (1988).

Note: ML, mean length; Muscle 65 and Muscle 85, grams of muscle at shell height 65 and 85 mm, respectively; stock size estimated considering only scallops >65 mm; Depth MD, depth interval with maximum density of scallops; Dredging, dredging conditions; *, partially within protected area; **, whole bed within protected area; $^{+}$, stock size outside protected area.

4.1.2. Bjørnøya and Svalbard

The first searches for scallop beds at Bjørnøya were conducted in 1968 and 1969. Scallops were found between 80 and 135 m depth, but marketable-sized scallops were below 105-m depth. Recruitment was very scarce, as most individuals were 6–11 yr old, and 91% of scallops were in the length interval 50–85 mm SH. The study concluded that a commercial fishery would be difficult to develop, partly because of rough bottom conditions (Wiborg, 1970). The next survey was carried out in 1973 and investigated both Bjørnøya and Svalbard. Scallops at Svalbard were also around 80 mm SH and 10–14 yr old, but muscle and gonad weight were larger in Bjørnøya. This second study was more optimistic about commercial use of the stock, but it indicated that only a small percentage of the total stock biomass should be taken, given the poor recruitment and low growth rates (Wiborg *et al.*, 1974).

Rubach and Sundet (1987) surveyed Jan Mayen, Bjørnøya and Svalbard in 1986 and estimated the stock size at about 418,000 tonnes, Bjørnøya and the Klovningen bed excluded (Table 7, Figure 17). However, the beds from Nordgat and Klovningen and part of the beds in Sjubreflaket and Breibogen were within the National Park north of Spitsbergen, so the fishable stock would be several thousand tonnes less. These new fishing grounds were considered suitable for dredging, but the scarcity of recruitment and the long time needed for scallops to recruit to the fishery (Table 7) required a very careful management. Just >2000 tonnes were taken in 1986 from the Svalbard area (Anonymous, 1988).

The survey carried out in 1987 estimated the stock off Bjørnøya at 23,000 tonnes of scallops >65 mm SH, and the Sørkapp bed was found to be 3000 tonnes larger than previously thought. On the other hand, the Moffen bed had decreased greatly (Anonymous, 1988) (Table 7).

During 1987, total catches rose to 41,000 tonnes despite the recommendations from the scientists. Nineteen vessels took part in the fishery, but 72% of the catches were taken by only four of them. In 1988, the fleet dredging in Svalbard dropped to seven boats and catches fell to 7000 tonnes according to Anonymous (1988), although these data do not agree with those from the Norwegian Directorate of Fisheries (Table 6). Comparison between scientific and official sources is difficult because of the different spatial divisions of the study area.

Catches between 1989 and 1995 have been between 2000 and 7000 tonnes, most of them from Bjørnøya, except for 1989, when most catches were from Moffen and Breibogen (Anonymous, 1990b) (Table 6). Bjørnøya opened in 1991 with a fixed quota of 2000 tonnes, and both Jan Mayen and Svalbard areas were surveyed again. Moffen was found to have been overfished in 1990.

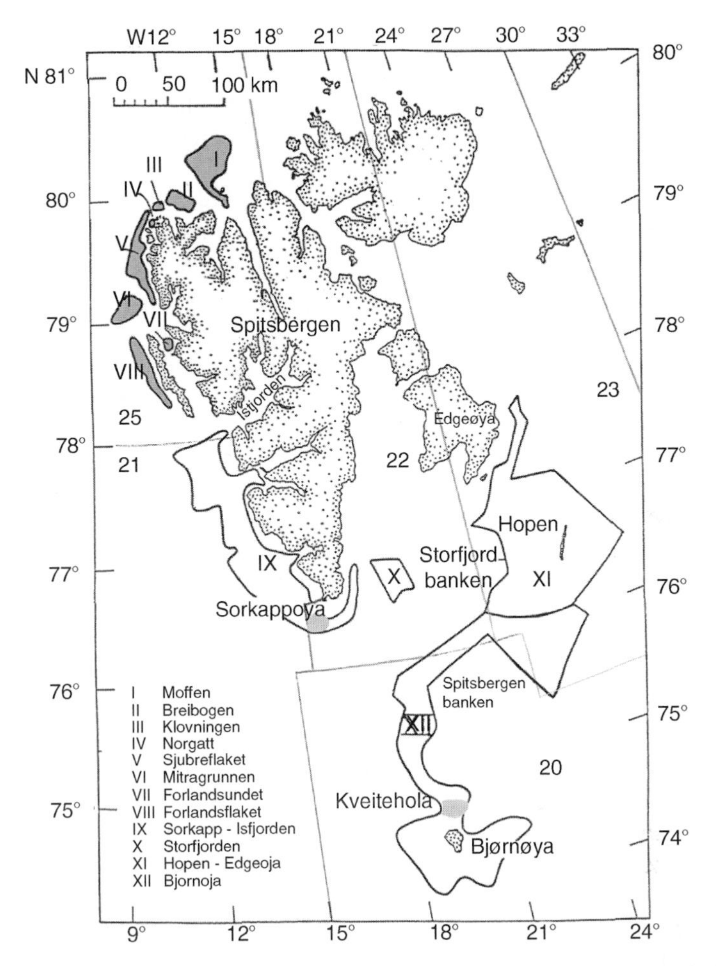

Figure 17 Svalbard and Bjørnøya, showing the areas surveyed, marked with Roman numerals, and the spatial division used by the Directorate of Fisheries, in Arabic numbers. The scallop beds surveyed in 1986 are shaded in grey. (Data from Rubach and Sundet, 1987.)

Besides, these beds seemed to be self-recruiting, meaning that recovery time would be very long (Anonymous, 1992).

Nevertheless, commercial and experimental dredging continued until 1995, mostly in Bjørnøya and Moffen. There were juvenile scallops, but they would need 7–8 yr to recruit to the fishery. A new scallop bed, Parryflaket, was found near Moffen, but the information available about commercial dredging there is unclear (Anonymous, 1992, 1995, 1996, 1997, 1998b).

4.2. Management

By mid 1987 it was obvious that the offshore scallop fisheries needed regulation and several management options were considered (Anonymous, 1988):

1. Introduction of quotas (for total catch, area or vessel)
2. Control fishing effort
3. Fixing a minimal size at capture
4. Opening/closing of scallop beds
5. Seasonal regulation of the fishery; based on the seasonal variations of muscle weight due to food availability and physiological changes associated with the spawning cycles (Sundet and Vahl, 1981; Sundet and Lee, 1984).

However, each of these options has certain requirements that the scallop fishery lacked. Thus, catches must be known accurately for quotas to be effective. In this case, only the number of processed meats was known, and total catch could not be estimated because the percentage of scallops broken and discarded during dredging and processing was unknown. Therefore, the total mortality was not accurately known and estimation of quotas would not be practical.

Similarly, regulating the fishery by limiting fishing effort required a model incorporating fishing mortality in order to estimate how much fishing mortality is generated by a certain effort (e.g., expressed as number of ships, or number of fishing days). This was not possible because catches were not known with accuracy. Nevertheless, limiting participation in the fishery could be used to prevent the fishing capacity of the fleet becoming larger than the resource could sustain. In fact, comparison of catches in 1987 and 1988 in relation to the total size of the stock suggests that fishing effort was above the maximum sustainable yield (MSY).

Minimal landing size can be used as a criterion to close or open an area to the fishery. It can be estimated from the effort or yield per recruit and spawning stock biomass. The problem is that dredges are not very selective and sorting is necessary on board. The result of this measure depends on whether scallops survive the process of sorting and to what extent the fleet can avoid dredging over areas with many small shells.

The closure/opening of areas, combined with monitoring, does not demand as much accuracy in the estimation of stock as the other methods, but it demands monitoring of the stock in the selected area. This option raises the following questions:

1. *Criteria to close/open an area.* This decision could be based on size and/or variability in the harvested stock. Both changes in size of fishable stock and the proportion below the minimal landing size need to be monitored.

Data from the commercial fishery plus survey data can be used to monitor the stock. It is possible to have observers on board to close or divide the fishing area if a large proportion of scallops is under the minimal landing size. Areas can be closed regardless of length distribution if there are indications of drastic reductions in stock size.

2. *Which scale is appropriate considering the size of the fishing area*? There are two possibilities to choose the scale: using natural demarcations for management or dividing the beds into smaller units and managing them independently. This last option can be applied in the short or long term to ensure that the stock can recover after fishing, without overly affecting the fleet. Data on stock size and spawning time are needed. This imposes a practical limit on the size of units used in the regulations, as too small units can reduce further the effect of regulations on the stock.

3. Seasonal closures based on the spawning cycle. This option was not very reliable because timing of spawning can be affected by environmental conditions and vary among years. Besides, there was not enough information available on the biology of this stock to investigate spawning patterns and select the optimal closure period.

It was, therefore, suggested to lower the minimal landing size to 60 mm SH, as this would have least impact on the biomass per recruit obtained but would increase the fishable stock from 200,000 to 290,000 tonnes. A fishing mortality of 0.2 would correspond to catches of around 50,000 tonnes whole weight, or 4–5 tonnes of muscle. In addition, Jan Mayen, Moffen and Bear Island were closed to the fishery at least until surveying the areas again in 1988 (Anonymous, 1988).

Dredging continued in Bjørnøya until 1995 and in the beds northwest of Spitsbergen until 1992 and again in 1995. The Moffen bed was also dredged in 1992 and 1994 (Table 6), but there is hardly any information concerning the scallop stocks in Jan Mayen, Bjørnøya and Svalbard and their management after 1988.

4.3. Research

The reports from the Danish Ingolf expedition (Jensen, 1912) mentioned the presence of scallops in Jan Mayen in 1882, but the first surveys to assess the offshore scallop resources estimate stock size and collect biological information took place around Bjørnøya during 1968/69 (Wiborg, 1970). Svalbard was surveyed in 1973 but not thoroughly. Nevertheless, some experimental fishing was carried out during the same survey (Wiborg *et al.*, 1974) and again in 1979 (Venvik, 1979) in a scallop bed of 69 km^2 detected north of the island, although density and average size of the scallops were lower than in

Bjørnøya. The Norwegian Directorate of Fisheries funded a 3-yr project to assess the potential of the shellfish fisheries. The main conclusion concerning the offshore scallop fishery was that CPUE and scallop density off Bjørnøya were lower than in coastal scallop beds, but the offshore beds covered such a large area that they could be profitable if appropriate scallop dredges and onboard processing facilities could be developed (Venvik, 1982).

Several Faeroese fishing vessels conducted two surveys in 1985 and 1986 that investigated mostly the area between Bjørnøya and Hopen. The data collected were used to investigate the distribution of scallops by the kriging method and compare the results with those obtained by the traditionally used swept area method. Regular surveys usually have a random design and they cover only the area in which scallops are most likely to be found, leading to overestimation of average density and underestimation of variance by comparing only areas of high–low scallop density. However, sessile animals are not randomly distributed, and this factor should be taken into account to improve biomass estimates. The kriging method assumes that scallop distribution is stationary in time, and it gives best results when sampling is distributed according to a regular grid, including as much as possible of the scallop distribution area. The average density is a weighted average of all sampling units. The weight is a function of the distance from the location sampled to each point within the study area. In this study, the kriging method resulted in lower biomass estimates, but variances were very high because of the unsuitable sampling design (Nicolajsen and Conan, 1987). In addition, thorough surveys to map, estimate the magnitude of scallop resources and investigate their population dynamics in Jan Mayen and Svalbard were conducted in 1986 and 1987 (Rubach and Sundet, 1987; Anonymous, 1988).

Additional research was carried out to find an appropriate storage method that would ensure maximum quality of the scallop muscle. This was a priority, as at the beginning of the fishery some vessels lacked onboard mechanical processing and the catch had to be stored for several days (Anonymous, 1988; Larsen, 1991).

The effect of dredging on the benthic communities was also investigated. No changes were detected in Jan Mayen after the closure of the fishery, apart from an increase in *Ophiura robusta*. In the Svalbard area, although the dominant species were the same in 1987 and 1988, the number of species and diversity decreased after dredging started. Mortality of by-catch might have been higher than in other scallop fisheries since onboard mechanical processing included heating to 80°C and everything in the length range of 45–70 mm went through the process (Aschan, 1988, 1991; Aschan and Sundet, 1990; Anonymous, 1995).

5. OVERVIEW

The main Icelandic scallop beds in Breiðafjörður have shown signs of decline since the early 1980s, but stock indices stabilised somewhat during the 1990s before the decline became more pronounced from 1998 onwards. Catches remained at 8000–9000 tonnes until 2000 because of technological improvements and increased fishing effort. However, natural mortality, estimated from the percentage of cluckers found in samples from several locations, rocketed from 2000 to 2003 and the stock decreased by 70% in Breiðafjörður and by 45–80% in smaller beds in Northwest Iceland (Jónasson *et al.*, 2004).

This increase in natural mortality was thought to be linked to the increasing sea temperature registered in Icelandic waters over recent years and especially during the period 2000–2003, as upper tolerance limits for *C. islandica* were thought to be around 10 °C and enhanced natural mortality of scallops following rising sea temperatures had been recorded in Norway and Canada (Dickie and Medcof, 1963; Wiborg, 1963a). However, research on scallops from several locations in Iceland has shown that temperature tolerance of *C. islandica* in summer is a few degrees above the temperatures registered over the last few years in Breiðafjörður (Jónasson *et al.*, 2004), leaving overfishing one of the major factors responsible for the observed decline in the scallop stock.

The Greenland Home Rule effort to develop a new fishery based on a scarce resource must be understood in the social and economical context of Greenland at that time. Greenland got its 200-mile EEZ in 1977, as the annual cod catches off West Greenland fell from 400,000 tonnes to around 50,000 tonnes. The closure of the fishery to outsiders, however, allowed Greenlanders to increase their own catch from 15,000 to 50,000 tonnes. Nevertheless, by this time it was evident that temperature change was affecting the distribution of cod, which was not found as far north as earlier in the century. On the other hand, most people already lived in the bigger settlements, close to cod fishing grounds, with landing and processing facilities. This was the result of a governmental plan from 1948 to reorganise the fishing industry. Extremely severe winters in 1982 and 1984 caused cod biomass to decrease by 70%, and although the stock recovered somewhat between 1988 and 1990, it finally collapsed in 1991. Many cod-based settlements found themselves without fishing resources, especially for inshore fishing (Hamilton *et al.*, 2000). With this background, all possibilities to create employment and maintain the fishing industry had to be explored.

It was estimated that a catch of 100 tonnes of scallop could generate 500 working days to distribute among 12–15 people, assuming that each fisherman would process 300 kg of scallops a day to produce 60 kg of meats and that one scallop vessel with a four-man crew could catch roughly 1500 kg of scallops a day (Pedersen, 1988e,f).

The surveys in West Greenland found scallop beds in most of the communities investigated, some of them apparently with some potential for a small-scale commercial fishery. The main problem was the impossibility of accurately estimating stock size and the appropriate fishing mortality, because of the lack of background data (Pedersen, 1988c,d) and resources to implement a monitoring programme to collect biological data prior to exploitation. Nevertheless, and despite the closure of some of the beds found in the 1980s, the fishery continues and catches are rather stable. The scallop fishery in Jan Mayen, Svalbard and Bjørnøya started with a very different goal. It was a private enterprise. Encouraged by the discovery of scallop grounds, the very high prices of scallop meats in the market and probably the lack of regulation, fishers started dredging the offshore beds without previous sound research of the stock. In fact, Sundet (1985) acknowledged that the scallop fisheries in Bjørnøya and Hopen Island were promising, adding that this assumption was based on the availability of "bottom area sufficient for scallop growth and not on the knowledge of an actual standing stock." By the time the first reports were published stressing that assessment was difficult and very careful management was required, it was too late. Catches had reached 45,000 tonnes and the stocks collapsed shortly afterwards.

The pelagic dispersal of scallop larvae may suggest that recruitment to a given scallop bed depends on influx of larvae from other beds. However, many scallop aggregations are far apart and have a precise location that in some cases has been known for centuries. In addition, scallops from different locations show variability in spawning time and growth rates. Moreover, most scallop beds are found in areas influenced by tides, gyres and water circulation regimes that favour larval retention in relatively fixed geographical areas, and genetic research performed in some scallop stocks has detected differentiation among scallops from different locations. All this evidence suggests that scallop beds might be self-recruiting, and therefore, too high a fishing effort will lead to recruitment overfishing, which takes place when recruitment is jeopardised due to depletion of the spawning stock. This poses a difficult problem for management, as the minimum spawning stock biomass required for a successful recruitment is unknown (Sinclair *et al.*, 1985). Recruitment overfishing might have been the case in some scallop beds in Breiðafjörður that became depleted after intensive effort. The assumption of self-recruitment adopted in Greenland is, therefore, a conservative management approach that seems to have been overlooked in Iceland and Norway.

Inadequate management may also lead to growth overfishing, or fishing above the MSY of the stock. This seems to have been the main reason for the collapse of the Iceland and Norway stocks.

It would have been desirable for Greenland, Iceland and Norway to establish a joint monitoring programme to study recruitment. In the case of Norway, it would have been useful to assess the possibilities of recovery, apart from the obvious scientific interest. For Iceland and Greenland, however, better knowledge of the dynamics of scallop populations would have enabled them to plan better management of their resources. Scallop beds in Greenland might be relic populations from much larger beds, as found by Pedersen (1988a). Some beds around Disko Island were surrounded by an extensive area covered with empty shells. Carbon^{14} analysis indicated that the shells were between 60 and 450 yr old, and some of them even 6300 yr old. This stresses further the need for careful monitoring and management of scallop stocks in the Arctic. It may be argued that political and economical constraints do not always allow for the expensive and time-consuming research required for understanding the population dynamics of a new resource. In addition, conservative management requires a data collection from surveys and commercial fishing spanning at least over a few years. The required fisheries data are age composition of the catch, annual CPUE and fishing effort. This information is fed into a model to estimate the removal and weight at age of scallops for any given fishing mortality. The goal is to optimise yield per recruit. The model can be used to regulate the fishery by regulating the age and size of capture (gear selectivity) and control fishing effort by regulating fishing mortality under the assumption of self-recruiting populations. Basically, this consists of fishing only the older scallops so each year-class has the chance to spawn over several years. It has been shown that this strategy increases yield and egg production significantly (Sinclair *et al.*, 1985).

This strategy could have been adopted in Greenland, Iceland and in Bjørnøya a few years after the fishery started. However, the management in Iceland seems to have focused on maximising landings each year instead of favouring a conservative approach for a long-term use of the resources. Catches have been kept stable in Greenland so far with a fixed quota based on landings during previous years, but no research has been carried out since the 1980s that could lead to improvements in management. Norway is a different case because the offshore stocks were wiped out so quickly that hardly any data had been collected before the fishery collapsed, but the conservative approach could have been tried in Bjørnøya, where scallop dredging has taken place occasionally over the past 15 years. Survey data from the Svalbard area indicate that the stock may be recovering (J. Sundet, personal communication).

Other scallop fisheries have followed a similar pattern to those reviewed here and yet the fishery has recovered. The largest Canadian *C. islandica* fishery started in 1993 on the Grand Banks off Newfoundland, and by 1996 catches had reached 21,000 tonnes. Fishing mortality proved to be too high for a stock that consisted mostly of old scallops and some beds became depleted in a short time with a consequent decline in landings (Naidu *et al.*, 1997).

The fishery for the weathervane scallop, *Patinopecten caurinus*, started in southeast Alaska in the late 1960s and depleted the stock in $<$10 yr. There was no dredging for several years from 1978 onwards. However, the stock recovered in the 1980s, and after dredging resumed, new management plans were implemented to ensure resource conservation. Catches have varied greatly among years ever since, with production between 90 and 770 tonnes of muscle, but dredging has not been interrupted again (Kruse *et al.*, 2000).

Similarly, in the Isle of Man fishery for *P. maximus*, the stock has recovered since 1991 after a long period of declining catches. The fishery relies nowadays on the strength of the recruited year-class instead of old scallops, and density is much lower than in the 1950s (9–20 scallops 100 m^{-2} versus $<$3 scallops 100 m^{-2} since 1990. Nevertheless, CPUE has increased on many grounds since 1991 (Brand, 2000; Beukers-Stewart *et al.*, 2003).

Scallops have very variable recruitment, influenced by several abiotic and biotic factors, not yet fully known (Young *et al.*, 1989). This characteristic makes scallop stocks more sensitive to recruitment failure, especially if the fishery depends on the strength of the recruited year-class (Beukers-Stewart *et al.*, 2003). It is interesting to note that a model developed by Smith and Rago (2004) showed that unexploited scallop populations in equilibrium had a large biomass and very low recruitment. When a fishery is introduced into the model, recruitment shows increased variability as the population oscillates towards a new equilibrium point with a much smaller biomass. As in this model, the Iceland scallop populations reviewed here had large average stock sizes and low recruitment levels. When the fishery started in Breiðafjörður, the mean density was 20–30 shells m^{-2} and most of them were $>$70 mm. Length distribution graphs from Breiðafjörður and Northwest Iceland show how the population consisted mostly of adult shells (Eiríksson, 1970a, 1986). It can be argued that dredges do not sample adequately the smaller shells and they are underrepresented, but since juvenile Iceland scallops are often attached to larger empty shells that are caught with the dredge, it is probably safe to assume that even allowing for underestimation of the smaller length groups, recruitment was very low. The same can be said of the West Greenland populations, made up mostly of individuals $>$75 mm and with very low recruitment levels (Pedersen, 1994), and the Svalbard area, where all individuals seemed to be sexually mature, and it was suspected that spawning did not take place every year (Rubach and Sundet, 1987). Data from Greenland and Svalbard are not available, but in Breiðafjörður the stock index has oscillated with a decreased trend and recruitment has been very variable in spatial and temporal terms during 1993–2003, being very low in the late 1990s (Jónasson, 2005).

Variability in stock size may be due to spatial variation in habitat quality, spatial concentration of the fishery, larvae dispersal and trends in mean size. Within the area occupied by a scallop population, there are differences in the habitat that make some sites more favourable, influencing growth rates and the potential for egg production and larval survival (Dickie, 1955; Claereboudt and Himmelman, 1996; Tammi *et al.*, 1997; Robinson *et al.*, 1999; Smith and Rago, 2004). Concentration of fishing effort on these areas can have a negative effect on spawning and, thus, recruitment by dispersing noncaught shells and thus affecting fertilisation (Oresanz *et al.*, 1991). Furthermore, diminished scallop abundance also leads to lower larval densities (Tremblay and Sinclair, 1988), making scallop populations vulnerable to fishing.

Smith and Rago (2004) proposed that fisheries should focus on avoiding recruitment overfishing rather than optimising yields. They proposed closure to fishing of the most productive areas as a strategy to reduce recruitment variability and increase yields. If egg production within the closed area is larger than the ratio between yield from the whole area and yield from the open area, recruitment would increase and compensate for yield losses after the closure. The more productive areas would export larvae to the less productive. This strategy assumes that the excess of effort is not redistributed over the remaining open areas.

The present situation of the scallop stocks of Iceland and Norway allows for experimenting with management strategies different from those used in the past. Both fisheries are closed nowadays, and considering their trends over the past 30 yr, it is very likely that fishermen will be willing to cooperate with management actions geared to rebuild the stock and make the fishery sustainable in the future. A detailed description of the alternative management options is out of the scope of this chapter, so only a brief suggestion for future action follows. The first step could be redesigning the stock assessment surveys. Iceland, Norway and Greenland use the swept area method for scallop stock assessment. This is a widely used method based on random sampling theory, and it assumes that scallop density is randomly distributed over the study area. However, this assumption does not apply to sedentary species such as scallops, with aggregated distribution patterns that are persistent in time and space and depend largely on environmental conditions. For these reasons, geostatistical sampling techniques such as kriging constitute a better approach. These model-based methods take into account the spatial organisation of the resource and their location on regular grids or fixed stations. Kriging consists on modelling the covariance between sampling units as a function of the distance between them and assigning to each sampling unit an optimal weight to estimate the average density of the resource within the study area (Nicolajsen and Conan, 1987; Caddy and Defeo, 2003).

Secondly, indicators of fishing pressure and reference points should be monitored, as noted in the FAO Code of Conduct for Responsible Fisheries. As scallop populations within a given location are structured as metapopulations, these variables should be estimated for the different scallop beds. Within a metapopulation, there are scallop beds acting as larval sources and beds acting as larval sinks. These should be identified. According to the source-sink hypothesis, all year-classes are present in source areas, whereas in sink areas, recruitment does not occur regularly and only a few year-classes will be present (Caddy and Defeo, 2003).

The next step would be closing permanently a large percentage of the source areas and implement rotational closures in the remaining scallop beds. Permanent closure of at least some source areas would ensure larval supply and it would also protect other benthic species that are vulnerable to dredging and are in some cases very important for scallop settlement, such as bryozoans (Harvey *et al.*, 1993; Hall-Spencer and Moore, 2000; Murawski *et al.* 2000; Caddy and Defeo, 2003). Rotational closures accompanied by decreased effort have been successfully implemented in the scallop fisheries in Georges Bank, which had reached a long-term low catch per vessel in the early 1980s (Sinclair *et al.*, 1985). However, following the closures, harvestable scallop biomass had increased by a factor of 15 according to survey data from 1994 to 1998 (Murawski *et al.* 2000). Closures have shown to be effective to increase yield per recruit and, what it is more important, to reduce the risk of recruitment and growth overfishing (Murawski *et al.* 2000; Hart, 2003), but they require a decrease of fishing effort. Rotational closures distribute fishing effort over the open areas, but at the same time, they are less restrictive for the fishery than permanent closures (Caddy and Defeo, 2003; Hart, 2003).

Finally, another measure that could be applied to prevent recruitment overfishing would be increasing the minimal capture size and regulating the fishing mortality at age to ensure that all year-classes are present in the population and that each individual has the chance to reproduce several years. Sinclair *et al.* (1985) estimated that if this approach would have been applied to the Georges Bank scallop stock, landings and egg production per recruit would have been 58% and 25% higher than they were. Note that Anonymous (1998) recommended decreasing the minimal landing size to increase the fishable stock in Svalbard.

Closures are difficult to implement because of their impact on social and economic issues, but they are necessary in view of the decreasing biomass trends of many target species. The main benefits of closures may not be increased catches in the future, as has been shown in the scallop beds of Georges Bank, but the knowledge we may gain from the monitoring and research carried out within their limits (Smith and Rago, 2004).

The fisheries examples cited here show us that management that focuses on long-term use of a resource, rather than on quick returns, can succeed if the stock is given a chance for recovery. Research and monitoring of the target species and its environment are necessary at all steps, but especially under anomalous circumstances. Good knowledge of the environment of a target species is probably the best tool for understanding the dynamics of its populations.

ACKNOWLEDGEMENTS

This study has been funded by the Nordic Council of Ministers as part of the project "Bottom trawling and scallop dredging in the Arctic," which is included in the Nordic Action Plan. The author wishes to thank the following: G. S. Bragason and H. Eiríksson, who have patiently answered an immense number of questions, L. Gunnarsdóttir for collecting all the papers and reports needed for this review, W. Vadseth for doing a manual search in the old files at the IMR Library in Bergen and E. Hopland for providing scallop catch data from the Norwegian Directorate of Fisheries. R. Frandsen read an earlier version of the manuscript on the Greenland fishery, H. Siegstad provided catch data from Greenland, J. Engelstoft supplied some information on management of the Greenlandic scallop stock, G. G. Þórarinsdóttir helped with some of the Danish reports and K. Gunnarsson gave good advice. A. A. Ramos and J. P. Jónasson read the manuscript, and V. Chosson helped me to produce the maps. Thanks are also due to the editor, Alan Southward, and three anonymous referees; their comments were very useful for improving the manuscript.

REFERENCES

Anonymous (1973). Rannsóknir á hörpudiski. *Hafrannsóknir* **4**, 23.
Anonymous (1974). Hörpudisksrannsóknir. *Hafrannsóknir* **5**, 19–20.
Anonymous (1981). Hörpudiskur. *Hafrannsóknir* **23**, 20.
Anonymous (1982). Hörpudiskur (*Pecten*). *Hafrannsóknir* **25**, 17–18.
Anonymous (1988). "Rapport fra Arbeidsgruppe Som har Vurdet Aktuelle Reguleringsmetoder for Haneskjellskraping i Svalbardsonen". JHY1–880718D Fiskeridepartementen, Oslo, Fiskeridirektoratet, Bergen.
Anonymous (1990). Nytjastofnar sjávar og umhverfisþættir 1990. Aflahörfur 1991. *Hafrannsóknastofnun Fjölrit* **21**, 57–59.
Anonymous (1990b). Haneskjell. *Fisken og Havet.* Ressursoversikt for 1990, 66–67.
Anonymous (1992). Haneskjell. *Fisken og Havet.* Ressursoversikt for 1992, 61.

Anonymous (1994). Hörpudiskur. *Hafrannsóknir* **46**, 31.
Anonymous (1995). Haneskjell. *Fisken og Havet.* Ressuroversikt for 1995, 65–66.
Anonymous (1996). Haneskjell. *Fisken og Havet.* Ressuroversikt for 1996, 92.
Anonymous (1997). Haneskjell. *Fisken og Havet.* Ressuroversikt for 1997, 94.
Anonymous (1998). "Grøndlandske Fisk, Rejer, Krabber og Muslinger—En Status over Ristige Ressourcer." Teknisk rapport nr 17. Grønlands Naturinstitut, Nuuk.
Anonymous (1998b). Haneskjell. *Fisken og Havet.* Ressuroversikt for 1998, 115.
Anonymous (2003). Nytjastofnar sjávar 2002/2003. Aflahörfur fiskveiðiárið 2003/2004. State of marine stocks in Icelandic waters 2002/2003. Prospects for the quota year 2003/2004. *Hafrannsóknastofnunin Fjölrit* **97**, 173.
Ansell, A., Dao, J-C., Lucas, A., Mackie, L. A. and Morvan, C. (1988). Reproductive and genetic adaptation in natural and transplant populations of the scallop, *Pecten maximus*, in European waters. Report to the European Commission on research carried out under ECC Scientific Cooperation Contract No. ST2J-0058–1-UK(CD) (mimeo).
Ansell, A., Dao, J. C. and Mason, J. (1991). Three European scallops: *Pecten maximus, Chlamys (Aequipecten) opercularis* and *C. (Chlamys) varia. In* "Scallops: Biology, Ecology and Aquaculture" (S. Shumway, ed.), pp. 715–752. Elsevier, Amsterdam.
Arsenault, D. and Himmelman, J. (1998). Spawning of the Iceland scallop (*Chlamys islandica* Müller, 1776) in the Northern Gulf of St. Lawrence and its relationship to temperature and phytoplankton abundance. *The Veliger* **41**, 180–185.
Aschan, M. (1988). The effect of Iceland scallop (*Chlamys islandica*) dredging at Jan Mayen and in the Spitsbergen area. *International Council for the Exploration of the Sea C.M. 1988/K:16* Shellfish Committee (mimeo).
Aschan, M. (1991). "Effects of Iceland scallop dredging on benthic communities in the Northeast Atlantic." ICES Benthic Ecology Working Group. Halifax, Canada, 6–10 May.
Aschan, M. and Sundet, J. (1990). Growth and recruitment in different populations of the Iceland scallop *Chlamys islandica*, in the Northeast Atlantic. *ICES Shellfish Commitee Symposium* **25**, 1–23.
Bergeron, J. P. and Buestel, D. (1979). L'Aspartate transcarbamylase, indice de l'activite sexuelle de la coquille Saint-Jacques (*Pecten maximus* L.). Premier resultats. *In* "Cyclic Phenomena in Marine Plants and Animals. Proceedings of the 13th European Marine Biology Symposium" (E. Naylor and R. G. Hartnoll, eds), pp. 301–308. Pergammon Press, Oxford.
Bergh, Ø. and Strand, Ø. (2001). Great scallop, *Pecten maximus*, research and culture strategies in Norway: A review. *Aquaculture International* **9**, 305–318.
Bernard, F. R. (1983). Catalogue of the living bivalvia of the eastern Pacific Ocean: Bering Strait to Cape Horn. *Canadian Special Publications of Fisheries and Aquatic Sciences* **61**, 1–102.
Beukers-Stewart, B., Mosley, M. and Brand, A. (2003). Population dynamics and predictions in the Isle of Man fishery for the great scallop (*Pecten maximus*, L.). *ICES Journal of Marine Science* **60**, 224–242.
Brand, A. (1991). Scallop ecology: Distributions and behaviour. *In* "Scallops: Biology, Ecology and Aquaculture" (S. Shumway, ed.), pp. 517–584. Elsevier, Amsterdam.
Brand, A. (2000). North Irish Sea scallop fisheries: Effects of 60 years dredging on scallop populations and the environment. *In* "A Workshop Examining Potential Fishing Effects on Population Dynamics and Benthic Community Structure of Scallops with Emphasis on the Weathervane Scallop *Patinopecten caurinus* in

Alaskan Waters," pp. 37–43. Alaska Department of Fish and Game, Division of Commercial Fisheries, Special Publication 14, Juneau, Alaska.
Brand, A., Paul, J. D. and Hoogester, J. N. (1980). Spat settlement of the scallops *Chlamys opercularis* (L.) and *Pecten maximus* (L.) on artificial collectors. *Journal of the Marine Biological Association of the U.K.* **60**, 379–390.
Buch, E. (2000). A Monograph on the Physical Oceanography of the Greenland Waters. Scientific report 00–12. Danish Metereological Institute, Copenaghen.
Caddy, J. F. and Defeo, O. (2003). Enhancing or restoring the productivity of natural populations of shellfish and other marine invertebrate resources. *FAO Fisheries Technical Paper* **448**, 159.
Cano, J. and Garcia, T. (1985). Scallop fishery in the coast of Málaga, SE Spain. *In* "5th International Pectinid Workshop, La Coruña (Spain), 6–10 May 1985." (mimeo).
Cano, J., Campos, M. J., Román, G., Vazquez, M. C., Garcia, T., Fernandez, L. and Presas, M. C. (1999). Pectinid settlement on collectors in Fuengirola, Málaga, South Spain. *In* "12th International Pectinid Workshop, Bergen (Norway), 5–11 May 1999. Book of Abstracts," pp. 70–71.
Claereboudt, M. and Himmelman, J. (1996). Recruitment, growth and production of giant scallops (*Placopecten magellanicus*) along an environmental gradient in Baie des Chaleurs, eastern Canada. *Marine Biology* **124**, 661–670.
Cochard, J. C. (1985). Observations sur la viabilite des oeufs de la coquille Saint-Jaques en Rade de Brest. 5th Pectinid Workshop, La Coruña, Spain, 6–10 May, 1985. (mimeo).
Dall, W. H. (1898). Contributions to the Tertiary fauna of Florida. Part IV. I. Prionodesmacea, II. Teleodesmacea. *Transactions of the Wagner Free Institute of Science of Philadelphia* **3**, 571–947.
Dickie, L. (1955). Fluctuations in abundance of the giant scallop, *Placopecten magellanicus* (Gmelin), in the Digby area of the Bay of Fundy. *Journal of the Fisheries Research Board of Canada* **12**, 797–857.
Dickie, L. and Medcof, J. (1963). Causes of mass mortalities of scallops (*Placopecten magellanicus*) in the Southwestern Gulf of St Lawrence. *Journal of the Fisheries Research Board of Canada* **20**, 451–482.
Dupuoy, H., de Kergariou, G. and Latrouite, D. (1983). L'explotation de la coquille Saint-Jacques *Pecten maximus* (L.) en France. 2eme partie: Evaluation et gestion du stock de la baie de Saint-Brieuc. *Science et Peche, Bulletin de l'Institut des Peches Maritimes* **331**, 3–11.
Eiríksson, H. (1970a). Athuganir á Hörpuðiski, *Chlamys islandica*, Müller, árið 1969. *Hafrannsóknir* **2**, 57–68.
Eiríksson, H. (1970b). Hörpudiskaleit í Breiðafirði. *Ægir* **63**, 334–339.
Eiríksson, H. (1970c). Hörpudiskaleit í Húnaflóa. *Ægir* **63**, 392–394.
Eiríksson, H. (1971a). Hörpudisksrannsóknir 1970. *Hafrannsóknir* **3**, 65–67.
Eiríksson, H. (1971b). Hörpudiskaleit á Vestfjörðum. *Ægir* **64**, 116–119.
Eiríksson, H. (1986). Hörpudiskurinn. *Hafrannsóknir* **35**, 5–40.
Eiríksson, H. (1997). The molluscan fisheries of Iceland. *National Ocean and Atmospheric Administration Technical Report NMFS* **129**, 39–47.
Eiríksson, H. and Nicolajsen, A. (1984). Fiskeri efter Kammuslinger ved Grønland Undersøgelse i Nuuk-området, Grønlands Hjemmestyre, Konsulentformidlingen, KIS & Grønlands Fiskeri-og Miljøundersøgelser, Nuuk.
Engelstoft, J. (1995). Status for Kammuslingebestanden i Nuuk 1994. Pinngortitaleriffik (Grønlands Naturinstitut), Nuuk.

Engelstoft, J. (2000). Omplanting af Kammuslinger, *Chlamys islandica*, ved Nuuk. Pinngortitaleriffik (Grønlands Naturinstitut), Nuuk.
Eydal, A. (2003). Áhrif næringarefna á tegundasamsetningu of fjölda svifþörunga í Hvalfirði. (Impact of nutrients on phytoplankton sucession in Hvalfjörður). *Hafrannsoknastofnunin Fjölrit* **99**, 44.
Gibson, F. (1956). Escallops (*Pecten maximus*) in Irish waters. *Scientific Proceedings of the Royal Dublin Society* **27**, 253–270.
Grau, G. (1959). Pectinidae of the Eastern Pacific. Allan Hancock Pacific Expeditions **23**, p. 308, University of Southern California Press, Los Angeles.
Greve, L. and Samuelsen, T. (1970). A population of *Chlamys islandica* (O. F. Müller) found in western Norway. *Sarsia* **45**, 17–24.
Gruffydd, L. (1976). The development of the larva of *Chlamys islandica* in the plankton and its salinity tolerance in the laboratory (Lamellibranchia, Pectinidae). *Astarte* **8**, 60–67.
Hall-Spencer, J. and Moore, P. (2000). Scallop dredging has profound, long-term impacts on maerl habitats. *ICES Journal of Marine Science* **557**, 1407–1415.
Hamilton, L., Lyster, P. and Otterstad, O. (2000). Social change, ecology and climate in 20th century Greenland. *Climatic Change* **47**, 193–211.
Hansen, K. and Nedreaas, K. (1986). Measurements of Iceland scallop (*Chlamys islandica* Müller) in the Spitzbergen and Bear Island regions. *International Council for the Exploration of the Sea C.M. 1986/K:26* (mimeo).
Hart, D. (2003). Yield- and biomass-per-recruit analysis for rotational fisheries, with an application to the Atlantic sea scallop (*Placopecten magellanicus*). *Fisheries Bulletin* **101**, 44–57.
Harvey, M. and Bourget, E. (1995). Experimental evidence of massive accumulation of marine bivalve larvae on filamenteous epibenthic structures. *Limnology and Oceanography* **40**, 94–104.
Harvey, M., Bourget, E. and Miron, G. (1993). Settlement of Iceland scallop *Chlamys islandica* spat in response to hydroids and filamenteous red algae: Field observations and laboratory experiments. *Marine Ecology Progress Series* **99**, 283–292.
Hauksson, E. and Gunnarsson, K. (1973). Nokkrar athuganir á fjörum við norðaustanverðan Breiðafjörð. Óbirt skýrsla (unpublished report).
Hickson, J., Johnson, A., Heaton, T. and Balson, P. (1999). The shell of the Queen scallop *Aequipecten opercularis* (L.) as a promising tool for palaeoenvironmental reconstruction: Evidence and reasons for equilibrium stable-isotope incorporation. *Palaeogeography, palaeoclimatology, palaeoecology* **154**, 325–337.
Hickson, J., Johnson, A., Heaton, T. and Balson, P. (2000). Late Holocene environment of the southern North Sea from the stable isotopic composition of Queen scallop shells. *Palaeontologia Electronica* **3**, 11.
Hopkins, D. M. (1967). The Cenozoic history of Beringia—A synthesis. *In* "The Bering Land Bridge" (D. M. Hopkins, ed.), pp. 451–484. Standford University Press, Standford, California.
Jensen, A. S. (1912). Lamellibranchiata. *In* "The Danish Ingolf Expedition," Vol. II, pp. 15–17. Bianco Luno, Copenhagen.
Johnson, A., Hickson, J., Swan, J., Brown, M., Heaton, T., Chenery, S., and Balson, P. (2000). The Queen scallop *Aequipecten opercularis*: A new source of information on late Cenozoic marine environments in Europe. *In* "The Evolutionary Biology of the Bivalvia" (E. Herper, J. Taylor and J. Crame, eds), pp. 425–439. Geological Society, Special Publications, London.

Jónasson, J. (2005). The effect of environment and fisheries on the status of Iceland scallop University of Iceland, Reykjavík, MSc thesis.

Jónasson, J., Thorarinsdottir, G., Eiríksson, H. and Marteinsdottir, G. (2004). Temperature tolerance of Iceland scallop, *Chlamys islandica* (O. F. Müller) under controlled experimental conditions. *Aquaculture Research* **35**, 1405–1414.

Kristmundsson, A., Eydal, M., Bambir, S. and Helgason, S. (2004). Sníkjudýr í hörpuskel (*Chlamys islandica*) við Ísland. *In* "Liffræði - vaxandi visindi. Afmælisráðstefna Líffræðifélags íslands og Liffræðistofnunar Háskolans," p. 68. öskju, 19. og 20. Nóvember 2004.

Kruse, G., Barnhart, J., Rosenkranz, G., Funk, F. and Pengilly, D. (2000). History and development of the scallop fishery in Alaska. *In* "A Workshop Examining Potential Fishing Effects on Population Dynamics and Benthic Community Structure of Scallops with Emphasis on the Weathervane Scallop *Patinopecten caurinus* in Alaskan Waters," pp. 6–12. Alaska Department of Fish and Game, Division of Commercial Fisheries, Special Publication 14, Juneau.

Larsen, S. (1991). Landing live Iceland scallops. *Canadian Translations of Fishery and Aquatic Sciences* **5530**, pp. 34.

Lubinsky, I. (1980). Marine bivalve molluscs of the Canadian central and eastern Arctic: Fauna composition and zoogeography. *Canadian Bulletin of Fisheries and Aquatic Sciences* **207**, 1–111.

Maack, M. (1987). Winter tolerance of *Pecten maximus* (L.). Bachelor Thesis, Department of Zoology, University of Göteborg.

MacDonald, B. and Thompson, R. (1985). Influence of temperature and food availability on the ecological energetics of the giant scallop *Placopecten magellanicus*. I. Growth rates of shell and somatic tissue. *Marine Ecology Progress Series* **25**, 279–294.

Mackie, L. A. (1986). Aspects of the reproductive biology of the scallop *Pecten maximus*. Ph.D. Thesis, Heriot-Watt University, Edinburgh.

MacNeil, F. S. (1967). Cenozoic Pectinids of Alaska, Iceland, and other northern regions. *U.S. Geological Survey Professional Paper* **553**, 1–57.

Madsen, F. (1949). Marine Bivalvia. *The Zoology of Iceland* **IV**, 1–116.

Mason, J. (1958). The breeding of the scallop, *Pecten maximus* (L.) in Manx waters. *Journal of the Marine Biological Association of the U.K* **37**, 653–671.

Mason, J. (1983). Scallop and Queen Fisheries on the British Isles. Fishing News Books Ltd., Surrey.

Masuda, K. (1986). Notes on origin and migration of Cenozoic pectinids in the Northern Pacific. *Palaeontological Society of Japan, Special Publications* **29**, 95–110.

Moore, E. (1984). Tertiary marine pelecypods of California and Baja California: Propeamusidae and Pectinidae. *U.S. Geological Survey Professional Paper* **1228-B**, 1–112.

Murawski, S. A., Brown, R., Lai, H.-L., Rago, P. J. and Hendrickson, L. (2000). Large-scale closed areas as a fishery-management tool in temperate marine systems: The Georges Bank experience. *Bulletin of Marine Science* **66**, 775–798.

Naidu, K. S. (1988). Estimating mortality rates in the Iceland scallop, *Chlamys islandica* (O. F. Müller). *Journal of Shellfish Research* **7**, 61–71.

Naidu, K. S., Cahill, F., Veitch, P. and Seward, E. (1997). Synopsis of the 1996 fishery for Iceland scallops in NAFO Div. 4R (Strait of Belle Isle). *Canadian Science Advisory Secretariat, Research Document* **97**, 6.

Nicolajsen, A. (1984). En populationdynamisk beskrivelse af kammuslingen *Chlamys islandica* i grønlandsk farvand, Nuuk området. Grønlands Hjemmestyre, Konsulentformidlingen, KIS & Grønlands Fiskeri-og Miljøundersøgelser, Nuuk.

Nicolajsen, A. (1985a). Kammuslingeundersøgelse ved Paamiut, Qarqortoq og Nanortalik i oktober 1984. Grønlands Hjemmestyre, Konsulentformidlingen, KIS & Grønlands Fiskeri-og Miljøundersøgelser, Nuuk.

Nicolajsen, A. (1985b). Kammuslingeundersøgelse ved Maniitsoq og Kangerdluarssugssuaq i juni-juli 1985. Grønlands Hjemmestyre, Konsulentformidlingen, KIS & Grønlands Fiskeri- og Miljøundersøgelser, Nuuk.

Nicolajsen, Á. and Conan, G. Y. (1987). Assessment by geostatistical techniques of populations of Iceland scallop (*Chlamys islandica*) in the Barents Sea. *International Council for the Exploration of the Sea c.m. 1987/K:14*, Shellfish Committee, (mimeo).

Ockelmann, W. (1958). The zoology of East Greenland. *In* "Meddelelser om Grønland" (M. Degerbøl, Ad. S. Jensen, R. Spärck and G. Thorson, eds), **122,** pp. 1–256. Bianco Luno, København.

Oganesyan, S. (1994). Biological characteristics and time of Icelandic scallop (*Chlamys islandica*) spawning on the Kanin Bank and off the Bear Island-Spitsbergen. *International Council for the Exploration of the Sea C.M. 1994/ K:45* (mimeo).

Oresanz, J., Parma, A. and Iribarne, O. (1991). Population dynamics and management of natural stocks. *In* "Scallops: Biology, Ecology and Aquaculture" (S. Shumway, ed.), pp. 625–714. Elsevier, Amsterdam.

Óskarsson, I. (1952). "Skeldýrafána Íslands I. Samlokur í sjó." Prentsmiðjan Leiftur, Hf., Reykjavík.

Oterhals, L. M. (1988). Erfaringar ved ombordproduksjon av haneskjell. *In* "Haneskjell Næringa-erfaringer of Framtidsutsikter," pp. 40–61.Tromsø 20–21 Januari 1988, Fiskeri Sjeferna i Nordnorge.

Ott, S. O., Mølgaard, J. and Christensen, T. (1987). Fiskeindustri, regional udvikling og beskæftigelse – har Grønland arvet en rød reje? Specialeopgave TEK-SAM Roskilde Universitets Center, April 1987.

Parsons, G., Dadswell, M. and Rodstrom, E. (1991). Scandinavia. *In* "Scallops: Biology, Ecology and Aquaculture" (S. Shumway, ed.), pp. 763–776. Elsevier, Amsterdam.

Paul, J. D. (1981). Natural settlement and early growth of spat of the queen scallop *Chlamys opercularis* (L.) with reference to formation of the first growth ring. *Journal of Mollusc Studies* **47**, 53–58.

Paulet, Y., Lucas, A. and Gerard, A. (1988). Reproduction and larval development in two *Pecten maximus* (L.) populations from Brittany. *Journal of Experimental Marine Biology and Ecology* **119**, 145–156.

Pedersen, S. (1986). "Status over Kammuslingefiskeriet i Nuuk-området, November 1986." Grønlands Hjemmestyre, Konsulentformidlingen, KIS & Grønlands Fiskeri-og Miljøundersøgelser, Nuuk.

Pedersen, S. (1987a). "Forsogsfiskeri efter Kammuslinger pa de Vestgrønlandske Banker, 1986." Grønlands Fiskeri-og Miljøundersøgelser, Kobenhavn.

Pedersen, S. (1987b). "Kammuslinger ved Vestgrønland." Grønlands Fiskeri og Miljøundersøgelser, Kobenhavn.

Pedersen, S. (1987c). "Kammuslinger ved Vestgrønland. Fiskeri efter Kammuslinger ved Attu 8–21. Juni 1986." Grønlands Fiskeri og Miljøundersøgelser, København.

Pedersen, S. (1988a). "Kammuslinger, *Chlamys islandica*, ved Vestgrønland." Grønlands Fiskeri- og Miljøundersøgelser, København.

Pedersen, S. (1988b). Inshore scallop resources, *Chlamys islandica*, in the Nuuk area West Greenland. *International Council for the Exploration of the Sea C.M. 1988/ K:17* (mimeo).
Pedersen, S. (1988c). "Kammuslingeressourcer ved Aasiaat-Kangaatsiaq-Attu, September 1988." Grønlands Fiskeriundersøgelser, København.
Pedersen, S. (1988d). "Kammuslingeressourcer ved Fiskenæsset, 1988." Grønlands Fiskeriundersøgelser, København.
Pedersen, S. (1988e). "Kammuslingeressourcer ved Sisimiut, 1988." Grønlands Fiskeriundersøgelser, København.
Pedersen, S. (1988f). "Kammuslinger ved Grønland. Kammuslingefiskeri i Maniitsoq-kommune. Status 1987." Grønlands Fiskeriundersøgelser, Kobenhavn.
Pedersen, S. (1989). Inshore scallop resources, *Chlamys islandica*, in the Nuuk area West Greenland. *North Atlantic Fisheries Organization SCR Doc. 89/20*, Serial no. N1596.
Pedersen, S. (1994). Population parameters of the Iceland scallop (*Chlamys islandica* [Müller]) from West Greenland. *Journal of Northwest Atlantic Fishery Science* **16**, 75–87.
Pedersen, S. and Bøje, J. (1987). Kammuslinger ved Vestgrønland. Underøgelse af Redskabeffektivitet for Kammuslingeskrabere. Grønlands Fiskeri-og Miljøundersøgelser, Kobenhavn.
Peña, J. B., Canales, J., Adsuara, J. M. and Sos, M. A. (1996). Study of seasonal settlements of five scallop species in the western Mediterranean. *Aquaculture International* **4**, 253–261.
Peña, J. B., Canales, J. and Mestre, S. (1993). Preliminary results on the reproductive cycle and the stored energy in *Aequipecten opercularis* L. off the coast of Castellon (E Spain). *In* "4th National Congress on Aquaculture. Book of Abstracts" (A. Cervino, A. Landin, A. de Coo, A. Guerra and M. Torre, eds), pp. 323–328. Vilanova de Arousa, Galicia (Spain), 21–24 September 1993. Centro de Investigaciones Marinas, Pontevedra (Spain).
Pérès, J. M. (1985). History of the Mediterranean biota and the colonization of the depths. *In* "The Western Mediterranean" (R. Margalef, ed.), pp. 198–232. Pergamon Press, Oxford.
Pérès, J. M. and Picard, J. (1964). Noveau Manuel de bionomie benthique de la mer Méditerranée. *Recueil des Travaux de la Station marine d'Endoume* **47**, 5–137.
Posselt, H. (1898). Grønlands Brachipoder og Bloddyr. *In* "Meddelelser om Grønland," 23, pp. 1–298. Bianco Luno, Kjøbenhavn.
Richards, H. (1962). Studies on the marine Pleistocene. Part I. The marine Pleistocene of the Americas and Europe. Part II. The marine Pleistocene mollusks of eastern North America. *Transactions of the American Philosophical Society* **52**, 1–141.
Robinson, S., Martin, J., Chandler, R. and Parsons, G. (1999). An examination of the linkage between the early life history processes of the sea scallop and local hydrographic characteristics. *Journal of Shellfish Research* **18**, 314.
Rubach, S. and Sundet, J. (1987). Ressurskartlegging av Haneskjell (*Chlamys islandica* [O. F. Müller]) ved Jan Mayen og i Svalbardsonen i 1986. Universitetet i Tromsø, Institutt for fiskerifag. Tromsø.
Siegstad, H. (2000). Denmark/Greenland research report for 1999. *North Atlantic Fisheries Organization SCS Doc. 00/22,* pp. 4.
Siegstad, H. (2001). Denmark/Greenland research report for 2000. *North Atlantic Fisheries Organization SCS Doc. 01/21,* pp. 6.

Simpson, J. (1910). Notes on some rare Mollusca from the North Sea and Shetland-Faeroe Channel. *Journal of Conchology* **12**, 109–115.
Sinclair, M., Mohn, R., Robert, G. and Roddick, D. (1985). Considerations for the effective management of Atlantic scallops. *Canadian Technical Report on Fisheries and Aquatic Sciences.* No. 1382.
Skreslet, S. (1973). Spawning in *Chlamys islandica* (O. F. Müller) in relation to temperature variations caused by vernal meltwater discharge. *Astarte* **6**, 9–14.
Skreslet, S. and Brun, E. (1969). On the reproduction of *Chlamys islandica* (O. F. Müller) and its relation to depth and temperature. *Astarte* **2**, 1–6.
Smith, S. and Rago, P. (2004). Biological reference points for sea scallops (*Placopecten magellanicus*): The benefits and costs of being nearly sessile. *Canadian Journal of Fisheries and Aquatic Sciences* **61**, 1338–1354.
Sundet, J. (1985). Are there potential resources of Iceland scallops (*Chlamys islandica*) in the Barents Sea? *International Council for the Exploration of the Sea C.M. 1985/K:43* (mimeo).
Sundet, J. (1996). Kartlegging av haneskjellforekomster i Sør-Varanger. Rapport 1/1996. Fiskeriforskning, Tromsø.
Sundet, J. (1997). Toktrapport fra haneskjellundersøkelser i Ytre Troms 24.–25. Juni 1997. Rapport 12/1997. Fiskeriforskning, Tromsø.
Sundet, J. (2003). Toktrapport fra haneskjellundersøkelser i Ytre Troms 1.–2. Juli 2003. Rapport 10/2003. Fiskeriforskning, Tromsø.
Sundet, J. and Vahl, O. (1981). Seasonal changes in dry weight and biochemical composition of the tissues of sexually mature and in mature Iceland scallops, *Chlamys islandica. Journal of the Marine Biological Association of the U.K.* **61**, 1001–1010.
Sundet, J. and Lee, J. (1984). Seasonal variation in gamete development in the Iceland scallop, *Chlamys islandica. Journal of the Marine Biological Association of the U.K.* **64**, 411–416.
Tammi, K., Turner, W. and Rice, M. (1997). The influence of temperature on spawning and spat collection of the bay scallop, *Argopecten irradians*, in southeastern Massachussets waters, USA. *Journal of Shellfish Research* **16**, 349.
Tang, S.-F. (1941). The breeding of the escallop (*Pecten maximus* (L.)) with a note on the growth rate. *Proceedings and Transactions of the Liverpool Biological Society* **59**, 9–28.
Taylor, A. and Venn, T. (1978). Growth of the queen scallop, *Chlamys opercularis*, from the Clyde Sea area. *Journal of the Marine Biological Association of the U.K.* **58**, 605–621.
Theroux, R. and Wigley, R. (1983). Distribution and abundance of East Coast bivalve mollusks based on specimens in the National Marine Fisheries Service Woods Hole collection. *National Oceanic and Atmospheric Administration Technical Report* NMFS SSRF-768.
Thorarinsdottir, G. (1991). The Iceland scallop, *Chlamys islandica* (O. F. Müller) in Breidafjordur, West Iceland. I. Spat collection and growth during the first year. *Aquaculture* **97**, 13–23.
Thorarinsdottir, G. (1993). The Iceland scallop, *Chlamys islandica* (O. F. Müller) in Breidafjordur, West Iceland. II. Gamete development and spawning. *Aquaculture* **110**, 87–96.
Thorarinsdottir, G. (1994). The Iceland scallop, *Chlamys islandica* (O. F. Müller) in Breidafjordur, West Iceland. III. Growth in suspended culture. *Aquaculture* **120**, 295–303.

Tremblay, M. and Sinclair, M. (1988). The vertical and horizontal distribution of sea scallop (*Placopecten magellanicus*) larvae in the Bay of Fundy in 1984 and 1985. *Journal of Northwest Atlantic Fishery Science* **8**, 43–53.

Ursin, E. (1956). Distribution and growth of the queen, *Chlamys opercularis* (Lamellibranchiata) in Danish and Faroese waters. *Meddelelser fra Danmarks Fiskeri- og Havundersøgelser* **13**, 1–32.

Vahl, O. (1978). Seasonal changes in oxygen consumption of the Iceland scallop (*Chlamys islandica* O. F. Müller) from 70° N. *Ophelia* **17**, 143–154.

Vahl, O. (1980). Seasonal variations in seston and in the growth rate of the Iceland scallop, *Chlamys islandica* (O. F. Müller) from Balsfjord, 70° N. *Journal of Experimental Marine Biology and Ecology* **48**, 195–204.

Vahl, O. (1982). Long-term variations in recruitment of the Iceland scallop, *Chlamys islandica* from northern Norway. *Netherlands Journal of Sea Research* **16**, 80–87.

Venvik, T. (1979). Fangstundersøkelser på haneskjellfeltene ved Bjørnøya. *Fiskets Gang* **65**, 650–651.

Venvik, T. and Vahl, O. (1979). "Muligheter og Begrensninger for Fangst og Produksjon av Haneskjell." Serie E: Fiskeriteknologi nr. 2/79. Institutt for fiskerifag, Universitetet i Tromsø.

Venvik, T. (1982). Muligheter for utnyttelse av skjell- og blötdyrressurser i Nord-Norge Sluttrapport fra Skjellprosjektet i Nord-Norge, 20 p. Fiskerisjefene i Nord-Norge.

Wagner, F. J. E. (1970). Faunas of the Pleistocene Champlain Sea. *Bulletin of the Geological Survey of Canada* **181**, 1–104.

Waller, T. (1991). Evolutionary relationships among commercial scallops (Mollusca: Bivalvia: Pectinidae). *In* "Scallops: Biology, Ecology and Aquaculture" (S. Shumway, ed.), pp. 1–74. Elsevier, Amsterdam.

Wiborg, K. (1962). Haneskjellet, *Chlamys islandica* (O. F. Müller) og dets utbredelse i noen nordnorske fjorder. *Fisken og Havet* **3**, 17–23.

Wiborg, K. (1963a). Some observations on the Iceland scallop *Chlamys islandica* (Müller) in Norwegian waters. *Fiskeridirektoratets Skrifter Serie Havundersøkelser* **13**, 38–53.

Wiborg, K. (1963b). Rapport om skjellundersøkelser med F/F "Asterias" 23. Mai–22. Juni 1963.

Wiborg, K. (1970). Utbredelse av haneskell (*Chlamys islandica* Müller) på Bjørnøybankene. *Fiskets Gang* **43**, 782–788.

Wiborg, K. and Bøhle, B. (1968). Forekomster av matnyttige skjell (muslinger i norske kystfarvann (med et tilleg om sjøsnegler). *Fiskets Gang* **54**, 149–161.

Wiborg, K., Hansen, K. and Olsen, H. (1974). Haneskjell (*Chlamys islandica* Müller) ved Spitsbergen og Bjørnøya–Undersølkelser i 1973. *Fiskets Gang* **60**, 209–217.

Wilson, J. (1987). Spawning of *Pecten maximus* (Pectinidae) and the artificial collection of juveniles in two bays in the west of Ireland. *Aquaculture* **61**, 99–111.

Young, P., Martin, R., McLoughlin, R. and West, G. (1989). Variability in spatfall and recruitment of commercial scallops (*Pecten fumatus*) in Bass Strait. *In* "Proceedings of the Australasian Scallop Workshop" (M. C. L. Dredge, W. F. Zacharin and L. M. Joll, eds), pp. 80–91.

Are Larvae of Demersal Fishes Plankton or Nekton?

Jeffrey M. Leis

Ichthyology, Australian Museum, Sydney, Australia

A pelagic larval stage is found in nearly all demersal marine teleost fishes, and it is during this pelagic stage that the geographic scale of dispersal is determined. Marine biologists have long made a simplifying assumption that behaviour of larvae—with the possible exception of vertical distribution—has negligible influence on larval dispersal. Because advection by currents can take place

ADVANCES IN MARINE BIOLOGY VOL 51

0065-2881/06 $35.00
DOI: 10.1016/S0065-2881(06)51002-8

over huge scales during a pelagic larval stage that typically lasts for several days to several weeks, this simplifying assumption leads to the conclusion that populations of marine demersal fishes operate over, and are connected over, similar huge scales. This conclusion has major implications for our perception of how marine fish populations operate and for our management of them. Recent (and some older) behavioural research—reviewed here—reveals that for a substantial portion of the pelagic larval stage of perciform fishes, the simplifying assumption is invalid. Near settlement, and for a considerable portion of the pelagic stage prior to that, larvae of many fish species are capable of swimming at speeds faster than mean ambient currents over long periods, travelling tens of kilometres. Only the smallest larvae of perciform fishes swim in an energetically costly viscous hydrodynamic environment (i.e., low Reynolds number). Vertical distribution is under strong behavioural control from the time of hatching, if not before, and can have a decisive, if indirect, influence on dispersal trajectories. Larvae of some species avoid currents by occupying the epibenthic boundary layer. Larvae are able to swim directionally in the pelagic environment, with some species apparently orientating relative to the sun and others to settlement sites. These abilities develop relatively early, and ontogenetic changes in orientation are seemingly common. Larvae of some species can use sound to navigate, and others can use odour to find settlement habitat, at least over small scales. Other senses may also be important to orientation. Larvae are highly aware of their environment and of potential predators, and some school during the pelagic larval stage. Larvae are selective about where they settle at both meso and micro scales, and settlement is strongly influenced by interactions with resident fishes. Most of these behaviours are flexible; for example, swimming speeds and depth may vary among locations, and speed may vary with swimming direction. In direct tests, these behaviours result in dispersal different from that predicted by currents alone. Work with both tropical and temperate species shows that these behaviours begin to be significant relatively early in larval development, but much more needs to be learned about the ontogeny of behaviour and sensory abilities in larvae of marine fishes. As a preliminary rule of thumb, behaviour must be taken into account in considerations of dispersal after the preflexion stage, and vertical distribution behaviour can influence dispersal from hatching. Larvae of perciform fishes are close to being planktonic at the start of the pelagic period and are clearly nektonic at its end, and for a substantial period prior to that. All these things differ among species. Larvae of clupeiform, gadiform and pleuronectiform fishes may be less capable behaviourally than perciform fishes, but this remains to be confirmed. Clearly, these behaviours, along with hydrography, must be included in modelling dispersal and retention and may provide explanations for recent demonstrations of self-recruitment in marine fish populations. Current work is directed at understanding the ontogeny of the gradual transition from planktonic to nektonic behaviour. Although it is clear that larvae of perciform fishes have the ability to strongly influence their dispersal trajectories, it is less clear whether or how these abilities are applied.

1. INTRODUCTION

The vast majority of demersal teleost fishes have a pelagic larval stage (Moser *et al.*, 1984; Leis, 1991), and this has major implications for the way these fish populations operate and for human management of them. Marine demersal fishes are generally considered to have open populations, that is, the young recruiting at any place will be the offspring, not of the adults living there, but of those from some other location, perhaps many kilometres away (Sale, 1991b; Caley *et al.*, 1996; Johnson, 2005). Thus, populations of marine demersal fishes in different locations may be connected by dispersal between them, and the extent to which these populations are linked is termed *connectivity* (Palumbi, 2003). It has long been thought likely that most dispersal in demersal teleosts takes place during the pelagic larval stage before it ends rather dramatically by settlement into a demersal way of life (Armsworth *et al.*, 2001; Kinlan and Gaines, 2003). Therefore, it is the pelagic larval stage, rather than the demersal adult stage, that sets the scale for population connectivity and for the geographic size of fish populations (Cowen, 2002; Kinlan and Gaines, 2003; Sale, 2004). Further, most mortality is thought to take place during this pelagic stage (e.g., Cushing, 1990; Bailey *et al.*, 2005), which also has implications for the spatial scales over which demographically meaningful dispersal can take place (Cowen *et al.*, 2000; Palumbi, 2001).

It is important to distinguish between genetic connectivity and demographic connectivity (Leis, 2002; Palumbi, 2003). Genetic connectivity is the movement of genes between populations. The genomes of genetically connected populations will differ little, if at all, but only a few individuals per generation moving between populations will prevent genetic differences from forming through drift (Shulman, 1998; Palumbi, 2003). A handful of recruits per generation will not, however, be demographically significant; it will not maintain a fishery, for example. In contrast, demographic connectivity is the movement of individuals between populations in numbers large enough to be demographically significant. What constitutes "significance" is context dependent and may be larger for fishery managers than for ecologists. In short, genetic connectivity is of evolutionary and biogeographic significance, whereas demographic connectivity is of ecological and management significance. Genetic connectivity is expected to operate over larger geographic scales than demographic connectivity (Palumbi, 2001; Swearer *et al.*, 2002). It is demographic connectivity that is relevant to this review.

Understanding the scale of dispersal during the pelagic larval stage is one of the major challenges facing marine biologists (Cowen, 2002; Warner and Cowen, 2002; Sale, 2004), as it is increasingly evident that management of marine populations, which has so often failed in the recent past, must take into account the scales over which these populations are demographically connected (Palumbi, 2001; Cowen *et al.*, 2003). A wrong guess about the appropriate scale is likely to doom management to failure. Generally, dispersal has been thought to operate over very large spatial scales (hundreds of kilometres), with management scaled accordingly.

The effectiveness of marine-protected areas (MPAs) as management tools depends on the scale of demographic connectivity (Palumbi, 2001, 2003). MPAs are frequently expected to fulfil two roles: biodiversity conservation and fishery replenishment. If the demersal fish populations of an MPA are indeed open, then the MPA will be dependent on other areas for its new recruits. This MPA may, therefore, be very vulnerable to events outside its borders, and as a result, might not fulfil its biodiversity conservation role. Because this MPA exports the propagules of its open fish populations, it may be successful in its fishery replenishment role, providing recruits to fished areas outside its borders. A major question then arises about the geographic scale over which this replenishment (connectivity) takes place. Further, assuming that fish populations of this MPA can be maintained by dispersal from elsewhere, it is important to know the sources of these recruits (i.e., the geographic scale of connectivity). In contrast, if the fish populations of a second MPA are more toward the closed end of the open–closed continuum, the second MPA may be largely self-recruiting, and because it supplies most of its own young, it should be able to fulfil its biodiversity role. Because this second MPA exports few propagules, it would not fulfil a fishery replenishment role; the scale of connectivity is too small. A third MPA may fulfil both roles by having a moderate degree of self-recruitment, yet still be exporting large numbers of propagules to fished areas. As with the first MPA, a major question will arise about the geographic scale over which this replenishment will occur and about the sources of exogenous recruits.

Evidence that dispersal and demographic connectivity might be at much smaller scales than often assumed was provided for herring by Iles and Sinclair, although their synthesis proved controversial (Iles and Sinclair, 1982; Sinclair, 1988), particularly among marine scientists and managers outside Canada. Increasing evidence, however, now supports their view that populations of marine fishes are demographically structured at more modest spatial scales than has been assumed in the past, in some cases as little as tens to hundreds of metres (Marliave, 1986; Jones *et al.*, 1999, 2005; Swearer *et al.*, 1999, 2002; Paris and Cowen, 2004). Although many researchers remain sceptical (e.g., Mora and Sale, 2002), little is to be gained by polarised debate that posits populations are either strictly open or strictly closed. Rather, it is appropriate to recognise that a continuum of connectivity is closer to the truth, with populations occupying differing, time- and location-dependent positions on the continuum (Morgan, 2001; Leis, 2002). The challenge is to identify and quantify the factors that contribute to that positioning and to quantify the spatial scale of connectivity.

Traditionally, based on limited evidence, it was thought that pelagic larvae of demersal fishes had swimming abilities so limited as to be irrelevant to dispersal, and that larvae could, therefore, be treated as passive particles. With this perspective, larval dispersal was presumed to be governed by hydrographic advection; if the hydrography were known with sufficient accuracy, all that was needed to determine a dispersal trajectory was the duration of the pelagic stage. What several authors (e.g., Frank *et al.*, 1993; Roberts, 1997) called the "simplifying assumption"—that behaviour does not matter and that fish larvae are planktonic animals with little control over their trajectories, rather than strongly swimming nektonic animals—has been used by modellers to predict pelagic dispersal, and by fishery and conservation managers to set geographic boundaries and scales of management. Typically, the implicit components of the simplifying assumption have included the following:

- Larvae are poor swimmers that can only drift passively with currents.
- The only biological variable of interest during dispersal is the pelagic larval-stage duration.
- To the extent that larvae have behaviour of relevance to dispersal, all larvae behave the same, whether within or among species and regardless of location.
- Larvae disperse with currents until they are sufficiently developed (i.e., competent) to settle, and then settle onto the first bit of suitable habitat they are pushed into by the current.

In short, what one needs to know to model dispersal is limited to the currents and the pelagic larval duration. In fact, relatively little has been known about behaviour of fish larvae during their pelagic sojourn away from demersal adult habitat or during the remarkable ecological and morphological transition from pelagic animal to benthic animal that is known as settlement. In the past 10 yr, however, research on the presettlement stages of reef fishes has revealed remarkable behavioural abilities that make this simplifying assumption and its components untenable. This contribution reviews these and other findings and discusses the implications for our understanding of the structure and operation of marine fish populations and for our management of them. To many workers, acceptance of the simplifying assumption has as a corollary that populations of marine fishes are demographically open over very large spatial scales, but in fact, acceptance does allow strictly physical means of attaining and maintaining population structure at smaller scales. Rejection of the simplifying assumption does not have as a corollary that populations of marine demersal fishes are necessarily structured over small scales, but it clears the way for introducing biological inputs into considerations of dispersal and provides potential explanations for demonstrations of smaller scale population structure. Many of the conclusions of this chapter should also apply to the larvae of benthic marine invertebrates, especially decapod crustaceans, as they occupy the same pelagic environments and may have similar behavioural capabilities (Bradbury and Snelgrove, 2001; Kingsford *et al.*, 2002; Sale and Kritzer, 2003; Queiroga and Blanton, 2005).

The term *plankton* has two connotations: (1) that of small size, typically <1 cm, but up to a few centimetres; and (2) that of insignificant swimming and orientation abilities relative to ambient currents ("all the floating or feebly swimming animals or plants that are swept about by the prevailing currents" [Johnson, 1957]). Because they typically hatch from small eggs (at most a few millimetres in diameter) at a small size (a few millimetres) in a relatively undeveloped state, and settle from the water column at one to a few centimetres in length, the larvae of demersal fishes have commonly been considered planktonic, with reference to the first connotation. The second connotation has normally been assumed either in the absence of information on the behavioural capabilities of the larvae or through guilt by association. The literature is full of references to the "planktonic larval stage"—references that carry both connotations, usually in an implicit, rather than an explicit, way. This has quite naturally led to the application of the simplifying assumption to dispersal models, resulting in numerical models of hydrography with little biological content. The more neutral term "pelagic" (*sensu*, Hedgpeth, 1957), meaning resident in the water column (as opposed to benthic), can be applied to the whole of the presettlement phase of the life history of benthic teleost fishes, as it does not carry the connotations of small size and passivity. Indeed, the information reviewed here shows that a significant portion of the pelagic early life-history phase of many fishes can correctly be considered nektonic rather than planktonic.

This review focuses on the pelagic early life-history stages of demersal teleost fishes. In part, this is to avoid sterile nomenclatural debates about when the pelagic larva of a pelagic species becomes a juvenile. But, in addition, the focus here is on dispersal and behaviour. In most demersal species, the pelagic early life-history stage is usually thought to be the one that disperses, whereas in pelagic species there is no reason to believe that the larvae are more dispersive than the adults. In pelagic species, there is little doubt that behaviour is the overriding factor in dispersal by adults, so behaviour is already recognised as a factor in dispersal of pelagic fishes, but this is not the case in dispersal of many demersal fishes, which are often relatively site-attached as adults. Amongst ichthyologists, there is no generally accepted definition of larva, or when a larva becomes a juvenile (e.g., Kendall *et al.*, 1984). For the purposes of this review, a broad, ecological (rather than morphological) definition of what constitutes a larva will be used that is equivalent to the presettlement stage (*sensu*, Kingsford and Milicich, 1987): the post-hatching pelagic life-history stage of demersal fishes.

Much of the work reviewed here is based on tropical species, many of them pomacentrid damselfishes, of which there are some 340 species (Allen, 1991). This work on the behaviour of tropical fish larvae is largely a result of a 25-yr focus on recruitment issues in coral-reef systems, and questions about the means by which fish larvae locate settlement habitat in often widely scattered coral reefs (reviews in Sale, 1991a, 2002, 2004). There is, however, no reason to believe that tropical systems are fundamentally different from those in cooler waters or that pomacentrid larvae are different from larvae of other perciform fishes (although, as shown later in this chapter, differences in behavioural capabilities amongst families do exist). Although pomacentrids, unlike the majority of marine fishes, hatch from demersal, not pelagic, eggs, the available evidence indicates that larvae of other perciform fishes have similar behavioural capabilities at any given size. The emerging perspective of the importance of larval fish behaviour to dispersal based on demersal perciform fishes living at low to moderate latitudes is in contrast to that originally developed from study of higher latitude northern hemisphere fishes of other orders (e.g., Clupeiformes, Gadiformes and Pleuronectiformes). I will attempt to examine whether there really is good evidence of differences in behavioural abilities among larvae of different orders.

Size, rather than age, is the most widely used ontogenetic metric. The reason is straightforward: "a given ontogenetic state is usually reached at a uniform size for a species, regardless of how long it takes to achieve it" (Fuiman and Higgs, 1997). In most work on fish larvae, size is expressed as length, either total length (TL) or standard length (SL), but at times body length (BL). For larvae of most perciform fishes, once the caudal fin is formed, SL is about 80% of TL (J. M. Leis, unpublished). Most often, reported size is of preserved specimens rather than live ones, although shrinkage from live length is variable, being dependent on species, size of the larva, fixative and preservative, handling protocol and length of time since fixation. Within a species, size is broadly correlated with age, but growth rate varies widely. In this chapter, size will be the ontogenetic metric of choice.

Speeds are expressed in cm s^{-1}; for perspective, 28 cm s^{-1} = 1 km h^{-1}. Speed may also be scaled to the size of the larva as SL s^{-1} or TL s^{-1} (or BL s^{-1} if the length measurement was unspecified); for perspective, human Olympic record holders swim at about 1 BL s^{-1}. Speed scaled to body size (BL s^{-1}, or "length-specific speed," but sometimes called "relative locomotor speed" [Bellwood and Fisher, 2001]) is frequently used in an attempt to create a common currency to allow comparisons of speed among species or ontogenetic stages. To some extent, this can be successful, as length-specific speed does not always increase with size, whereas absolute speed nearly always does. But, in many cases, length-specific speed does increase with size, particularly over the size range when fins are forming and muscle mass is increasing markedly. Length-specific speed is inversely proportional to body size when the comparison is restricted to "individuals with fully developed locomotor systems" (Bellwood and Fisher, 2001), so that larvae that settle at the smallest sizes are likely to have the greatest length-specific speed. This is a clear indication that the use of length-specific speed for comparisons will not remove the effects of size.

2. BEHAVIOUR AS A FACTOR IN LARVAL FISH BIOLOGY

Behaviour of larval fishes has largely been ignored in considerations of dispersal until very recently, with behavioural research focusing on feeding, particularly in the very early larval stages, and on vertical distribution (e.g., Blaxter, 1986). Feeding behaviour has received considerable attention, but it operates over millimetre scales and will not generally influence dispersal. Vertical distribution, which also has received considerable attention, can have a decisive indirect influence on dispersal where current velocity varies with depth (velocity is used herein as in physics: a vector that includes both speed and direction) (Forward and Tankersley, 2001; Sponaugle *et al.*, 2002; Paris and Cowen, 2004). Yet, relatively few dispersal models (other than those involving estuaries) explicitly include vertical distribution behaviour of larvae, and when included, it is usually very generalised, often including only the vertical "centre of gravity" of the population without variation about it, and with scant consideration of ontogenetic, diel or spatial changes in vertical distribution behaviour.

When behaviour of fish larvae has been considered in the context of dispersal, the behavioural information has most often been based on published laboratory studies of larvae of a few species of clupeiform, gadiform and pleuronectiform fishes of cooler waters of the northern hemisphere, rather than on studies on the particular species whose dispersal is of interest. Many of these heavily studied exemplar species are not demersal as adults but are pelagic in coastal or even oceanic environments. Reviews that have touched upon behaviour of fish larvae (few have focused on behaviour as such) have generally concluded that the swimming speed of the larvae is very low: BL s^{-1} (Theilacker and Dorsey, 1980; Blaxter, 1986; Miller *et al.*, 1988). Based upon this, most workers have concluded that swimming by larvae could make little difference to dispersal, and these low speeds would render other behaviours, such as orientation, irrelevant. Given the reported average larval speed of about 1 TL s^{-1}, one major review concluded, not unjustifiably, that "sustained swimming at almost constant speed ... is used primarily in searching for food" (Miller *et al.*, 1988).

Several factors make this application of information from the literature, and the conclusions based upon it, suspect. There is no reason to believe that either the morphological or the behavioural ontogenies of larvae of clupeiform, gadiform or pleuronectiform fishes are representative of the ontogenies of perciform fishes (here considered to include scorpaeniforms; Johnson and Patterson, 1993; Mooi and Gill, 1995) that dominate demersal communities in warmer water. These four orders of fishes have been separate lineages for at least 50–60 million yr, and the Clupeiformes for much longer (Carroll, 1988), so assuming that information on clupeiform fishes would apply to perciform fishes would be equivalent to assuming information on rodents or bats would apply to seals or camels. Comparisons across such wide taxonomic gaps should be undertaken only with great caution (Billman and Pyron, 2005). Secondly, behaviour of fish larvae in the laboratory may not be representative of behaviour in the sea (von Westernhagen and Rosenthal, 1979; Theilacker and Dorsey, 1980; Hecht *et al.*, 1996; Utne-Palm, 2004). Further, methodological differences among studies may have contributed to the perceived differences in behavioural capabilities amongst these different ecological and taxonomic groups, and this is considered below. Finally, there are sound physical and physiological reasons to believe that swimming should be both easier and more efficient in warm water than in cold water (Fuiman and Batty, 1997; Hunt von Herbing, 2002). It is clearly important to control for phylogenetic and methodological differences when making comparisons (Billman and Pyron, 2005), and the physiological and hydrodynamic effects of differences in water temperature may also be important.

Fishes swim more efficiently in an inertial rather than in a viscous hydrodynamic environment (Webb and Weihs, 1986). At a given water viscosity, the larger and the faster a larva is, the more likely it is to be swimming in an inertial environment. A hydrodynamic measure called the Reynolds number ($Re = U^*L/v$, where U is speed, L is total length, and v is the kinematic viscosity of sea water; Webb and Weihs, 1986) is used to indicate in which environment swimming is taking place. Although it was originally thought that a viscous environment applies at $Re < 30$ and an inertial environment at $Re > 200$ (Webb and Weihs, 1986), recent experiments indicate that the viscous environment extends to Re 300, and a fully inertial environment does not come into play until $Re > 1000$ (Fuiman and Batty, 1997; McHenry and Lauder, 2005; Sarkisian, 2005), with an intervening intermediate zone where both viscous and inertial forces are important. It can be seen from Figure 1A that very small larvae must swim at speeds of many BL s^{-1} to attain $Re > 300$, and that slow larvae will not reach $Re > 300$ until they reach a large size. The speed required to reach $Re = 300$ decreases rapidly with growth so that (at 20°C), by about 5 mm TL, speeds of only 10 BL s^{-1}

A

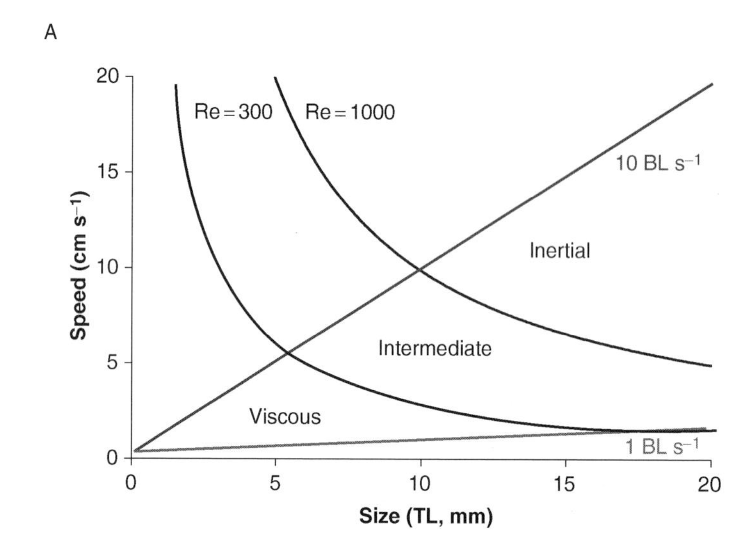

B

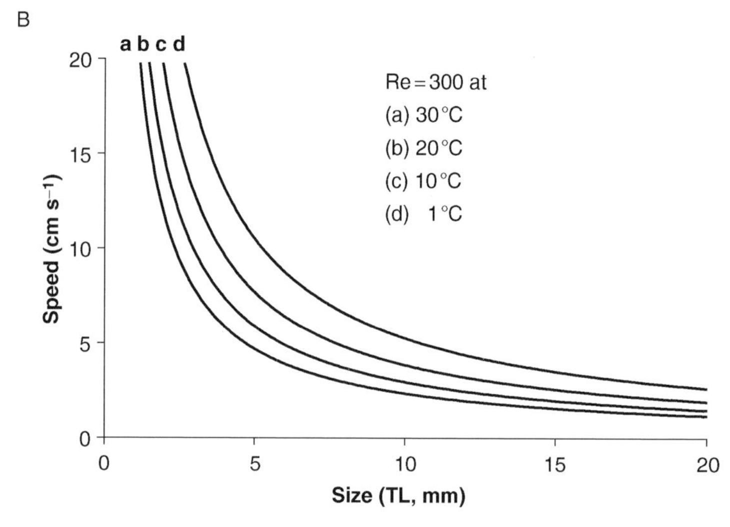

Figure 1 Speed versus length in seawater, showing the combination of speed and size at which Reynolds numbers (Re) of different magnitudes are achieved. (A) Re of 300 and 1000 at 20°C. Inertial hydrodynamic forces predominate at Re > 1000, whereas viscous hydrodynamic forces are most important at Re < 300. Between Re values of 300 and 1000 lies a transition area between the two hydrodynamic zones. The straight lines portray speeds of 1 and 10 BL s^{-1}. (B) Re of 300 (marking the upper boundary of the viscous hydrodynamic zone) at different water temperatures. In cooler water, the curves shift away from the origin toward the upper right, because of the increased dynamic viscosity of water. Thus, for a larva of a fixed size, greater speeds are required to achieve any given Re value in cooler water. The effect is greater for smaller larvae and in cooler temperatures. A similar shift applies to the lower boundary of the inertial hydrodynamic zone (Re = 1000).

(5 cm s^{-1}) are necessary to reach Re > 300, and for larvae of 10 mm TL, speeds of only 2 BL s^{-1} (2 cm s^{-1}) are needed. Therefore, (at 20 °C) a larva swimming at 10 BL s^{-1} leaves the fully viscous zone at about 5 mm and emerges into the fully inertial zone at about 10 mm TL (Figure 1A). A larva that swims at 20 BL s^{-1} leaves the viscous zone at about 4 mm TL and enters the inertial zone at about 7 mm. Fish larvae are clearly small, and as the speeds reported for larvae in major reviews were low (1–5 BL s^{-1}), it was generally thought larvae were swimming in a viscous environment, and that, therefore, swimming was energetically very costly (Osse and Drost, 1989; Osse and van den Boogaart, 2000). For example, larvae that swim at 1 BL s^{-1}, which published reviews ensured readers was a "typical" swimming speed, would not reach Re = 300 until about 20 mm TL. Because of this, it was assumed that swimming was not something larvae were likely to do other than to search for food (slowly, and over limited distances). But, as shown below, many larvae swim much faster than was previously believed.

Temperature has two effects on swimming performance in fish larvae (Hunt von Herbing, 2002). First, fish larvae are ectotherms, and fish muscle cells operate more efficiently at higher temperatures (Hunt von Herbing, 2002). Secondly, the viscosity of seawater increases as temperature decreases, which means that either higher speed or greater size is required to reach a given Reynolds number at cooler temperatures (Figure 1B). This viscosity effect of temperature is greatest for smaller larvae and at lower water temperatures. At cooler temperatures, greater speeds are required to reach an inertial environment and the increased swimming efficiency an inertial hydrodynamic environment offers. Both of these effects contribute to increased swimming efficiency at higher temperatures, although within natural temperature ranges, the effects of viscosity may outweigh the effects of muscle efficiency (Fuiman and Batty, 1997). Although it would be meaningless to measure swimming performance at temperatures outside the range at which the larvae of a particular species are found in nature, within the natural range, it is clear that larvae can swim faster and more efficiently in warmer water (Hunt von Herbing, 2002). This has implications for the possible influence that larval fish behaviour can have on dispersal in different thermal environments.

Distributional studies of larval fishes frequently show that larvae of different species originating from the same sort of habitat (often, a coral reef) have differing distributions (reviews in Leis, 1991; Boehlert, 1996; Cowen and Sponaugle, 1997), which should not be the case if the simplifying assumption of passive larval behaviour applies. Several authors commented on this, and some speculated that behaviour might be responsible (Leis and Miller, 1976; Richardson and Pearcy, 1977; Leis, 1982; Suthers and Frank, 1991), but there was little or no hard information on behaviours of the larvae, other than vertical distribution. Work on temperate systems in the 1980s led several authors to conclude that behaviour of larvae would be important to dispersal (e.g., Marliave, 1986; Sinclair, 1988; Kingsford, 1988; Frank *et al.*, 1993), although demonstrations of relevant behavioural abilities in larvae were again lacking.

In the mid 1990s, work on larvae of coral-reef fishes began to show that the swimming abilities of perciform larvae approaching settlement were considerable, in terms of both speed and endurance, and in the case of speed, this was based on observations on larvae in both the laboratory and the sea (Stobutzki and Bellwood, 1994, 1997; Leis *et al.*, 1996; Leis and Carson-Ewart, 1997). This prompted a reassessment of the influence that behaviour of fish larvae might have on dispersal and, in turn, an interest in the behavioural capabilities and ontogeny of behaviour in marine fish larvae. At about the same time, renewed focus began to be applied to the question of demographic connectivity in populations of marine fishes, and evidence started to emerge that self-recruitment, far from being the rare exception to demographic panmixia, was in fact common (Jones *et al.*, 1999; Swearer *et al.*, 1999, 2002), supporting views by other researchers (Iles and Sinclair, 1982; Sinclair, 1988). This has led to a search for the mechanisms that allow self-recruitment and restricted dispersal and, inevitably, to more interest in behaviour of pelagic larvae as a potential influence. It should be noted that the behaviours reviewed here may be important in contexts other than dispersal, and my focus on dispersal should not be taken to indicate otherwise.

3. METHODS OF STUDYING BEHAVIOUR IN FISH LARVAE

Because of their small size and fragility, as well as the difficulty of laboratory rearing, larval fishes present a number of difficulties to those attempting to study their behaviour. A mixture of traditional and innovative means has been developed to overcome these difficulties. Five broad approaches have been used to investigate behaviour in larval fishes. In the first method, used particularly to study vertical distribution behaviour, larvae are quantitatively captured with towed nets and behaviour inferred by the distribution of abundance or size (e.g., Pearre, 1979). The remaining approaches rely on observations of live larvae. The second approach can be called "undisturbed observation" in the laboratory, wherein larvae are observed (often filmed) going about their business, undisturbed by any direct intervention by the experimenter, in laboratory tanks of various sizes and shapes (e.g., Fuiman *et al.*, 1999; Fisher and Bellwood, 2003). The third is "manipulative observation" in the laboratory, wherein larvae are placed in various experimental devices such as raceways where flow can be manipulated or choice chambers where the larvae are presented with sensory choices (e.g., Stobutzki and

Bellwood, 1994; Arvedlund *et al.*, 1999; Atema *et al.*, 2002). The fourth approach is "*in situ* observation," wherein behaviour of larvae is observed in the sea by divers. Observations of larvae that "just happen to swim by" (von Westernhagen and Rosenthal, 1979; Potts and McGuigan, 1986) are rare because larvae are dilute and difficult to detect in the sea, so more common are observations made on larvae released into the sea by a diver. Most of this *in situ* work has been done in tropical or warm-temperate waters for obvious reasons, although some of the earliest *in situ* observations of larval fish behaviour were made in cold waters (Bainbridge, 1952; Marliave, 1977b; von Westernhagen and Rosenthal, 1979; Potts and McGuigan, 1986). A fifth approach is "*in situ* experimentation," wherein a variety of field experiments are performed using larvae in various cages or selection chambers (Stobutzki and Bellwood, 1998; McCormick, 1999; Tolimieri *et al.*, 2004) placed in the sea, by determining how many and what types of larvae are attracted to different sensory cues (Tolimieri *et al.*, 2000; Simpson *et al.*, 2004, 2005a; Leis and Lockett, 2005), or manipulating various aspects of settlement habitat to determine the effects on settling larvae (e.g., Sweatman, 1988; Leis *et al.*, 2002b; Almany, 2003). Finally, it may be possible to gain information about behavioural capabilities of fish larvae as a by-product of other studies; comparisons of different sampling methods, for example, can provide insight about swimming abilities or attraction to light (Choat *et al.*, 1993; Anderson *et al.*, 2002).

The live larvae under study might be wild individuals coming on a field experiment by chance or captured in the sea and introduced into an experiment or released for *in situ* observation, or they might be larvae reared in captivity and used in similar ways. Light attraction, particularly light traps (Doherty, 1987; Stobutzki and Bellwood, 1997), has provided wild larvae for many *in situ* and laboratory studies, but other methods, including reef-crest nets (Dufour, 1991; Doherty and McIlwain, 1996) or seines, have also been used. But these methods capture primarily older larvae nearing settlement, so studies of younger individuals tend to use reared larvae. Laboratory studies tend to rely more on reared larvae, but much work on swimming behaviour in the laboratory has used wild settlement-stage larvae from light traps. An obvious question is whether reared larvae have behaviour, or specifically, swimming performance, that is similar to wild larvae. The answers are mixed. It is generally agreed that behaviours with a learned component, such as avoiding predators, will differ between wild and reared individuals because larvae from the two sources have very different experience (Olla *et al.*, 1998; Kellison *et al.*, 2000; Brown and Laland, 2001). Other behaviours may or may not differ. For example, reared fishes may swim faster or slower than wild fishes, or they may have equivalent performance, or these can vary ontogenetically, and the same is true of other behaviours (Blaxter, 1976; von Westernhagen and Rosenthal, 1979; Danilowicz, 1996;

Smith and Fuiman, 2004; Leis *et al.*, 2006). So, it is desirable to compare the behaviour of reared larvae to that of wild ones when possible.

4. SWIMMING SPEED

Swimming speed is central to assessment of the simplifying assumption, as without speeds on the order of the currents of the water in which they swim, larvae are unlikely to be able to influence their dispersal trajectories directly. In contrast, indirect influence of dispersal trajectories via vertical distribution behaviour would still be possible even if horizontal swimming speeds were very limited (vertical swimming speeds are considered in the section on vertical distribution behaviour, below). Of course, current speeds vary spatially and temporally, so the influence that swimming by larvae can have is context dependent. The concept of "effective swimming," referring to the ability to swim faster than the mean ambient current speed (Leis and Stobutzki, 1999), was introduced with this in mind. "Effective swimming" is obviously easier to achieve in an area where mean current speed is 10 cm s^{-1} (e.g., many coastal locations) than where it is 100 cm^{-1} (e.g., the Gulf Stream) (e.g., Leis and McCormick, 2002; Fisher, 2005). Horizontal swimming speeds less than "effective speeds" can strongly influence dispersal trajectory if swimming direction is normal to the current direction. For example, in coastal waters, nontidal currents are largely parallel to depth contours, so larvae over the continental shelf that are seeking to enter nursery grounds in shallow water near shore or in estuaries would only be required to swim at right angles to the flow.

Yet, the question remains, how great must swimming performance be before it can significantly influence trajectories? Numerical models of circulation indicate that only modest speeds (0.3–10 cm s^{-1}) are required to have large effects on dispersal. A vertical swimming speed of >5 cm s^{-1} was considered necessary "to overcome vertical mixing" in a tidal channel (Smith and Stoner, 1993). Near Georges Bank, on-bank swimming by larvae of 0.3–1.0 cm s^{-1} "would substantially enhance shoalward displacement" and result in modelled distributions consistent with field observations (Werner *et al.*, 1993). On the Newfoundland shelf, directed horizontal swimming of 1–3 cm s^{-1} by cod larvae was considered able to "greatly increase their retention on the shelf (and on banks, too)" (Pepin and Helbig, 1997). In a numerical model of the Florida coast, simulated larvae that swam at only 1 cm s^{-1} had settlement 36 to 300% greater than passive larvae, whereas larvae that swam at 10 cm s^{-1} had settlement rates "many times" greater (Porch, 1998). In a numerical model of an Australian coral reef, a swimming speed of 10 cm s^{-1} by simulated settlement-stage larvae resulted in a duplication of measured distributions of larvae that was impossible to achieve with passively drifting model larvae (Wolanski *et al.*, 1997).

A confusing factor in attempting to assess the importance of swimming by fish larvae is the number of different ways in which swimming speed is measured. When speeds are cited, it is often not obvious by what method they were measured, thus adding to the confusion. It is clear that the methodology used to measure swimming speed influences the resulting speed estimate. The least appropriate measure in the context of dispersal is burst speed (the highest speed of which a fish is capable), as this is measured and can be maintained (almost always in the laboratory) only over very short periods and is considered to be fuelled anaerobically (Plaut, 2001). Because burst speeds can be maintained over very short periods (typically <20 s; Plaut, 2001), they are inappropriate for considerations of dispersal, although they have been used for this purpose (e.g., Bradbury *et al.*, 2003). It would, however, be appropriate to use burst speed when examining predator escape or avoidance of plankton nets.

Many reports of swimming speed of fish larvae are based on a laboratory measure usually called "routine speed," in which larvae are observed in laboratory containers, usually in still water. Thus, the larva chooses the speed, but in an artificial laboratory context that replaces the natural conditions, stimuli and cues of the sea with those, both intentional and unintentional, provided by the experimenter. Measures of routine speed usually include periods when the larva did not move, so they include a behavioural component other than the larva's ability to reach and maintain a particular speed. Size of the container in which the larvae swim varies but is frequently very small (e.g., 20 cm diameter, 1 cm deep, in the study of Fuiman *et al.*, 1999), and often the area over which measurements are made may be smaller still, particularly when observations are made by filming (e.g., equivalent to a square 3.3–8.0 cm on a side in the study of Fisher and Bellwood, 2003). The time over which routine speed is measured varies from a few seconds to a few minutes, so it is not clear how long it might be maintained. Few studies have considered the influence container size and other laboratory conditions may have on routine speed measurements, but at least two have noted that swimming speed of larvae is lower in small containers than in large ones (von Westernhagen and Rosenthal, 1979; Theilacker and Dorsey, 1980), albeit with larvae of pelagic, not demersal, species (e.g., anchovy larvae swam three to four times faster in large than in small containers). In the laboratory, the density of larvae and differences in food availability can influence the percentage of time that larvae spend swimming (Hecht *et al.*, 1996), with higher densities leading to less swimming. The fact that density of both larvae and food in the laboratory usually are orders of magnitude greater than in the field has obvious implications for swimming behaviour measured in the culture conditions. Unfed larvae of two sparid species had routine speeds 20–100% faster than fed larvae (Fukuhara, 1985, 1987), which adds a further complication to consideration of routine speeds. Turbidity may also influence the activity levels of larvae in the laboratory; herring larvae swam a higher proportion of the time, and the duration of swimming was greater in turbid than in non-turbid water (Utne-Palm, 2004). One review concluded that "laboratory studies may have underestimated swim speeds by up to 20% due to sensory deprivation" (Miller *et al.*, 1988). Most of the swimming speed values included in well-cited reviews (e.g., Blaxter, 1986; Miller *et al.*, 1988) are routine speed. Published larval fish routine speeds are in the vicinity of 1–5 BL s^{-1}, and they seldom exceed 5 cm s^{-1} except in the Pomacentridae (Table 1) (Miller *et al.*, 1988; Fuiman *et al.*, 1999; Masuda and Tsukamoto, 1999). Routine speeds of larvae of perciform fishes are greater than those of clupeiform, gadiform or pleuronectiform fishes, especially when expressed as BL s^{-1} (see Bailey *et al.*, 2005) (Table 1).

A second laboratory measure of swimming speed is critical speed (Brett, 1964), wherein larvae in a raceway are made to swim against an incrementally increasing flow until they can no longer maintain position and are swept onto a downstream mesh. The greatest speed maintained is termed "critical speed" (U_{crit}). Although U_{crit} is frequently considered a measure of prolonged or sustained speed (Plaut, 2001), in fact, U_{crit} in fish larvae is maintained for a maximum of 2–5 min (albeit, following a variable period swimming at lesser speeds), and it is probably better termed a measure of potential speed. Thus, U_{crit} is a measure of the swimming capacity of larvae, and that potential may or may not be sustainable and may or may not actually be used in the sea. U_{crit} is not directly applicable to field situations but is a useful measure of *relative* speed, for comparisons among taxa or developmental stages. Published U_{crit} values for settlement-stage larvae at the end of their pelagic phase often exceed 20 BL s^{-1} and 50 cm s^{-1} (Figure 2A). Many species have mean U_{crit} values in excess of mean ambient current speeds (Fisher, 2005; Fisher *et al.*, 2005) (Figure 3).

Relatively few measures of U_{crit} are available in larvae smaller than settlement stage, but the available information on seven tropical and warm-temperate species (families Apogonidae, Percichthyidae, Pomacentridae, Sciaenidae, Sparidae) as small as 3 mm TL shows that speed increases relatively steadily with growth (Figure 4A), and that relationships between U_{crit} and size were linear or close to linear. Speeds great enough to influence dispersal (>5 cm s^{-1}) occurred after notochord flexion and concomitant formation of the caudal fin; this developmental milestone takes place at about 5 mm SL in the seven marine and estuarine perciform species studied (Fisher *et al.*, 2000; Clark *et al.*, 2005). By 10 mm SL, a size at which settlement takes place in many species, mean U_{crit} values of 8–32 cm s^{-1} were found in the seven study species. These are equivalent to about 7–20 SL s^{-1} near to the completion of notochord flexion and 9–20 SL s^{-1} at 10 mm SL. Whereas

Table 1 Routine speeds of larvae of marine fishes compared to U_{crit} and *in situ* speeds of related taxa

	Routine speed		U_{crit}		*In situ* speed	
	cm s^{-1}	BL s^{-1}	cm s^{-1}	BL s^{-1}	cm s^{-1}	BL s^{-1}
Clupeiformes						
Clupeidae–Clupeinae	0.6–2.5	1–1.2			0.4–3.0	0.5–4.0
Clupeidae–Dussumieriinae, Spratellodini			4–41	2–11		
Engraulidae	0.1–2.0	~1				
Gadiformes						
Gadidae	0.1–0.3	<1				
Pleuronectiformes						
Paralichthyidae	0.5–3.0	~1				
Pleuronectidae	0.6–1.6	0.9–1.5				
Soleidae	0.5–1.3	~1				
Perciformes						
Apogonidae	1–2	1.5–4.0	3–25	5–15	5–12	1–12
Carangidae	0.8–2.5	0.8–2.5	12–40	15–28	4–20	4.5–13.0
Lutjanidae–*Lutjanus* spp.	0.3–2.0	0.5–2.5	3–40	3–14	14–30	9–20
Pomacentridae						
Pomacentrus amboinensis	1–7	6–11	3–36	6–27	11	9
Amphiprion spp.	2–4	4–10	4–35	7–35		
Sciaenidae	0–5	0–3	2–16	4–10	1–9	2–6
Sparidae	0.3–5.0	1–4	2–27	2–22	1–12	6–10

Note: Numbers, ages and sizes of larvae measured all vary among studies. In most taxa, there is a substantial range of values because larvae of a range of sizes and developmental states are included.

Values are drawn primarily from reviews (Blaxter, 1986; Miller *et al.*, 1988; Bailey *et al.*, 2005) but with additional values from the primary literature (Fukuhara, 1985, 1987; Sakakura and Tsukamoto, 1996; Hunt von Herbing and Boutilier, 1996; Doi *et al.*, 1998; Fuiman *et al.*, 1999; Fisher and Bellwood, 2003; Fisher *et al.*, 2005; Clark *et al.*, 2005; Leis *et al.*, 2006; Leis and Fisher, 2006; J. M. Leis, unpublished data).

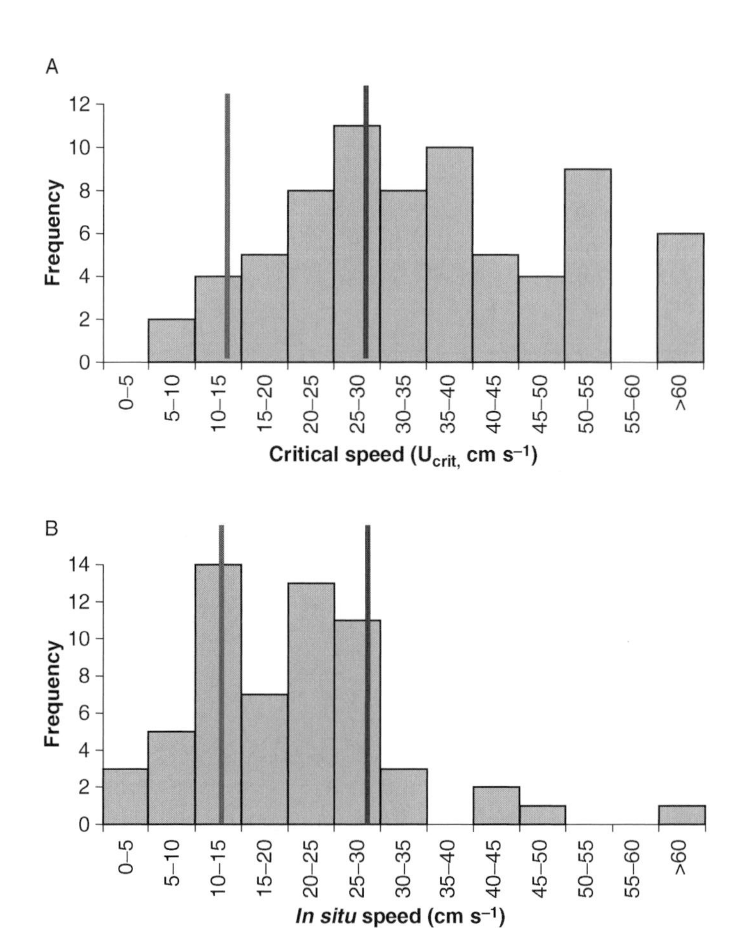

Figure 2 Frequency distribution of species-mean values of swimming speed (cm s^{-1}) in settlement-stage reef-fish larvae measured by two methods (data from Leis and Fisher, 2006). Thirty-six species (11 families) are common to both data sets. To provide context, vertical lines show average current speeds for the Lizard Island area (left, 13 cm s^{-1}), where many of these measurements were made, and, 1 km h^{-1} (right, equal to 28 cm s^{-1}). These measurements were made at temperatures of 28–30°C. (A) U_{crit} of 70 species (17 families). (B) *In situ* speed of 60 species (16 families).

absolute speed increases with growth of larvae, length-specific speed (BL s^{-1}) may not. Nearly all reported U_{crit} values across a range of larval sizes and states of development are 10–20 SL s^{-1} (Figure 4A).

Very limited data on U_{crit} in the youngest larval stages are available. Just hatched larvae (1.5–4.5 mm TL) of 10 species of coral-reef fishes had U_{crit} values as high as 4 cm s^{-1} (mean ~2 cm s^{-1}; up to 14 TL s^{-1}) (Fisher, 2005).

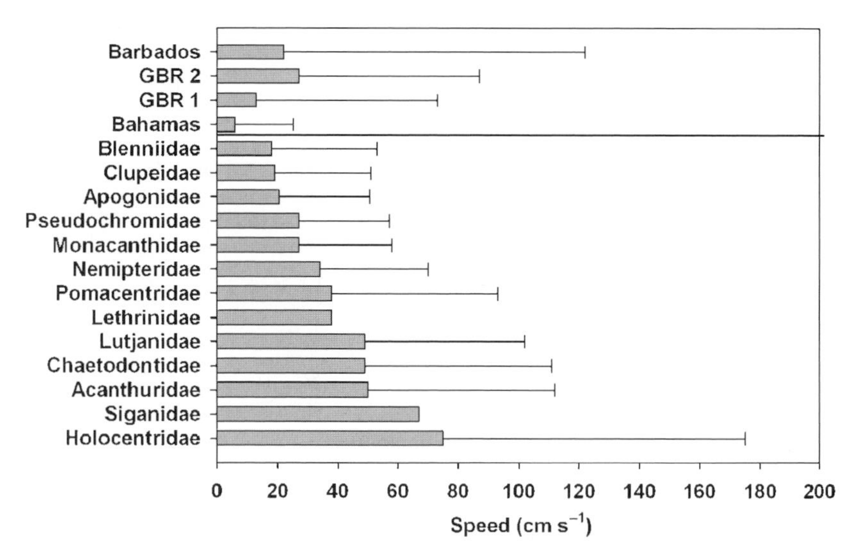

Figure 3 Swimming speeds (U_{crit}) of settlement-stage reef-fish larvae of 12 families from the Indo-Pacific and Caribbean (plus clupeids of the reef-associated, tropical tribe, Spratellodini) compared to current speeds at four tropical locations. Above the line (data from Fisher, 2005) are mean and maximum current speeds (GBR 1 and 2 are different locations on the Great Barrier Reef, Australia). Below the line are mean U_{crit} for the family and for the fastest species in each family (data from Fisher *et al.*, 2005). At settlement, average species of most families can swim faster than average currents at all locations. The fastest species can swim roughly twice as fast as the family average, and some of the fastest species are faster than the greatest currents of most locations. In the Lethrinidae and Siganidae, values for several unidentified species were used to calculate a family mean, but values for individual species were not provided by Fisher *et al.* (2005).

U_{crit} of yolk-sac larvae of a temperate moronid species increased from 0.5 to 2.6 cm s^{-1} (1.3–4.3 TL s^{-1}) over the size range of 4–6 mm TL (Peterson and Harmon, 2001). U_{crit} speeds great enough to have a important influence on dispersal and great enough to be "effective" in many situations occur over a large proportion of the pelagic larval stage of the species that have been studied. The perciform larvae in which ontogeny of U_{crit} has been examined swim in a viscous hydrodynamic environment until they are about 5 mm long and enter the inertial hydrodynamic environment at about 7–10 mm, depending on species (Figure 4A).

A laboratory method of measuring swimming abilities in fish larvae that shares characteristics of both the routine and the critical speed methods provides a result intermediate between those two methods (Hogan and Mora, 2005). Larvae placed in a long (280 cm) relatively broad (13 cm)

A

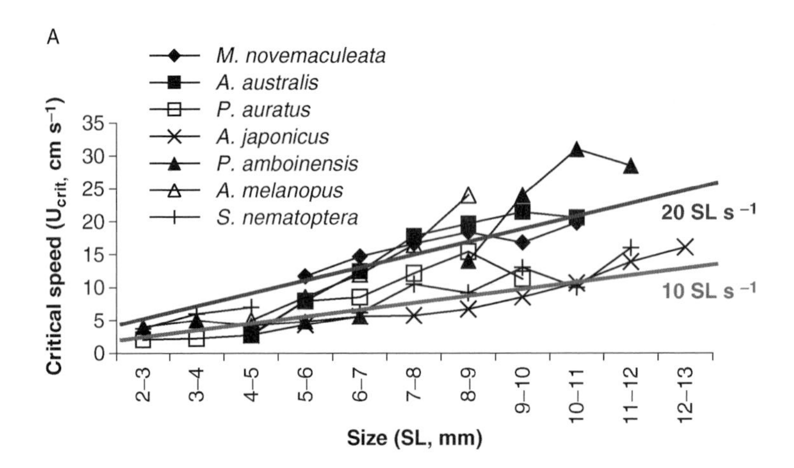

B

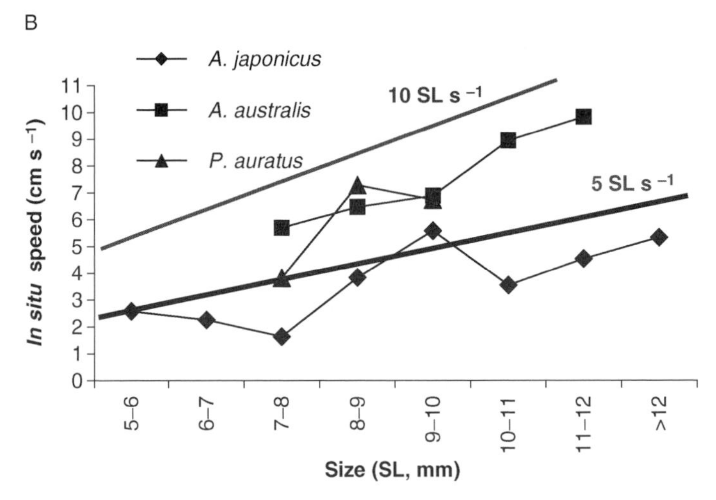

Figure 4 Ontogeny of swimming speed in reared larvae measured by two methods based on mean performance in each 1-mm increment of size. Species with the same shape of symbol are from the same family. All seven species are perciform taxa with similar, relatively generalised developmental morphology. Swimming speed increases approximately linearly with increasing size. These measurements were taken at approximately 20°C for the warm-temperate species and 28–30°C for the tropical species. (A) U_{crit} for the larvae of four warm-temperate species (*Macquaria novemaculeata* [Percichthyidae], *Acanthopagrus australis* [Sparidae], *Pagrus auratus* [Sparidae], *Argyrosomus japonicus* [Sciaenidae]) and three tropical species (*Amphiprion melanopus* [Pomacentridae], *Pomacentrus amboinensis* [Pomacentridae] and *Sphaeramia nematoptera* [Apogonidae]). Lower and upper lines indicate swimming speeds of 10 and 20 SL s^{-1}, respectively. The relationships in Figure 1 are based on TL, rather than the SL used here, and must be modified accordingly, but it appears that larvae smaller than about 3–4 mm SL would swim in a viscous hydrodynamic environment, and those larger than 6 mm (at 20 SL s^{-1}) to 8 mm

swimming channel were observed for up to 140 s. In the pomacentrid species studied, mean swimming speed kept pace with increasing mean current until about 20 cm s^{-1} (~12 TL s^{-1}) and remained at about this speed at faster current speeds. In this species, asymptotic swimming speed was about 60% of the U_{crit}, so larvae chose to swim slower than they were able. So, this method that combines elements of routine speed and critical speed returns a swimming speed higher than routine speed, but lower than critical speed, a characteristic shared with *in situ* speed (see below). Because the larvae choose not to swim as fast as the current, even though they were capable of doing so, it was concluded that larvae of this pomacentrid species "do not use the entire potential of their swimming capabilities" (Hogan and Mora, 2005). This is supported by similar conclusions reached from comparisons of U_{crit} and *in situ* speed (see below) in >70 species (Leis and Fisher, 2006) and for "maximum sustainable speed" of Fisher and Wilson (2004) in 9 species (see below).

Field measurements of the swimming speed of larvae made by divers using flowmeters, called *in situ* speed, have the advantage that they take place in the sea, and that the larva chooses the swimming speed. The behaviour of larvae may be influenced by the presence of the divers, and at present, there is no direct way of quantifying this possible influence. In addition, the possible influence on behaviour of the release of larvae into the pelagic environment from a small container after spending 1–10 h in captivity is very difficult to assess. Indirect considerations indicate, however, that *in situ* behaviours constitute the best available measure of what larvae actually do in the field for the following reasons: (1) *in situ* speeds are considerably less than maximum speed measurements such as U_{crit} (Fisher and Wilson, 2004; Clark *et al.*, 2005; Leis and Fisher, 2006); (2) *in situ* behaviours are repeatable but can vary among locations, times and swimming directions in spite of the constant presence of observer divers (Shanks, 1995; Leis and Carson-Ewart, 2001, 2002, 2003; Leis *et al.*, 2006); (3) larvae observed settling by divers do not necessarily disappear into the first coral head they encounter; rather, they are selective about where they settle,

(at 10 SL s^{-1}) would swim in an inertial environment, whereas larvae of sizes between these would swim in an intermediate hydrodynamic environment. (B) *In situ* speed for the larvae of three warm-temperate species (*Acanthopagrus australis, Pagrus auratus* and *Argyrosomus japonicus*). Lower and upper lines indicate swimming speeds of 5 and 10 SL s^{-1}, respectively. Larvae of the sparids would swim in an intermediate hydrodynamic environment in most cases, but only the larger sciaenids (>8 mm SL) would reach this intermediate environment. The smaller (<8 mm SL) sciaenids would swim in a viscous environment. (A, From data in Fisher *et al.*, 2000, and Clark *et al.*, 2005; note gaps in data for *P. amboinensis* and *S. nematoptera* for 7–8 mm and 5–6 mm SL increments, respectively. B, From data in Leis *et al.*, 2006b.)

frequently rejecting potential settlement sites following close inspection (Leis and Carson-Ewart, 1998, 2002; Leis and McCormick, 2002); (4) observed larvae frequently shelter with observer divers when predators are present, and, as opposed to larvae approached by divers, seem unconcerned by a diver who maintains a constant distance (Leis and Carson-Ewart, 1998; personal observations); and (5) observed larvae frequently feed (Leis and Carson-Ewart, 1998; Trnski, 2002; Hindell *et al.*, 2003). So, it is clear that larvae are not swimming unnaturally fast in order to escape from the observing divers. On the contrary, they cruise at speeds far below their potential (about one-third to one-half [Leis *et al.*, 2006; Leis and Fisher, 2006]). Two disadvantages of the *in situ* technique of making observations of behaviour must be pointed out. Firstly, the technique cannot be applied at night, and secondly, not all species are amenable to it (Leis *et al.*, 1996). Some species will simply sink to the bottom when released, some will simply float in midwater, some will attempt to settle on, or otherwise closely associate with, the diver and a few (notably tetraodontids) will simply hover in the water and closely watch the observer diver.

Larvae do not necessarily swim constantly in the sea. They do feed, typically by "eating on the run" with little change in speed or direction, but larvae have been observed to slow or stop to feed (Leis and Carson-Ewart, 1998; Trnski, 2002; Hindell *et al.*, 2003). Some individuals will slow considerably or simply stop swimming and "hang" in the water column before resuming their prior swimming speed. The reason for this behaviour is not always obvious, although, in some cases, this behaviour is apparently a reaction to the presence of larger fishes (Leis and Carson-Ewart, 1998, 2001). Published *in situ* speed measurements include such pauses.

In situ speed values are available for wild settlement-stage larvae of 60 species of 16 families of tropical reef fishes (Leis and Carson-Ewart, 1997, 1999; Leis and Fisher, 2006) (Figure 2B), and of wild settlement-stage larvae of four species (two families) of warm-temperate coastal fishes (Trnski, 2002). In addition, *in situ* speed has been measured in reared larvae of three species of warm-temperate benthic fishes (two families) and a reef-associated tropical carangid (Leis *et al.*, 2006) of a range of sizes (5–18 mm SL). Species mean *in situ* speeds of settlement-stage tropical reef fishes ranged from 2 to 66 cm s^{-1} (2–34 SL s^{-1}) (Figure 2B), whereas those of settlement-stage wild temperate larvae were 5–15 cm s^{-1} (4–11 SL s^{-1}), at the slow end of the range of tropical species.

Two studies of the ontogeny of swimming speed *in situ* are available. Larvae of three warm-temperate species (Sciaenidae, Sparidae) increased in speed from about 2 cm s^{-1} at 5–6 mm to 5–10 cm s^{-1} at 10–12 mm SL (settlement size) at 0.4–2.0 cm s^{-1} mm increase in SL (Figure 4B). Speed scaled to body length (length-specific speed) was not strongly correlated with size and ranged from 4–10 SL s^{-1}. In a reef-associated tropical carangid, *in situ* speed increased at about 1.7–2.6 cm s^{-1} mm increase in SL over a size range of 8–18 mm SL, but the actual speed varied among locations (see below) and ranged from 3 to 20 cm s^{-1}. In this carangid, mean length-specific *in situ* speed was 9.5 (0.9 SE) SL s^{-1}, and did not increase with size (Leis *et al.*, 2006b). The few perciform species in which ontogeny of *in situ* speed has been examined would swim in a viscous hydrodynamic environment when smaller than about 5–6 mm and an intermediate environment thereafter, with only some individuals reaching a fully inertial environment just before settlement at about 10–12 mm (Figure 4B). Settlement-stage reef-fish larvae of nearly all species, in contrast, would be well into the inertial hydrodynamic environment at their mean *in situ* speeds.

It may be only the best performers that survive the pelagic larval phase and are of relevance to considerations of dispersal. The mortality rates of fish larvae in the ocean are very high, so high, in fact, that the average fish larva will die well before the end of the pelagic larval stage (e.g., Cushing, 1990; Bailey *et al.*, 2005). Work on other aspects of larval fish biology has frequently shown that the best performers—for example, the fastest growers (e.g., Searcy and Sponaugle, 2001; Vigliola and Meekan, 2002)—are the individuals most likely to survive (Fuiman and Cowan, 2003). Therefore, it may be most relevant to use the values of the best performers rather than mean values when incorporating behaviour into considerations of dispersal. In settlement-stage reef-fish larvae, the maximum speed was on average 1.48 and 1.54 times the mean for U_{crit} and *in situ* speed, respectively (Leis and Fisher, 2006). The fastest *in situ* speeds of larvae of three warm-temperate species were 6–8 cm s^{-1} faster (depending on species) than the slowest larvae at any size (Leis *et al.*, 2006), whereas U_{crit} for the "best performers" (the fastest 20–30% of individuals of any size) in four species was 1–5 cm s^{-1} greater than the mean speed, depending on species and size (Clark *et al.*, 2005). Given the magnitude of the differences between them in swimming speed, best performers would probably have very different dispersal trajectories than mean performers if swimming were a factor in dispersal. Researchers should report both mean and maximum speeds for these reasons.

There are consistent taxonomic influences on swimming speed. Comparisons of swimming speed among taxa are confounded by the different methods used to measure speed and, in some cases, the fact that different developmental stages were measured. Further, the different modes of development and differing durations of the larval phase complicate comparisons and have led to proposals for methods of standardising developmental intervals, for example, as a percentage of the length differential or time between hatching and the end of the larval period (Fuiman and Higgs, 1997; Ditty *et al.*, 2003). These morphology-based proposals founder on the lack of a generally agreed definition of the end of the larval period, which is unlikely to be forthcoming for the simple reason that larval development is so varied, and that any definition based on morphological criteria is ultimately arbitrary. An ecologically based (as opposed to morphologically based) measure based on size at settlement might be a more practical standardisation, particularly in the context of dispersal, which depends on events during the pelagic portion of the life history, regardless of the morphological state of development. However, this, too, presents difficulties given that size at settlement varies widely amongst species (from at least 8 to 200 mm) (Leis, 1991; Moser, 1996; Leis and Carson-Ewart, 2004). With these caveats, some comparisons among taxa will be attempted.

As noted above, routine speeds of larval perciform fishes are greater than those of clupeiform, gadiform or pleuronectiform fishes. A comparison of U_{crit} of settlement-stage larvae of 89 species in 21 families of primarily perciform reef-fishes revealed a broad range of speeds 6–100 cm s^{-1}, mean 37.3 cm s^{-1} (3–46 TL s^{-1}, mean 19.0 TL s^{-1}) (Figures 2 and 3) (Fisher, 2005; Fisher *et al.*, 2005; Clark *et al.*, 2005) and some clear taxonomic associations: the slowest species were in the families Apogonidae, Blenniidae, Nemipteridae and Pomacanthidae, whereas the fastest species were in the families Acanthuridae, Holocentridae and Siganidae. Most species had U_{crit} that was greater than the mean current speeds in the study area (Fisher, 2005; Fisher *et al.*, 2005; Leis and Fisher, 2006). Intraspecific variation in swimming ability was not explained by differences in size or condition factor within each species (Fisher *et al.*, 2005). However, interspecific variation in swimming ability (at both the family and species levels) was significantly influenced by the size and morphology of larvae (Fisher *et al.*, 2005). Speed was loosely correlated with size, whereas length-specific speed (BL s^{-1}) was negatively correlated with size. Comparisons of *in situ* speeds of settlement-stage larvae of 60 species in 16 families of primarily perciform reef fishes revealed speeds lower than U_{crit}, a similar order of families based on swimming speed (Figure 2B), and again a loose correlation with size (Leis and Carson-Ewart, 1997; Leis and Fisher, 2006). Most species could swim faster *in situ* than the mean currents in the study area (Leis and Carson-Ewart, 1997; Leis and Stobutzki, 1999); that is, they were "effective swimmers." Among the species common to both data sets, the two speed measures were significantly correlated, with U_{crit} on average being twice the *in situ* speed.

Spatial differences in *in situ* swimming speed by larvae have been documented in several cases. Settlement-stage larvae of a pomacentrid swam 20% faster in an atoll lagoon in the Tuamotu Islands than in the adjacent deep ocean, whereas a microdesmid was 35% faster in the ocean (Leis and Carson-Ewart, 2001). Settlement-stage larvae of another pomacentrid swam up to 27% faster on the leeward than windward side of Lizard Island, Great Barrier Reef, although three other species had no such difference (Leis and Carson-Ewart, 2003). Larvae of a sciaenid had a greater increase in speed with size in a bay than in the Tasman Sea off the warm-temperate east coast of Australia (Leis *et al.*, 2006a,b), and swam up to 4 cm s^{-1} faster in the bay. Larvae of a reef-associated carangid swam at different speeds in three coastal locations off southern Taiwan, with differences among locations of up to 10 cm s^{-1} at a given size (Leis *et al.*, 2006b), although the rate of increase in speed with growth was similar in all three areas. Settlement-stage sparid larvae swam slower near the bottom than in midwater in an Australian estuary (Trnski, 2002). At present, we do not know the reasons for these spatial differences in speed, but they are something dispersal models must take into account.

Both the locations and the direction of swimming can influence swimming speed. Settlement-stage larvae released adjacent to a coral reef swam faster when they were going away from the reef than when they were approaching the reef or swimming over it (Leis and Carson-Ewart, 1999, 2002; Leis and McCormick, 2002; J. M. Leis, unpublished). These differences in speed were up to 12 cm s^{-1}. This behaviour has been shown in several families including Pomacentridae, Chaetodontidae, Lutjanidae and Serranidae.

The *in situ* speeds mentioned above are speeds through the water that was normally moving itself; they are not speeds over the bottom. Arguably, speed over the bottom (or net speed), which combines both swimming and water movement, is more relevant to dispersal than swimming alone. Of course, swimming direction of larvae relative to the current plays a major role in the resultant net speed over the bottom. This is discussed in the section on direct tests of the simplifying assumption, later in this chapter.

All the observations of swimming speed reviewed thus far were made during daylight, which provides only part of the story. The only study to address swimming speeds of fish larvae at night has revealed substantial differences between day and night (Fisher and Bellwood, 2003). Larvae of five species (families Apogonidae and Pomacentridae) were filmed in a laboratory tank swimming without external stimulus—similar to the methods used to measure routine speed. The larvae swam constantly during the day, but at night all species initially swam only about 15% of the time, although this increased to about 60–80% at the end of the pelagic period, with an overall mean throughout the larval duration of about 34%. During the night periods of observation, individuals seemed to be either "active" (i.e., swimming constantly as during the day) or "inactive" (i.e., hanging in the water at an angle without swimming). Late in the pelagic phase of the anemonefish species under study, the swimming speed of "active" individuals at night was nearly twice that during the day, although this difference would be less if the inactive individuals were included in the calculation of the mean value. These day–night differences in speed and proportion of time spent swimming indicate day and night swimming behaviours are unlikely to be equivalent, and that more work on behaviour of larvae at night is required.

A comparison of swimming speeds of fish larvae made by three methods and on different species allows several conclusions. Although there are few comparisons of routine speed to other measures of speed that involve the same species, or even closely related ones, there are systematic differences among methods (Clark *et al.*, 2005; Leis *et al.*, 2006) (Table 1). Unfortunately, broad comparisons of methods are only possible for perciform fishes, but critical speeds (U_{crit}) are much greater than routine speeds, often by a factor of 5–10. What small overlap in values there might seem to be between the two methods is due to differences in the size or stage of development of the larvae studied. One study comparing swimming abilities in larvae of three species (Apogonidae and Pomacentridae) concluded that routine speed was only 19% of U_{crit} (Fisher and Bellwood, 2003). Third, *in situ* speeds are also greater than routine speeds (except, perhaps, for clupeids, although the authors of the comparisons stated that the *in situ* speeds were higher), with twofold differences being the norm and some differences reaching 10-fold (Table 1). Finally, U_{crit} values are nearly always greater than *in situ* speeds. *In situ* speed was about half of U_{crit} in settlement-stage larvae of 83 species of coral-reef fishes of 11 families (Leis and Fisher, 2006), and 35–50% over a range of developmental stages (5–12 mm SL) in larvae of three species of warm-temperate demersal fishes (Leis *et al.*, 2006b). U_{crit} and *in situ* speed are correlated, however (Leis *et al.*, 2006b; Leis and Fisher, 2006), making it possible to predict *in situ* speed from the more easily measured U_{crit}.

Which speed is the most relevant for considerations of dispersal? Routine speed has the advantage of being a measure of swimming speed undisturbed by divers or any overt forcing by the investigator, but it carries the disadvantage of being measured in artificial laboratory conditions. U_{crit} is most relevant for comparisons of relative performance, but it is not a performance measure that can be directly included in dispersal models and is almost certainly faster than larvae actually swim in the sea. *In situ* speed has the clear advantage of being measured in the sea, but with the unknown influence of the observing divers. In spite of this, *in situ* speed is the best existing measure of how fast larvae actually swim in the sea, and therefore, the most relevant for dispersal models. But, speed alone is not the only consideration (see Section 5).

In summary, upon hatching, larvae of most perciform fishes have limited swimming abilities, but by the time the caudal fin is formed at about 5 mm SL, swimming speeds can be attained that numerical models indicate are able to greatly influence dispersal trajectories, and by the time of settlement, larvae of most species are fast enough to be "effective swimmers" in many environments (e.g., Fisher, 2005). Within species, larger larvae will be able to swim faster than smaller larvae, but there are large differences among taxa in swimming ability that are unrelated to size. Swimming speeds of tens of cm s^{-1} by settlement are achieved *in situ* by most perciform species studied. This means that larvae of 1–2 cm in length can swim at impressive speeds of 0.5–1.5 km h^{-1}. On a length-specific basis (BL s^{-1}), clupeiform, gadiform and pleuronectiform larvae are poorer swimmers than perciform larvae, but whether this is due to phylogenetic differences (Billman and Pyron, 2005) or to differences in water temperature or to a combination is unclear. The relatively high speeds (U_{crit}) found in tropical clupeids of the tribe Spratello-dini (Fisher *et al.*, 2005) (Table 1) are difficult to interpret, as they were from relatively large individuals (30–38 mm TL) swimming in high temperatures (28–30 °C), and these tropical species are in a different subfamily than the temperate herrings from which the much lower routine and *in situ* speeds were obtained. There does not seem to be a strong difference in swimming speeds between warm-temperate and tropical perciform species when larvae of similar size are compared. Many tropical species are larger at settlement, however, and therefore likely to be faster at this stage than temperate species, but even species that are of similar length at settlement may differ greatly in morphology, propulsive area and muscle mass, and this probably contributes to differences in performance (Fisher *et al.*, 2005) (Figure 5). In addition, because of higher water temperatures, tropical species are likely to have faster growth rates and, therefore, are likely to become "effective swimmers" sooner than temperate species. Theoretically, warmer water should favour faster swimming due to both physiological and physical factors (Fuiman and Batty, 1997; Hunt von Herbing, 2002), but there is little evidence of this in the performance of the larvae, although the range of temperatures over which related species have been compared with similar methods is limited. Swimming speed is flexible, and larvae may alter their swimming speeds in different locations or when swimming in different directions, at least near settlement sites. At night, the limited evidence indicates that swimming by fish larvae is initially less frequent than during the day, but this changes with growth, and near the end of the pelagic phase, swimming at night is nearly as frequent as during the day, and nocturnal swimming speeds may be greater. Such swimming capabilities are a clear violation of the simplifying assumption. The fact that swimming speed varies among species during the pelagic larval phase also violates the simplifying assumption (of behavioural equivalence among species). Behaviour in larvae of most species remains unstudied, however, so there is clearly still much work to be done on swimming speed alone.

Figure 5 Preserved settlement-stage larvae of (top) a warm-temperate sparid (*Acanthopagrus australis*) and (bottom) a tropical pomacentrid (*Chromis atripectoralis*). Both are about 8.5 mm SL but differ substantially in morphology, propulsive area and muscle mass, although larvae of tropical sparids are similar to the temperate species and larvae of temperate pomacentrids are similar to the tropical species. The two species also differ substantially in swimming performance. The mean *in situ* speed and unfed endurance of the sparid are 8–10 cm s^{-1}(Leis *et al.*, 2006a) and 8 km (Clark *et al.*, 2005) whereas those of the pomacentrid are 20–25 cm s^{-1} (Leis and Fisher, 2006) and 18 km (Stobutzki and Bellwood, 1997). For them to be valid, comparisons of behavioural performance must control for phylogeny, temperature and method, something few studies have accomplished. (Photo courtesy A. C. Hay.)

5. SWIMMING ENDURANCE

All the measures of swimming speed in the previous section were determined over short periods—a few to 10 min. Therefore, there may be doubt about the length of time over which such speeds can be maintained. Swimming endurance has also been called *sustained swimming ability*, but given that U_{crit} is frequently referred to as a measure of sustained swimming speed, these similar terms are avoided here. Three methods of measuring swimming endurance over periods of time that are relevant to dispersal have been devised.

A modification of critical speed called "maximum sustainable speed" (Fisher and Wilson, 2004) is designed to determine in a laboratory raceway what speeds can be maintained for a 24-h period. Like U_{crit}, this is a measure of potential performance of unfed larvae, but over considerably longer periods. Estimates of maximum sustainable speed are available for the settlement-stage larvae of nine species of three families of tropical reef fishes (Apogonidae, Lethrinidae and Pomacentridae) and range from 8 to 24 cm s^{-1} (5–14 TL s^{-1}) (Fisher and Wilson, 2004). This was about one-half of U_{crit} measurements of the same species. These maximum sustainable speed values were similar to values of *in situ* speed reported for settlement-stage larvae of the same or related species.

Similar to maximum sustainable speed is a measure called "50% fatigue velocity" (50% FV) (Meng, 1993; Jenkins and Welsford, 2002), which aims to identify the speed that 50% of individuals could maintain for a given period, usually 1–2 h. This speed is likely to be somewhat greater than "maximum sustainable speed" but lower than U_{crit}, although there are no direct comparisons of these measures within species. One of the two species of marine/estuarine fishes for which this measure is available—a temperate sillaginid—was studied only at settlement (20–25 mm SL) and had a low 50% FV of 6 cm s^{-1} (~2.4–3.0 SL s^{-1}) (Jenkins and Welsford, 2002). The other, a temperate moronid, was studied at sizes between 6 and 9 mm (Meng, 1993) and had 50% FVs ranging from 1.7 to 3.0 cm s^{-1} (3–4 BL s^{-1}), and these increased with growth.

Settlement-stage larvae of coral-reef fishes are capable of swimming for days and for 10s of km at a fixed speed (Figure 6). Stobutzki and Bellwood (1997) developed a multichamber raceway in which larvae were stimulated by rheotaxis to swim without food or rest until they could no longer keep pace with the constant flow, usually 13.5 cm s^{-1}. The speed of 13.5 cm s^{-1} was chosen by Stobutzki and Bellwood because it was the mean current speed in the vicinity of Lizard Island, on the Great Barrier Reef, where their work took place (Frith *et al.*, 1986). Wild, settlement-stage larvae of >20 species from nine families (Acanthuridae, Apogonidae, Chaetodontidae, Lethrinidae, Lutjanidae, Monacanthidae, Nemipteridae, Pomacanthidae, Pomacentridae) were tested in the raceway, with a range in family mean values of 7.4–194 h and 3.6–94 km, the means being 84 h and 41 km, respectively. The maximum endurance was achieved by a 29 mm TL lutjanid that swam for 289 h and covered 140 km. Endurance was positively correlated with both size and pelagic larval duration. Within-family (among species) differences in endurance of settlement-stage larvae of up to 7.5-fold are known (Stobutzki, 1998) in the families Chaetodontidae (10 species) and Pomacentridae (24 species). Endurance swimming ability was correlated with size (TL) and weight of the larvae in pomacentrids, but not in chaetodontids.

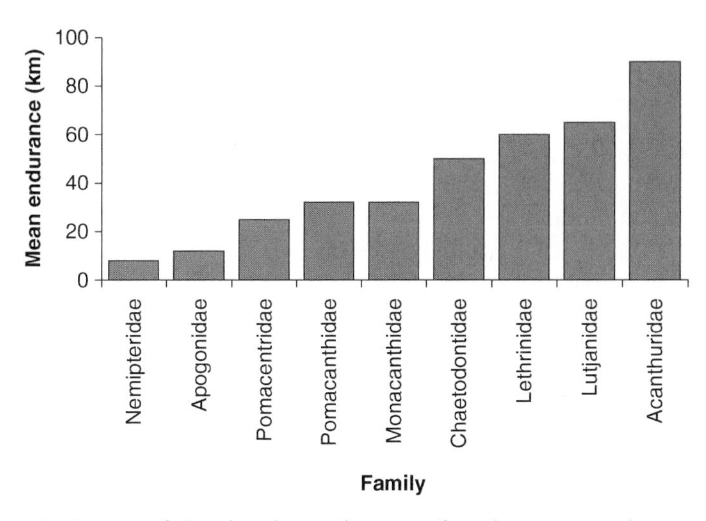

Figure 6 Mean unfed swimming endurance of settlement-stage larvae of nine families of coral-reef fishes (data from Stobutzki and Bellwood, 1997). Data are from larvae (8–25 mm, SL, depending on species) that swam in a flume without food or rest at 13.5 cm s^{-1} until exhaustion. To provide context, the cross-shelf width of the Great Barrier Reef in the Lizard Island region, where these measurements were made, is about 50 km. Temperatures were not reported but were probably about 30°C, based on the seasons and procedures used (Leis and Clark, 2005).

The reaction to this research showing that larvae 1–2 cm long could swim 10s of km was interesting. Stobutzki and Bellwood's (1997) results were greeted with disbelief by reviewers; because their data so conflicted with the embedded simplifying assumption, reviewers simply would not believe them. The reviewers could, however, not find any fault with the work, and it was indirectly supported by appearance of some of the first *in situ* observations of behaviour of larval reef fishes (Leis *et al.*, 1996), so the paper was finally accepted for publication. Stobutzki must also be the first larval-fish researcher to run afoul of animal ethics issues when an article criticising her work on the baseless grounds that she used a cruel method of "electrical shocking of larval fishes in experimental races to keep them swimming" appeared in the newsletter of a learned society following a presentation of the work at an international conference, prior to publication. The criticism, by a prominent ichthyologist, was subsequently withdrawn, but it was probably at least partly inspired by the belief that so gross a violation of the simplifying assumption could only be achieved by extreme cruelty to the larvae. Both the results and the methodology are now firmly accepted and highly cited, so much so, that other investigators (e.g., Dudley *et al.*, 2000) have persisted with the use of 13.5 cm s^{-1} as a test speed for measuring endurance, even though they work where the average current speed is almost certainly different. There is, however, merit in using 13.5 cm s^{-1} as a standard speed upon which to base comparisons of endurance among species (B.S. Danilowicz, personal communication), but, as with speed, such comparisons must control for phylogeny and temperature if they are to be useful.

Settlement-stage larvae of three species of temperate-reef fishes (20–36 mm SL, families Monacanthidae, Mullidae and Scorpididae) were tested using the Stobutzki-Bellwood methodology and had endurance at least as impressive as the tropical species (Dudley *et al.*, 2000). Other work on endurance in fish larvae has focused on ontogeny or on methodological issues (Fisher *et al.*, 2000; Fisher and Bellwood, 2002a; Clark *et al.*, 2005; Leis and Clark, 2005).

Not surprisingly, swimming speed influences endurance (Fisher and Bellwood, 2002a), at least in settlement-stage larvae of one pomacentrid species. Over the range of 4–16 cm s^{-1}, decreasing swimming speed increased endurance in *Amphiprion melanopus* at a rate of about 500 m for each 1 cm s^{-1}. It has also been noted that endurance and speed are correlated in settlement-stage larvae, leading to the suggestion that swimming abilities, no matter how defined, are related (Leis and Stobutzki, 1999). Endurance may be correlated with speed because the fixed speed at which endurance is typically measured (Stobutzki and Bellwood, 1997) will constitute a smaller proportion of the speed capabilities of faster species, in effect, allowing them to swim at a slower relative speed, thus allowing greater endurance (R. Fisher, personal communication).

Stobutzki and Bellwood (1997) predicted that larvae allowed to feed would have substantially greater endurance values, and unsurprisingly, this prediction has subsequently been verified (Fisher and Bellwood, 2001; Leis and Clark, 2005). An increase of 1.8–2.4 fold in endurance was found in a single pomacentrid species (*Amphiprion melanopus*) when fed (Fisher and Bellwood, 2001), with some individuals not reaching exhaustion, and growing as fast as undisturbed larvae. In six other pomacentrid species, the experiment was terminated after a two to fivefold increase in endurance over unfed values was attained with the larvae still swimming strongly (Leis and Clark, 2005), indicating that endurance is less limited by fatigue than by energy supplies. Larvae *in situ* do feed while swimming (Leis and Carson-Ewart, 1998; Trnski, 2002; Hindell *et al.*, 2003), so assuming that food is available in the sea, it seems unlikely that settlement-stage larvae would fail to reach a settlement site due to fatigue alone (but see Mora and Sale, 2002). Although all the estimates of unfed endurance reviewed here may, in reality, be measuring the condition of the larvae, rather than any realistic indication of how long or far larvae would actually swim in the sea, it is clear that the endurance capabilities of larvae at settlement are considerable (measured in kilometres or tens of kilometres) and appear to be related in predictable ways to their swimming speed. Measurement of endurance in a manner that would be directly applicable to field situations and dispersal models would be valuable.

Attempts to investigate the ontogeny of endurance are few but are consistent in their results. They show that endurance swimming abilities are very limited (seldom >100 m) until after notochord flexion is complete at about 5 mm SL, after which they increase rapidly with size until they reach 10s of km at settlement (Fisher *et al.*, 2000; Clark *et al.*, 2005) (Figure 7). Study of ontogeny of endurance is limited to only six species of four families of demersal fishes (Apogonidae, Pomacentridae, Sciaenidae and Sparidae), equally split between warm-temperate and tropical taxa. There were large differences among species in the postflexion size at which endurance began to improve, in the rate of that improvement with growth, and in the ultimate endurance at settlement (Figure 7). This may be due to the confounding factors of speed capability and experimental speed (R. Fisher, personal

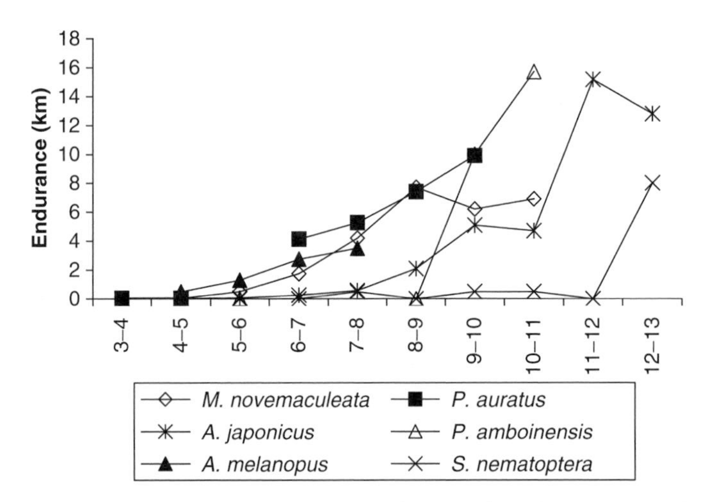

Figure 7 Ontogeny of unfed endurance swimming based on mean performance in each 1-mm size increment for the larvae of the three warm-temperate species (*Macquaria novemaculeata, Pagrus auratus, Argyrosomus japonicus*) and three tropical species (*Amphiprion melanopus, Pomacentrus amboinensis, Sphaeramia nematoptera*). Data from Fisher *et al.*, 2000, who used a speed of 11 TL s^{-1}, and Clark *et al.*, 2005, who used a speed of 10 cm s^{-1}. A value of 55 km for the 12- to 13-mm size interval of *P. amboinensis* is off scale to the top. Note gaps in data for *P. amboinensis* for 6–8 and 11–12 mm, and *P. auratus* for 5–6 mm SL size increments. Species with the same shape of symbol are from the same family. The shapes of the size–endurance relationships vary, with some clearly nonlinear. See Figure 4 for temperature information.

communication). There was, however, no indication of any consistent warm-temperate versus tropical difference in these variables. Both studies of ontogeny of endurance documented large among-individual variation in performance at any size. This may have been an artefact of the use of reared larvae in both studies, but it is equally likely to have been an expression of the natural variation in performance to be expected in wild populations, as shown by the within-species variation in endurance of wild settlement-stage larvae (Stobutzki, 1998). Many species have greater endurance at settlement than the six species studied thus far, and examination of the ontogeny of endurance in better-performing taxa would be interesting. It is noteworthy that unfed endurance of settlement-stage larvae of one pomacentrid species was greater in reared than in wild individuals (Leis and Clark, 2005), and perhaps an indication that reared larvae can be "fatter" than wild larvae (Blaxter, 1976).

Endurance swimming abilities as measured in laboratory swimming chambers are highly artificial but are thus far the only data available. On one hand, it seems unlikely that a larva in the field would actually swim until exhaustion, but on the other hand, it also seems unlikely that a larva would swim continuously for hours or days without feeding. In view of the fact that swimming speed influences the distance swum, at least in unfed trials, the results of laboratory endurance trials at an arbitrarily fixed speed are difficult to apply to the field with any validity. In the laboratory, larvae that are fed seem to be likely to reach the end of their period of competency to settle and to metamorphose before they become so fatigued that they stop swimming (Leis and Clark, 2005). Both maximum sustainable speed and 50% FV are based on arbitrary durations (1–24 h) that are very short compared to the pelagic larval duration. Although this helps to minimise the influence of withholding food during the trial, including a range of durations would provide more insight as to what is possible in the sea. The few studies on the ontogeny of endurance swimming indicate that prior to notochord flexion, the endurance capabilities of perciform larvae are very limited (Fisher *et al.*, 2000; Clark *et al.*, 2005; Leis *et al.*, 2006), but different results might be obtained if slower test speeds were used for smaller larvae (R. Fisher, personal communication). Endurance abilities increase rapidly after notochord flexion, and by the time larvae are competent to settle, they are capable of swimming 10s of km when unfed and much more when they have access to food. In the later portions of the pelagic period of larvae attempting to remain nearby or swim to particular settlement habitats, it seems likely that the ability of the larva to detect distant sensory stimuli that might serve as orientation cues will be more limiting than the ability to swim long distances. The endurance swimming abilities and variations among species that have been documented thus far are certainly inconsistent with the simplifying assumption.

6. AVOIDING THE FLOW INSTEAD OF SWIMMING

Most workers and dispersal models assume that larvae of demersal fishes are found in the water column and are thus subject to the flows measured by current meters or predicted by numerical hydrodynamic models. Near the bottom, however, is a boundary layer wherein the water column current speeds drop rapidly to zero. The height of this boundary layer varies depending on the current speed, the roughness of the bottom and the length, density and type of vegetation that is growing on the bottom (Nowell and Jumars, 1984; Folkard, 2005). Larvae that occupy this boundary layer will be subject to much reduced advection compared to larvae that remain above it (Black and Moran, 1991; Black *et al.*, 1991), and the typically measured or predicted water-column flow will not be relevant to their dispersal.

Several studies show that fish larvae do indeed occupy the boundary layer, at least for a portion of their pelagic period. In shallow (<30 m) water off the southern California coast, net tows revealed that postflexion larvae of gobiids and two sciaenids left the middle portions of the water column and were found almost exclusively in the epibenthic layer, and a clinid moved into the epibenthos at night (Barnett *et al.*, 1984; Jahn and Lavenberg, 1986). In shallow water over oyster reefs in Chesapeake Bay, schooling goby larvae nearing settlement were observed by divers low over the bottom and sheltering in the lee of objects (Breitburg, 1991). In shallow inshore waters near rocky reefs in New Zealand, divers observed larvae of gobiesocids and tripterygiids in the benthic boundary layer (Kingsford and Choat, 1989). In shallow inshore waters near reefs off southern England, larvae of several families (labrids, gadoids and ammodytids) were present in the epibenthic layer and around the reefs, but gobiid larvae were particularly abundant and seemed to occupy this environment throughout their larval stage (Potts and McGuigan, 1986). Including the four studies noted above, larvae of at least 13 families from a wide variety of mostly temperate locations are known to enter the benthic boundary layer (Marliave, 1977b; Grout, 1984; Lindeman, 1986; Leis *et al.*, 1989; Steffe, 1990; Kaufman *et al.*, 1992; Beyst *et al.*, 1999). Judging from the available literature, larvae of gobiids and sciaenids are the taxa found most frequently in the epibenthos. In some cases, only a relatively small proportion of larvae present in the water column occupy the benthic boundary layer, meaning that different dispersal trajectories will apply to different life-history stages. Other areas of reduced flow, including within seagrass or kelp beds (e.g., Olney and Boehlert, 1988) and along reef edges (e.g., Marliave, 1989; Taylor *et al.*, 2005), offer similar refuge from currents. All the examples noted above occurred in shallow water, and most have been discovered by divers. Similar boundary layers exist in water too deep for SCUBA divers but have not been examined for larvae. Another similarity of most of the cited studies is that only more advanced (usually postflexion-stage) larvae were found in the boundary layer. Younger larvae were either shown, or presumed, to be higher in the water column, meaning that an ontogenetic vertical migration was involved. In only one case—the gobiids off southern England—was it demonstrated that the entire larval stage of a species was spent in the benthic boundary layer (Potts and McGuigan, 1986).

Larvae that occupy the boundary layer and thereby avoid passive advection with currents do so behaviourally, and this behaviour seems to occur primarily in larger, more developed larvae. Such behaviour, which might be regarded as a special case of vertical distribution behaviour (see below), certainly violates the simplifying assumption. It is unclear how taxonomically widespread this sort of behaviour is, as the epibenthos is normally ignored in studies of both vertical and horizontal distribution of fish larvae. Where occupancy of the epibenthic boundary layer has been looked for, it has generally been found, but only in a minority of the species present in the area.

7. SCHOOLING BY LARVAL FISHES

Late larvae of a few species of benthic fishes are known to school prior to settlement. This has been noted in some species of the taxa Aulorhynchidae, Blennioidei, Cottidae, Gobiesocidae, Gobiidae, Lutjanidae, Mullidae, Pomacentridae and Trichodontidae (Marliave, 1977a; Potts and McGuigan, 1986; Breitburg, 1989; Leis and Carson-Ewart, 1998; Leis and McCormick, 2002). Some, but not all, of these species also school following settlement from the pelagic environment. Marliave (1977a) argues that "schooling behaviour is perhaps typical of more species in the larval stage than in the adult stage." There is limited information on when these species first begin to school, but larvae of species of the demersal taxa Aulorhynchidae and Trichodontidae and pelagic taxa Clupeiformes and Osmeridae begin to school almost from the time of hatching, whereas a gobiesocid begins to school "midway during larval development" (Marliave, 1977a,b; Morgan *et al.*, 1995). Most larvae of demersal fishes observed to school are well advanced in their larval development. Schooling is normally thought of as antipredator behaviour, and in this role, it is clear that the large number of sensory systems provided by the school enhance the possibility of survival (Pitcher and Parrish, 1993). In the same way, it is possible that the increased number of sensory systems present in a school of larvae may enhance the detection of or response to cues that may aid in orientation in the pelagic environment and the arrival at appropriate settlement sites (Simons, 2004).

This, of course, applies once the school is formed, but an intriguing question is how do the larvae find each other to form a school in the first place? The concentration of fish larvae per unit volume of ocean is low, and it is not at all obvious how larvae, many of which are transparent in whole or in part, discern each other or gather together to begin to school. Perhaps at a particular stage of development, they increase their local concentration by gathering at interfaces such as the surface, the bottom or convergence fronts. In the Great Barrier Reef Lagoon, one study concluded that patches of late-stage larvae were due to "active aggregation," perhaps to take advantage of higher productivity, although the active behaviour that accomplished this was not identified (Thorrold and Williams, 1996). Larvae of a gadid aggregate in response to light and remain aggregated within food patches, which may offer a means of gathering together to initiate schooling (Davis and Olla, 1995). The experimental work on the ontogeny of schooling in pelagic fishes like carangids has been done in relatively small laboratory tanks and offers little insight into this aspect of the initiation of schooling (e.g., Masuda and Tsukamoto, 1998).

Schooling by fish larvae violates the simplifying assumption because this behaviour leads to movement different from passive drift, as schools actively move, following the lead of a few individuals—they do not simply drift. Schooling is another demonstration that fish larvae are indeed active. In addition, the fact that not all individuals or species school is another violation of the simplifying assumption. Finally, the simplifying assumption does not allow for possible group effects in a school that might result in better survival or better ability to locate settlement habitat.

8. VERTICAL DISTRIBUTION BEHAVIOUR

Vertical distribution was the first behaviour of marine larval fishes to be studied, primarily by towing fine-mesh nets, and such studies continue today. There are probably hundreds of published studies on vertical distribution behaviour of fish larvae using this method, and it is well established that larvae are not uniformly distributed vertically, and that temporal differences in vertical distribution—at least diel ones—are common. It is also well established that vertical distribution behaviour can indirectly modify dispersal trajectories (Sponaugle *et al.*, 2002). A rich and long-standing literature also exists on selective tidal stream transport of larvae of many species in estuaries (reviewed by Forward and Tankersley, 2001). This transport relies on vertical distribution behaviour of the larvae interacting with reversing tidal flows, and the resulting trajectories of larvae are very different from what would be expected from the simplifying assumption alone.

Given this background, it is perhaps surprising how many attempts—even contemporary ones—to model dispersal ignore vertical distribution by explicit or implicit application of the simplifying assumption (e.g. Roberts, 1997; James *et al.*, 2002). When vertical distribution is taken into account in a three-dimensional model, it is frequently only the centre of gravity of the distribution that is used to model dispersal (e.g., Bartsch, 1988). Vertical distributions are seldom very narrow, so this approach ignores a large proportion, perhaps the majority, of individuals, and makes the implicit assumption that individual larvae remain at a given depth, rather than allowing for the possibility that individuals move between strata even if the vertical distribution pattern remains static at the population level. More realistic means of including in dispersal models relatively complex vertical distribution that varies temporally and ontogenetically are a major step forward (e.g., Bartsch and Knust, 1994; Hare *et al.*, 1999). One model, for example, included five diel periods with three size groups of larvae and used data on the percentage of the larvae that occupied five depth bins rather than simply the centre of gravity of the population (Bartsch and Knust, 1994). The other model included ascent velocity of eggs plus vertical swimming behaviour of larvae that differed in an age-dependent manner, including excursions to the surface for swim bladder inflation, and age-dependent sinking rates (Hare *et al.*, 1999). Dispersal models should use depth–frequency data that provides information on the relative period of time that larvae spend at different depth intervals (Figure 8), combined with information on the vertical structure of current velocity at the same scales.

The inability to discern behaviour of individual larvae is inherent in the normal towed-net study (Pearre, 1979, 2003), which provides information on population vertical distribution only, as the sampling is destructive and without replacement. Acoustic studies of vertical distribution in fish larvae avoid some of these limitations and can sample large volumes of water, yet are rare, possibly because of difficulties in distinguishing species. This lack of information on behaviour of individuals can result in misleading dispersal predictions, as a simple example will show. Imagine a stratified system with a flow of X in an upper layer equal, but opposite, to that in a lower layer, and with the larvae equally distributed vertically between the two layers. If there is no movement by individual larvae between layers, at the end of time Z the larvae in the upper layer will be advected a horizontal distance of 2XZ relative to those in the lower layer. If movement of larvae between layers is constant and individuals spend an equal amount of time in each layer, then the larvae in the two layers will not become horizontally separated at all. Depending on the proportion of time an individual spends in each layer, any other result intermediate between these extremes is possible. Application of the simplifying assumption to a three-dimensional model of this system would provide only the first result, that is, relative displacement of 2XZ.

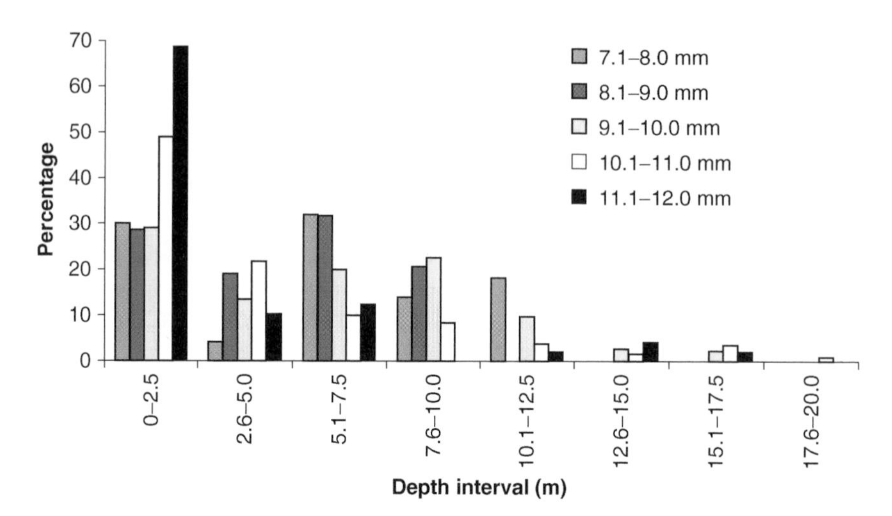

Figure 8 Ontogenetic change in vertical distribution as shown in size-specific depth–frequency data obtained from *in situ* observation of 38 larvae of *Acanthopagrus australis* (Sparidae) off the southeastern Australian coast. The vertical axis represents the percentage of observations (taken every 30 s) that were present in each depth interval. In this species, smaller larvae (<10 mm SL) avoided the 2.5–5.0 m depth interval, resulting in a bimodal depth distribution that was significantly different from that of larger larvae (>10 mm SL), which were strongly surface orientated. (Data from Leis *et al.*, 2006b.)

This review concentrates on vertical distribution information acquired by methods other than towed nets. This should not be interpreted as a denigration of towed-net studies but should be taken as an indication that they are a well-recognised and reviewed means of providing information on larval behaviour (e.g., Pearre, 1979, 2003; Neilson and Perry, 1990; Cowen, 2002). One limitation of towed-set studies, however, is that they tend to contain little information on larger larval stages. This is because, due to mortality, they are rarer than smaller younger stages, and because larger larvae can avoid many towed nets due to their superior sensory and burst-swimming abilities. Gear avoidance has been recognised for many years and perhaps should have provided an early warning that behaviour of fish larvae was important.

The vertical distribution behaviour of pelagic fish eggs has received far less attention than that of larvae, but perhaps surprisingly, the vertical distribution of eggs is not static. Pelagic eggs are generally slightly positively buoyant, and where spawning takes place away from the surface, ontogenetic ascents are common (Sundby, 1991) and may take place over tens of metres. These could be argued to be passive as they are not under behavioural control by the egg but are merely the inevitable result of differences in density between egg and seawater. In addition, pelagic fish eggs may experience ontogenetic changes in density, and most frequently, this constitutes a sharp increase in density as hatching approaches (Olla *et al.*, 1996), resulting in the egg sinking at least to water of greater density. Egg density may also change in response to physical factors, such as exposure to light or changes in temperature or salinity (Olla and Davis, 1993). Like any change in vertical distribution, those of eggs, whether they result from ontogenetic ascents or from ontogenetic, or extrinsically induced changes in density, can have a profound influence on dispersal (Sponaugle *et al.*, 2002).

In situ observations of vertical distribution of fish larvae during the day offer the possibility of gaining data on the behaviour of individual larvae as well as on larvae outside the size range normally captured by towed nets (Leis *et al.*, 1996). *In situ* observation of settlement-stage larvae of coral-reef fishes has received the most attention, but a few studies on larvae of temperate fishes are available. *In situ* studies of vertical distribution have several limitations other than the usual concern with presence of the divers. First, they can only be done during the daytime. Second, because of safety concerns, they can only examine vertical distribution in the upper portions of the water column, usually the upper 20 m. And finally, individuals are usually observed for only relatively short periods, on the order of 10 minutes.

Few, if any, settlement-stage larvae of species of the seven families of coral-reef fishes (Acanthuridae, Chaetodontidae, Holocentridae, Lutjanidae, Microdesmidae, Pomacentridae, Serranidae) that have been observed *in situ* were not selective about the depths at which they swam (Leis and Carson-Ewart, 1999, 2001; Leis and McCormick, 2002; Leis, 2004). Aside from the data on individual species, a major outcome from these observations is that variation in depth-selection behaviour is great at small spatial and temporal scales and among and within individuals. As an example, settlement-stage larvae of the pomacentrid, *Chromis atripectoralis* (~10 mm SL), swam significantly deeper and more variably (both within and among individuals) off the windward than off the leeward side of Lizard Island, Great Barrier Reef (Figure 9). Further, off the leeward side, mean swimming depth and its variation increased significantly with distance offshore (and water-column depth), whereas off the windward side, mean depth did not increase with distance from shore (and water-column depth), but variation in depth did increase significantly, at least among individuals (Leis, 2004). Differences in depth among locations played a role in these differences, but it is clear that bottom depth was not solely responsible for the differences, as depth distributions were wider at the windward 100-m location than at the leeward 1000-m location in spite of similar depths. Similar differences among locations were found in other species, and in other locations, with the general pattern that

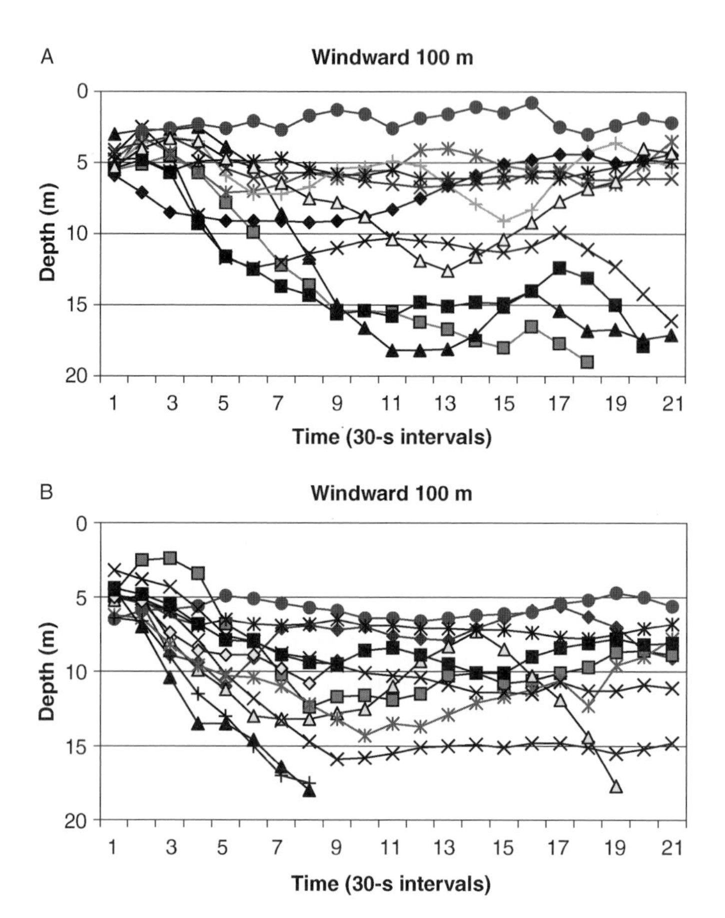

Figure 9 (Continued)

larvae swam deeper and more variably (both among and within individuals) in deeper water, although larvae rarely ventured near the bottom (Leis and Carson-Ewart, 1999, 2001; Leis and McCormick, 2002; Leis, 2004). There were also large differences in vertical distribution behaviour among individual larvae, with some individuals remaining within a very narrow depth range, whereas others occupied a wide depth range, some oscillating over much of the available water column depth. Thus, at least in coral-reef environments, vertical distribution behaviour of fish larvae can differ with location over relatively small horizontal scales (100s to 1000s of metres).

Settlement-stage larvae of a temperate sillaginid in southern Australia also selected different depths in different locations within an estuarine bay

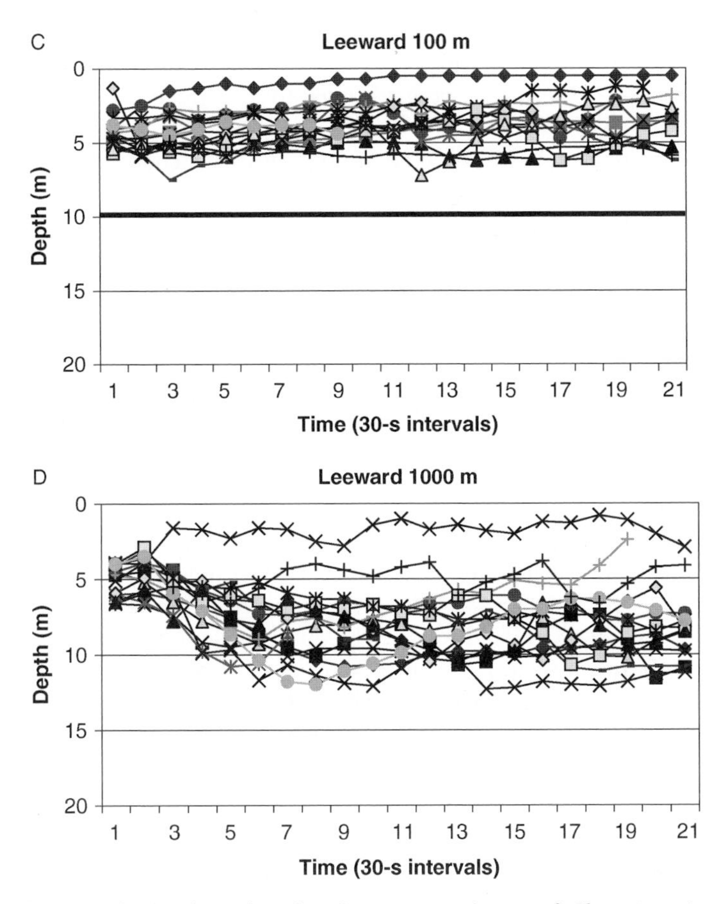

Figure 9 Vertical trajectories of settlement-stage larvae of *Chromis atripectoralis* (Pomacentridae, ~10 mm SL) released and observed by divers on two sides (windward and leeward) of Lizard Island, Great Barrier Reef, and 100 and 1,000 m from shore. Each line represents an individual larva. Bottom depths at each location were as follows: (A) windward 100 m, 22–25 m; (B) windward 1000 m, 33–36 m; (C) leeward 100 m, 9–10 m (as indicated by the bold line); (D) leeward 1000 m, 18–21 m. Some trajectories were for less than 10 min because larvae were not followed deeper than 18 m, and because some larvae were lost by the observer (see Leis, 2004, for further details). Note that no larvae swam deeper than 18 m at the leeward 1000-m location, whereas several did so at the windward locations (short trajectories that terminate at or near 18 m depth).

(Hindell *et al.*, 2003), in effect demonstrating an inverse relationship between average swimming depth and water-column depth. This species tended to swim rapidly toward the bottom in shallow (<1.5 m) water. In water 1.5–3.0 m deep, swimming depth was variable, and larvae were found throughout the water column. The observer diver could not see the bottom in deep water (7 m), and in this location, most fish swam to the surface, the implication being that larvae were attracted toward the bottom when it was visible.

Settlement-stage larvae observed in the entrance channel of a warm-temperate estuary in southeast Australia had consistent, species-specific vertical distributions (Trnski, 2002). Two species of sparids and a girellid—all of which settle into sea-grass beds—swam near the surface, and a third sparid, a species that does not settle in seagrass, swam in the lower portions of the water column, often very near the bottom. The species swimming near the surface seem to have selected the most unfavourable place for entry into the estuary, as nontidal surface flow in estuaries is typically oceanward; this would not constitute efficient swimming (*sensu,* Armsworth, 2001).

In situ studies of the ontogeny of vertical-distribution behaviour are uncommon, but they are available for four species in three families and show clear ontogenetic changes in swimming depth, something found in many studies based on towed-net samples (e.g., Paris and Cowen, 2004). In warm-temperate Australian waters, over size ranges of 5–14 mm SL, a sparid and a sciaenid undertook ontogenetic descents (e.g., Figure 10), whereas a second sparid showed a clear ontogenetic ascent (Figure 8; Leis *et al*., 2006b). In contrast, in the tropics of Taiwan, over a size range of 8–18 mm SL, a reef-associated carangid showed an initial ontogenetic ascent followed by an ontogenetic descent (Leis *et al*., 2006).

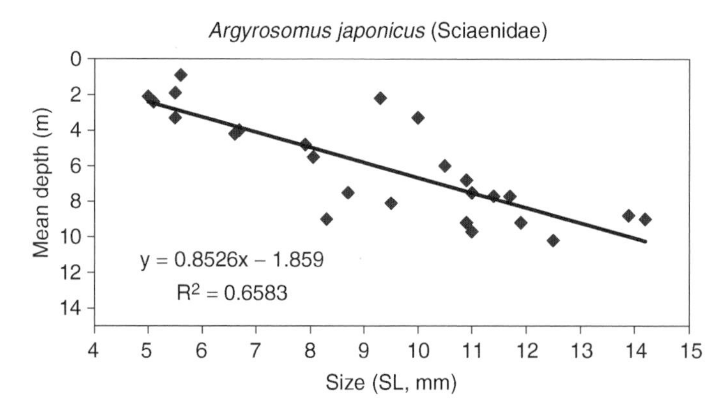

Figure 10 Ontogenetic descent of larvae of a warm-temperate sciaenid, *Argyrosomus japonicus*, in coastal waters off central New South Wales, Australia. Each point is the mean depth of a reared larva observed *in situ* and, in most cases, is based on 20 observations taken at 30-s intervals. Larvae had been released at 5 m depth. (Data from Leis *et al*., 2006b.)

In situ studies provide additional unconventional perspectives on the factors that larvae use to determine their depth. Some species seem to attempt to remain in the middle of the water column, regardless of its depth (Leis, 2004) (Figure 9), which indicates that vertical positioning may involve orientation relative to the bottom, not only to the surface, as is generally accepted. Bottom type may also play a role in depth selection, because in locations where the view downward was dark due to algal cover on the bottom (or due to great depth), larvae of some species swam deeper than in locations where the view downward was bright due to bare sand bottom (Leis and Carson-Ewart, 1999, 2001; Leis, 2004), even if the bottom itself was not visible.

Vertical distribution has also been studied using light traps, which give the same sort of information as towed nets (i.e., nothing on behaviour of individuals), but for fewer taxa due to light-trap selectivity, and only of settlement-stage larvae (which towed nets usually undersample) at night (Choat *et al.*, 1993). Nocturnal data from light traps are a useful complement to the daytime *in situ* data. Several studies in the Caribbean and Great Barrier Reef report a variety of species-specific vertical distribution patterns but report that larvae of many taxa were most abundant near the surface and that a few species preferred greater depths (Doherty and Carleton, 1997; Hendriks *et al.*, 2001; Fisher and Bellwood, 2002b; McKinnon *et al.*, 2003; Fisher, 2004). Fisher (2004) found consistent size-specific vertical distributions in several of the most abundant families, with larger individuals within each species near the surface and smaller individuals in the middle of the water column, and suggested that many larvae may migrate into surface waters at night prior to settlement.

Laboratory studies of vertical distribution are rare for the simple reason that the scale of vertical distributions in the sea is difficult to duplicate. There are, however, a number of laboratory studies that have helped to understand the proximal and ultimate causes for the patterns of vertical distribution that are found in the sea. Light, salinity, tide, temperature, flow and turbulence have all been shown to influence vertical distribution patterns and the influence of some variables changes ontogenetically (e.g., Davis, 2001). Further, some of these factors (e.g., light and temperature) can interact, leading to reversal of behaviour that would be initiated by one factor alone.

Vertical distribution behaviour has long been recognised as having the potential to indirectly influence dispersal and even result in retention, as reviewed by Sponaugle *et al.* (2002). For example, larvae that moved inshore or were not advected off the southeastern U.S. continental shelf were found deeper in the water column than larvae that were exported from the shelf. Combined with a review of other work, this led Hare and Govoni (2005) to hypothesise that in many marine systems fish larvae can achieve onshore transport by moving deeper in the water column. In another example, workers on pleuronectiform fishes have concluded that because the larvae seem to be such poor swimmers, adjustment of vertical distribution is the prime behaviour used by flatfish larvae to achieve retention or transport to nursery areas (Bailey *et al.*, 2005). But, aside from estuarine systems, there have been few actual demonstrations that such behaviour actually does result in retention of larvae near the natal area. One such demonstration has shown how larvae of a pomacentrid undertake an ontogenetic descent of 60 m, documented by towed-net samples, that interacts with vertically structured flow in the vicinity of Barbados, resulting in the larvae remaining in the vicinity of the island for over 18 days (Paris and Cowen, 2004). The retained pomacentrid larvae still have to return to settle on the Barbados reefs from the retention area 2–15 km offshore, but based on horizontal swimming abilities documented in other pomacentrid larvae, this would certainly be possible, as would enhancement of retention by directed horizontal swimming from about midway through the pelagic period of about 30 days.

Vertical distribution behaviour that results in retention or selective tidal stream transport is clearly a violation of the simplifying assumption. Whenever current velocity varies with depth, non-uniform vertical distribution of larvae violates the simplifying assumption.

9. ORIENTATION IN THE PELAGIC ENVIRONMENT

Very good swimming abilities will make little difference to dispersal trajectories if swimming in the sea is random (i.e., not orientated). Therefore, it is important to determine whether larvae in the ocean are swimming in a directionally random manner or if they have an orientation to their swimming. *In situ* studies have demonstrated that most individual larvae (66–100%, depending on species and location) can swim directionally in the pelagic environment (Figure 11) Leis *et al.*, 1996, 2006, in press; Leis and Carson-Ewart, 1999, 2001, 2003), and there are similar demonstrations for the larvae of decapod crustaceans and for jellyfish (Shanks and Graham, 1987; Shanks, 1995). Interestingly, the few studies that have addressed the ontogeny of orientation found no indication of an ontogenetic improvement in individual directionality—that is, that the percentage of directionally swimming individuals or the precision of the swimming direction increases with development—of larval fishes over a size range of 5–20 mm SL. Perhaps observations of smaller larvae will be necessary to detect this expected improvement in orientation. Individual fish larvae do not appear to be swimming randomly.

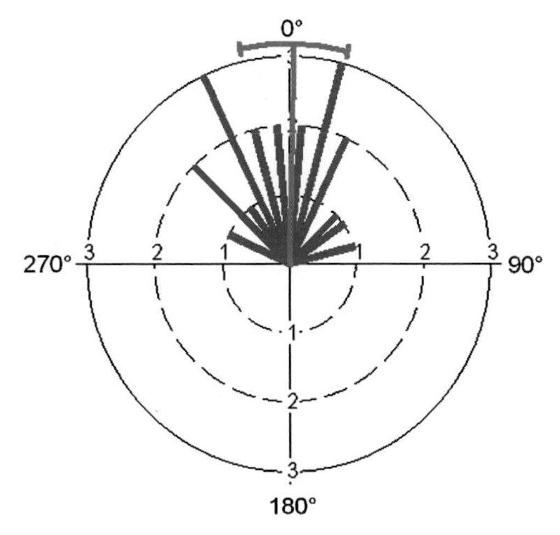

Figure 11 Frequency distribution of the swimming direction of a single larva to illustrate the data obtained from *in situ* observations of an individual larva. In this case, the observations were of a settlement-stage larva (~20 mm SL) of the lutjanid *Caesio cuning* and represent compass bearings of swimming direction taken every 30s over 10 min in open water in the vicinity of Lizard Island, Great Barrier Reef. The mean direction is the radius that pierces the outer circle, and the arc attached to it is the 95% confidence interval (CI) (shown in red). This larva swam in a highly directional manner over the 10-min observation period ($p < .001$, Rayleigh Test; J. M. Leis, unpublished).

In addition to directionality in swimming at an individual level, larvae of some species demonstrate directionality in swimming at a population level. For example, settlement-stage larvae of some pomacentrid species in the northern Great Barrier Reef swim to the southeast (Figure 12), and some of the pomacentrids swim more westerly in the afternoon than in the morning implying that a solar compass is involved (Leis and Carson-Ewart, 2003). In contrast, larvae of other species in this region seemingly orientate with respect to settlement habitat (i.e., a coral reef up to 1 km away) and swim away from the reef during the day (chaetodontids [Figure 13], perhaps others) and toward it at night (apogonids, pomacentrids) (Leis *et al.*, 1996; Leis and Carson-Ewart, 2003). In southern Taiwan, larvae of a reef-associated carangid swam away from shore off an open coast but had no overall direction within a U-shaped bay (Leis *et al.*, in press), perhaps suggesting that either motivation or ability to orientate was less in the bay. Population-level ontogenetic changes in directionality *in situ* have been demonstrated in warm-temperate Australian waters using reared larvae. In one sparid species, larvae smaller than 10 mm SL swam toward shore (northwest), whereas larger larvae swam parallel to shore (northeast) Leis *et al.*, 2006b; the authors suggested

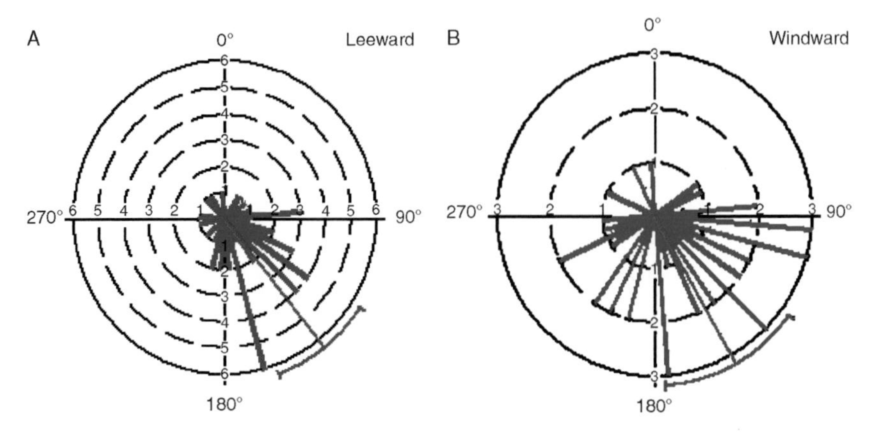

Figure 12 Frequency distribution of the mean swimming direction of settlement-stage larvae of the pomacentrid, *Chromis atripectoralis*, 100–1,000 m off the leeward and windward sides of Lizard Island, Great Barrier Reef. The plotted data are the mean swimming direction for each individual measured, usually over a 10-min period. See Figure 11 for explanation of format. (A) Leeward side of Lizard Island, based on 42 individuals. (B) Windward side of Lizard Island, based on 39 individuals. In both areas, the overall mean swimming direction was to the southeast on average ($p < .0001$, Rayleigh Test). These data are from the morning; in the late afternoon at the same time of year, larvae swam more westerly (mean, 56°), implying that they were using the sun to orientate their swimming (for more details, see Leis and Carson-Ewart, 2003, from which these data were obtained).

this behaviour could help larvae remain close to the southeastern Australian coast until they were ready, at about 10 mm, to find settlement habitat in estuarine seagrass beds, whereupon swimming parallel to shore might facilitate detection of plumes of estuarine water. In contrast, inside a bay, smaller larvae (<8 mm SL) of a second sparid species had no overall swimming directionality, whereas larger larvae swam to the east (Leis *et al.*, 2006b). All but one of the aforementioned studies were based on *in situ* observations of larvae by divers during the day (Stobutzki and Bellwood, 1998). In the only study of orientation of fish larvae at night, other than work that broadcasted recorded sounds (see below), orientation cages moored at least 30 m from reefs in the northern Great Barrier Reef were used to demonstrate that larvae of pomacentrids and apogonids were more likely to swim toward, rather than away from, the nearest coral reef (Stobutzki and Bellwood, 1998).

Swimming directionality of fish larvae may differ among locations, distances from shore or times of the day. *In situ* studies in the Great Barrier Reef using settlement-stage larvae (Leis and Carson-Ewart, 2003) showed that some pomacentrid species had the same directionality in each location (i.e., sides of an island or distances from shore). In contrast, other species

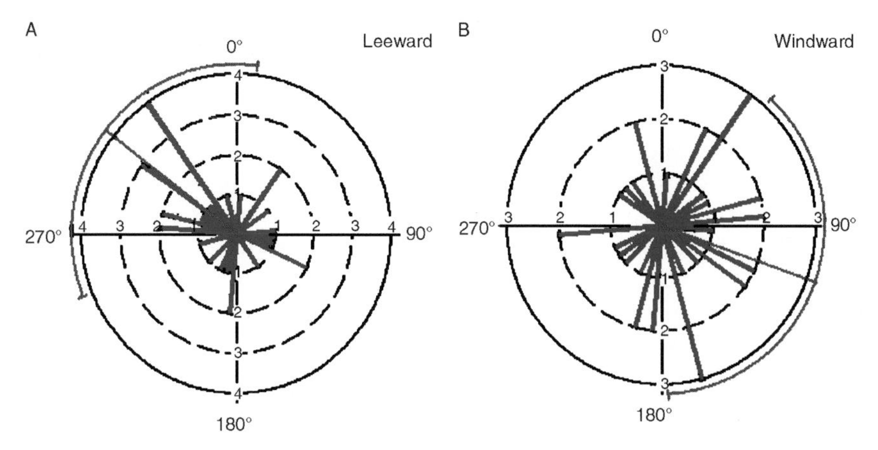

Figure 13 Frequency distribution of the mean swimming direction of settlement-stage larvae of the chaetodontid, *Chaetodon aureofasciatus*, 100–1,000 m off the leeward and windward sides of Lizard Island, Great Barrier Reef in the morning. The plotted data are the mean swimming direction for each individual measured, usually over a 10-min period. See Figure 11 for explanation of format. (A) Leeward side of Lizard Island, based on 33 individuals. (B) Windward side of Lizard Island, based on 44 individuals. In both areas, the larvae swam away from shore, on average, although this was to the northwest on the leeward side, and to the southeast off the windward side. The distributions on opposite sides of the island were significantly different (see Leis and Carson-Ewart, 2003, from which these data were obtained). This implies that the larvae were using the reefs of the island to orientate their swimming.

(chaetodontids and other pomacentrids) differed in the direction they swam on opposite sides of the island or with distance from shore. Finally, some species swam in significantly different directions in the late afternoon than in the morning. Combining these results with those of other studies (Leis *et al.*, 1996; Stobutzki and Bellwood, 1998) reveals that some taxa had differing orientation during the day than at night.

Directionality in swimming of larvae with respect to an underwater sound source has been shown in several experiments, both day and night, and with a variety of fish taxa. It is clear that settlement-stage reef-fish larvae of many families can hear and can localise a sound source at least at night (Tolimieri *et al.*, 2000, 2002, 2004; Leis *et al.*, 2002a, 2003; Leis and Lockett, 2005; Simpson *et al.*, 2005a,b; Wright *et al.*, 2005). In other words, settlement-stage reef-fish larvae can use sound for orientation. One field experiment showed that larvae of a pomacentrid species could distinguish between sounds recorded over a reef and sounds that were an artificial combination of pure tones (Leis *et al.*, 2002a). At least during the day, larvae could distinguish sound with probable biological relevance from sound without biological relevance. The distance over which sound sources might serve to orientate swimming is unclear. The influence of sound demonstrated in the field studies noted above was up to a few 100 m from the sound source. In contrast, Egner and Mann (2005) estimated from laboratory measurements that one pomacentrid species just after settlement might be able to hear reef sounds >2 km from a reef, although they doubted that sound plays a significant role in finding reefs >1 km away. Several studies have speculated that sound was the cue most likely to have given rise either to orientation in field experiments or to distributions of larvae in the field (Leis *et al.*, 1996; Doherty and Carleton, 1997; Wolanski *et al.*, 1997; Stobutzki and Bellwood, 1998).

Over smaller scales, settlement-stage apogonid and pomacentrid larvae can use smell to orientate their swimming and to locate the source of the odour. Odour sources have been conspecifics, heterospecific fishes, anemones (i.e., settlement habitat) and "lagoonal water" (Sweatman, 1988; Elliott *et al.*, 1995; Arvedlund *et al.*, 1999; Atema *et al.*, 2002). Of course, larvae orientating to an odour can do so only from downstream of the source (not true for sound), so, all else being equal, on average, larvae using a sense of smell would have to be better swimmers than those using sound to achieve the same result (e.g., number of individuals reaching settlement habitat [Armsworth, 2000]).

Rheotaxis is well documented in fish larvae, at least in laboratory situations where there are (usually visible) cues external to the moving water that can serve as a point of reference; indeed, rheotaxis is the basis for some laboratory measures of swimming abilities, including U_{crit}. Generally, orientation in relation to currents would not be expected in the pelagic environment unless an external reference, such as a view of the bottom, were available to the larva (Montgomery *et al.*, 1997, 2000, 2001; see also Chapter 3 in this volume). This is because the current itself is not detectable without such an outside reference, just as passengers cannot detect the motion of a airliner flying at a constant altitude and speed except by looking out the window. Although turbulence associated with currents could inform a larva that it was moving (Forward and Tankersley, 2001; Montgomery *et al.*, 2001), turbulence would not supply information about the direction or speed of movement (Montgomery *et al.*, 2000). External references such as magnetic fields, electric fields caused by the movement of sea water (a conductor) within the earth's magnetic field and celestial cues have been speculated upon, and although some are used for navigation by adult fishes (Kingsford *et al.*, 2002), there is no demonstration that any are used by larvae for orientation or rheotaxis. In addition, other possibilities exist. Small (1-mm s^{-1}) relative current velocities might be detected by a fish moving vertically in a stratified ocean due to the acceleration the fish would experience as it moved between strata that differed in current velocity

(Sand and Karlsen, 2000). An upstream odour source might be able to provide the cues needed for a larva to swim upstream, but this remains to be demonstrated in the pelagic environment. If a larva had the sensory ability to detect a fixed target, for example, a reef from sound produced on it, then the larva could theoretically determine it was in a current by not swimming and monitoring the relative motion of the target (Armsworth, 2001). Rheotaxis, or at least the influence of currents on swimming direction, has been looked for without success (e.g., Leis and Carson-Ewart, 2001, 2003), and there appears to be only one case in which orientation into the ambient pelagic currents by fish larvae has been shown in the field. Larvae of a sillaginid swimming in water up to 7 m deep in temperate Australia consistently orientated into the current regardless of the location or depth or whether the bottom was visible to the diver observing the larvae (Hindell *et al.*, 2003). The ability to orientate in relation to currents in the pelagic environment would be an important advantage to larvae attempting to resist dispersal, and perhaps further work would be worthwhile on this sillaginid to determine what means it used to orientate into the flow.

The above research on orientation was largely carried out during the day, and there are reasons to expect that orientation differs at night. In one field experiment, pomacentrid larvae were attracted to reef sounds at night, but not during the day (Tolimieri *et al.*, 2004), a result consistent with other observations of some species (Leis *et al.*, 1996; Stobutzki and Bellwood, 1998; Leis and Carson-Ewart, 2003). Larvae of both marine and freshwater fishes may lose the ability to maintain a position at night (Kobayashi, 1989; Copp *et al.*, 2002). Swimming frequency and speed of larval fishes may also differ on a diel basis, and this has obvious implications for orientation (Fisher and Bellwood, 2003). More work on orientation of fish larvae at night is needed.

10. SENSORY ABILITIES THAT ENABLE NONPASSIVE BEHAVIOUR (HOW DO LARVAE ORIENTATE?)

The senses that larvae might use for orientation either when swimming within the pelagic water column or when searching for settlement habitat have received increased attention, and the reader is referred to other reviews for details (Leis *et al.*, 1996; Montgomery *et al.*, 2001; Kingsford *et al.*, 2002; Leis and McCormick, 2002; Myrberg and Fuiman, 2002; Mora and Sale, 2002; see also Chapter 3 in this volume). My intention here is to mention only a few issues that bear directly on dispersal including the scales over which these senses might be useful and where there have been demonstrations that the senses were actually used for orientation.

Sensory abilities and linked behaviour of fish larvae are key factors in several aspects of dispersal. Armsworth (2001) provided modelled examples to demonstrate that "to swim efficiently in the sea, an organism should exploit current-driven movements ... favourable currents should be ridden and unfavourable currents avoided," and that larvae can substantially increase their influence on dispersal if they exploit the vertical structure of the water column. To do these things, the larvae must, at a minimum, be able to sense a target (e.g., a settlement site) or have some knowledge of the structure of the water column. A heuristic model of larval dispersal and recruitment showed that survival of larvae was "highly sensitive to both the sensing and orientating abilities of the larvae" (Codling *et al.*, 2004).

As mentioned in Section 9, several senses, most notably those of smell and hearing, are known to be useful for orientation. Much work has been done to document the morphological development of sense organs in larval fishes, and it is clear that most sense organs (eyes, ears, nose, lateral line) are well developed by the middle stages of larval development (Myrberg and Fuiman, 2002; Fuiman *et al.*, 2004). It is unclear whether other sense organs whose use is more speculative (e.g., magnetic and electric) are morphologically developed, primarily because the organs involved are not clearly understood, if they exist at all. It is difficult, however, to predict the performance of a sense organ from its structure alone (e.g., Pankhurst *et al.*, 1993; Job and Bellwood, 1996). Emphasis has switched from form to function of the sense organs, with studies now available showing that by the time of settlement the sense organs are fully functional (e.g., Poling and Fuiman, 1997, 1998), and increasingly, research is making it possible to say just what the larvae can see, hear or smell. Attempts are now underway to determine at just what stage of larval development the senses of smell and hearing are functional (S. D. Simpson and K. J. Wright, personal communications). The precision required for any sense to be effective in orientation varies with the cue and the situation. For example, if the goal is to find a small oceanic island, the accuracy and precision of the orientation required for success would necessarily be high. In contrast, if a larva in the Coral Sea is seeking the 2000-km-long Great Barrier Reef on the east coast of Australia, then only a relatively inaccurate and imprecise ability to swim to the west is required.

Vision can operate over only limited scales (tens of metres) for orientation in the ocean due to the attenuation of light, and over these scales, it is most likely to be useful for vertical distribution and settlement behaviour, and both have been demonstrated (Kingsford *et al.*, 2002; Myrberg and Fuiman, 2002). If celestial cues, such as solar compass, are used for orientation, the scale over which visual orientation can operate would be greatly extended, but the use of celestial cues is likely to be limited to the upper portions of the water column. During the day and on moon-lit nights, gobiid larvae in Hawaii were found in higher concentrations over the slope of coral reefs than in open water, but no difference was found on dark nights (Kobayashi, 1989). This led Kobayashi to conclude that the larvae could "visually orient to" reefs and resist dispersal, but when they were unable to see the reef on dark nights, they were advected away. Vision is very likely to play a key role in situations where rheotaxis exists (Vickers, 2000).

Hearing has the potential to operate over large distances (tens to hundreds of kilometres) as sound travels well in water with little attenuation (Popper and Carlson, 1998). Further, hearing is what has been called a current-independent cue because the stimulus is not distributed by currents (Armsworth, 2000). This increases the utility of sound as an orientation cue not only because it spreads in all directions, but also because weaker swimmers can utilise such a cue as they need not necessarily swim upstream when localising its source. There is increasing evidence that by the settlement stage, larval fishes can hear well and can localise sound sources (see Chapter 3 in this volume). What remains to be determined is which sound frequencies are used, at what stage in development this ability is first present, and over what spatial scales sound-based orientation actually operates. Electrophysiological measurements of hearing ability of pomacentrids at and just after settlement show that they hear well at frequencies between 100 and 1000 Hz, but that hearing sensitivity drops off rapidly at higher frequencies (Egner and Mann, 2005; Wright *et al.*, 2005). Reaction to sound was demonstrated in pomacentrids still unhatched in the demersal egg that would normally be located on a reef (Simpson *et al.*, 2005), raising the possibility that some species might be able to hear throughout the larval phase. Thus far, all demonstrations of sound being useful as an orientation cue have been at scales of <1 km.

The sense of smell can operate over moderate scales, but because odours are diluted (and their components perhaps degraded) with distance from the source, and because odour is a current-dependent cue, it is potentially less useful for orientation than sound. There is no question that settlement-stage fish larvae can detect odours, can change behaviour in response to them and can localise them, at least over scales tens of metres (Montgomery *et al.*, 2001; Kingsford *et al.*, 2002; Leis and McCormick, 2002; Myrberg and Fuiman, 2002). Further, even newly hatched larvae are capable of chemokinesis over millimetre scales that can be used for location of prey and perhaps for other orientation (Døving *et al.*, 1994), and before their demersal eggs hatch, anemone fish (Pomacentridae) young can be imprinted with the odour of the host anemone (Arvedlund *et al.*, 1999). These results indicate that the sense of smell may operate throughout the pelagic larval period, at least in some species. There has yet to be a demonstration, however, that olfactory abilities can be used by fish larvae for orientation in the pelagic environment (Vickers, 2000). As with hearing, what remains to be determined about olfaction is what substances can be detected, at what stage in development this ability is present and over what spatial scales such orientation can actually operate. Settlement-stage larvae can distinguish lagoon from ocean water (apogonids; Atema *et al.*, 2002) or river water from seawater (pleuronectids; Dando, 1984) and can detect conspecifics or amino acids (pomacentrids; Wright *et al.*, 2005) as well as settlement habitat such as anemones (pomacentrids; Arvedlund *et al.*, 1999). Most of what has been said about a sense of smell in larvae also applies to a salinity sense and sense of taste, although much less is known about them in larvae (Myrberg and Fuiman, 2002).

The use of other senses, such as magnetodetection and electrodetection, for orientation by larvae of marine fishes is largely speculative, although both are known to be used by adults of some species (Kingsford *et al.*, 2002; Myrberg and Fuiman, 2002; Wiltschko and Wiltschko, 2005). It is reasonably clear that gravity and pressure senses are used in vertical distribution behaviour, that they can be used over large scales, and that they operate at a very early stage in development. The lateral line system and its developmental precursors may be able to mediate rheotaxis, but orientation to currents cannot operate without an external reference (Montgomery *et al.*, 1997), as discussed in Section 9.

In summary, although much remains to be learned, it is clear that several classes of sensory cues may be used by fish larvae for orientation, and that this certainly violates the simplifying assumption. Several authors have argued that successful completion of the pelagic larval phase requires the detection of more than one orientation cue, involving more than one sense, either in sequence or simultaneously (e.g., Forward and Tankersley, 2001; Kingsford *et al.*, 2002). The need now is for more information about when during the pelagic larval phase the simplifying assumption stops being acceptable on the basis of sensory abilities. The research is at an early stage but, based on the evidence reviewed here, is starting to appear as if the larvae of perciform fishes have the sensory capabilities necessary to influence dispersal at about the same stage in development at which they have the locomotor capabilities necessary to make use of the sensory input.

11. INTERACTION WITH PREDATORS IN THE PELAGIC ENVIRONMENT

Mortality during the pelagic larval phase of benthic marine fishes is extremely high, although it is seldom taken into account in dispersal models. Dispersal models that do incorporate mortality estimates, however crude, seem to indicate that the high mortality rates make long-distance dispersal unlikely in demographically significant numbers (Cowen *et al.*, 2000, 2003; Hare *et al.*, 2002), even if the simplifying assumption is met. If this is true, it is likely that temporal variability in mortality will result in temporal variability in the spatial scale of dispersal. Although the simplifying assumption does not explicitly consider mortality, aspects of the interaction of larval fishes with their potential predators are relevant when considering its validity.

In situ studies of larval fish behaviour can result in unplanned observations of interactions between larvae and their predators. These show that larvae are far from passive in the presence of larger fishes, and that they usually react with behaviours that make encounters with the predator less likely ("predator avoidance," sensu Fuiman and Magurran, 1994). When approaching a reef, settlement-stage larvae that encounter a school of adult fishes will usually do one of four things that depend on the species of larva (Leis and Carson-Ewart, 1998, 2001, 2002): (1) stop swimming and hang in the water column; (2) swim back into open water away from the reef; (3) swim upward near or to the surface and over the adult school; or (4) attempt to shelter in the vicinity of the observer diver. Similar behaviour is observed in interactions with aggressive residents occupying potential settlement sites. These behaviours can be interpreted as making larvae less likely to be seen by larger fishes. Larvae are smaller and usually largely transparent or otherwise camouflaged and would not generally be visible to adult fishes at the time adults were detected by the larvae (usually several metres away). All the behaviours noted above would make it less likely that the larvae were seen than would unswerving continuance on their original course. They also result in trajectories different from that of the original course and, as such, violate the simplifying assumption. Nothing is known of the ontogeny of these behaviours in the field. Predation studies in the laboratory typically take place in small containers (i.e., smaller than the distance at which larvae in the field are observed to react to adult fishes) and thus are of limited relevance to considerations of dispersal.

12. FEEDING AS A FACTOR: FUELLING THE SWIMMER

The impressive swimming abilities of fish larvae must be fuelled. Stobutzki (1997) analyzed the body composition of settlement-stage reef-fish larvae (Pomacentridae, Chaetodontidae) both before and after unfed endurance swimming in the laboratory. Lipid appeared to be the compound most energetically important to endurance swimming. Some species lost 75% of their lipid, 66% of carbohydrate and 39% of their protein energy stores during endurance swimming. The unfed larvae had extensive internal reserves to power this swimming, but in the sea, it is very unlikely that larvae would swim until exhaustion without feeding. Indeed, *in situ* observations show that these perciform larvae feed frequently while swimming, usually "on the run" and without the slow and often inefficient feeding "C" strikes of very young larvae (Leis and Carson-Ewart, 1998; Trnski, 2002; Hindell *et al.*, 2003). Published estimates of *in situ* speed include the slight pauses necessitated by these "on-the-run meals" (Leis and Carson-Ewart, 1997).

If larvae can fuel their swimming by feeding while swimming, then laboratory values of endurance using unfed larvae will be underestimates (Fisher and Bellwood, 2001; Leis and Clark, 2005), but it is also possible that larvae swimming long distances may have a lower level of reserves or slower growth, as a result. The growth and physiological condition of fish larvae have received much attention, with most authors concluding that poor growth or condition is the result of poor food availability (e.g., Suthers, 1998). Little attention has been given to the possibility that a decrease in condition or growth may result from swimming or other behaviours that consume energy (Mora and Sale, 2002). If growth rates or physiological condition influence survival either before or after settlement (McCormick, 1998; Searcy and Sponaugle, 2001), then the amount of swimming that larvae undertake could also influence survival. It is not unreasonable to speculate that larvae that disperse long distances, at least in part by swimming, may have lower survival either before or after settlement as a result (Leis and McCormick, 2002). Similarly, larvae that achieve self-recruitment by prolonged swimming could have decreased survival for the same speculative reasons. Larvae of reef fishes do, however, appear to swim efficiently; pomacentrid larvae that were fed in endurance trials grew up to 14% in length and 40% in lateral area in spite of swimming >23 h d^{-1} for several days (Fisher and Bellwood, 2001; Leis and Clark, 2005). These larvae were relatively large and swam at high speeds in an inertial environment. In contrast, it is energetically costly for small larvae to swim slowly in a viscous environment (Hunt von Herbing *et al.*, 2001), as some taxa clearly do.

None of this would be expected under the simplifying assumption because larvae would not expend energy by swimming other than to find food.

13. SETTLEMENT BEHAVIOUR

The simplifying assumption asserts that once larvae become competent to settle, they settle onto the first bit of suitable habitat into which they are passively advected by the current. Some species seem to have a very narrow time window of competency, based on a narrow range of otolith ages found in settled individuals (e.g., Thresher *et al.*, 1989; Wellington and Victor, 1989). If this is so, it is pertinent to ask what happens to those individuals that have not found a suitable place to settle at the end of this window. Do larvae become increasingly less selective about where they settle, readily settling into places late in the competency period that would have been rejected earlier? Differences among pomacentrid larvae of different ages in the corals selected and settlement sites chosen suggest that selectivity at different scales does vary with age (Danilowicz, 1997). Do larvae that are still pelagic irresistibly metamorphose at the end of the competency period, with all the disadvantages a demersal phenotype would bring in a pelagic environment? In contrast, larvae of other species have a very wide window of competency (Victor, 1986b), although almost nothing is known about what controls the decision to settle or not. Do these taxa have an increased settlement rate as a result, or are they more selective about where they settle? Work on invertebrates suggests that delayed metamorphosis can reduce postmetamorphic growth and survival (Gebauer *et al.*, 1999), and although fish larvae with delayed metamorphosis frequently grow slower during the delay, there has been no demonstration that this leads to postsettlement disadvantages in fishes (Victor, 1986a; McCormick, 1999). Does a specific cue dictate attempts to settle and the concomitant morphological changes, as it does in some invertebrates (Hadfield, 1998)? Flexibility in the length of the pelagic period or the competency period might be expected to result in some benefit, perhaps better ultimate success in settlement. The length of the pelagic period is not, however, correlated with variability in population sizes (Eckert, 2003), but there has been no examination to determine whether the same applies to the length of the competency period. These questions remain unanswered but are important inputs to any attempts to predict dispersal.

As described earlier, in the pelagic environment, most individual larvae have directional trajectories, but in the vicinity of reefs, the trajectories are much less linear (Leis and McCormick, 2002). Presumably, this can be attributed to a search for suitable settlement habitat, but it is a clear difference in behaviour between locations that must be taken into account in modelling.

Fish larvae are clearly selective about where they settle (Marliave, 1977b; Booth and Wellington, 1998; Montgomery *et al.*, 2001), and this selectivity operates over a range of scales. When settlement-stage larvae are released near a reef, a high proportion of them reject the reef and swim back into the pelagic environment (Leis and Carson-Ewart, 1999, 2002; Leis and McCormick, 2002). Some of this rejection is due to mesoscale selectivity at the species level. For example, a lutjanid and a pomacanthid are known to reject leeward or windward reefs in favour of lagoonal reefs (Leis and

McCormick, 2002; unpublished data). A pomacentrid does the opposite (Doherty *et al.*, 1996). Some individuals, however, will reject a reef upon which other conspecifics have settled in some numbers (Leis and Carson-Ewart, 1999, 2002). At a microscale, selectivity of settlement location varies widely among species. Some pomacentrids, for example, seem to settle on the first hard substrate they encounter, and this is reflected in the distribution of their juveniles on the edges of the reef (Sale *et al.*, 1984; Leis and Carson-Ewart, 2002). Other pomacentrids seek out particular corals or reef morphology or schools of similar-sized (not necessarily conspecific) new recruits in which to settle (Danilowicz, 1996; Öhman *et al.*, 1998; Leis and Carson-Ewart, 2002), and some chaetodontids will settle only into live coral (Leis and McCormick, 2002). Warm-temperate sparids chose a variety of species-specific settlement habitats on the basis of depth, slope, type of substratum and biotic cover (Harmelin-Vivien *et al.*, 1995). Larvae of several families (Gobiesocidae, Cottidae, Agonidae, Stichaeidae, Pholidae) in a cool temperate environment displayed settlement preferences based on tactile cues and light transmission, but in some species, current speed and salinity also influenced settlement (Marliave, 1977b). Depth selectivity in settlement locations is common (Leis and Carson-Ewart, 1999, 2002; Leis *et al.*, 2002b; Srinivasan, 2003). Most of this knowledge has been gained through either observational or experimental work *in situ*, but a few studies have taken place in the laboratory (e.g., Marliave, 1977b; Danilowicz, 1996; Öhman *et al.*, 1998; Lecchini and Galzin, 2005).

In some species settlement takes place largely, or perhaps even exclusively, at night, although the number of species for which this has been clearly shown is small (Leis and Carson-Ewart, 1999; Kingsford, 2001). If these species arrive at a settlement site at the wrong time of day, then under the simplifying assumption, they would simply drift on with the current and miss the opportunity to settle. The precise timing in arrival at settlement sites by larvae of many reef fishes is a clear indication that this does not happen (Dufour, 1991, 1994; Dufour and Galzin, 1993; Doherty and McIlwain, 1996). Presumably, these settlement-stage larvae use their swimming and orientation abilities to remain in the vicinity of the settlement location and then move to it for settlement at their chosen time, and some *in situ* observations and experiments are consistent with this notion, at least with some species (Leis *et al.*, 1996; Stobutzki and Bellwood, 1998; Leis and Carson-Ewart, 2003).

In some cases, return to the pelagic environment seems to be attributable to avoidance of predators or aggressive attacks by reef residents, whereas in other cases, there is no obvious reason why the larva elected to remain pelagic (Leis and Carson-Ewart, 1999, 2002; Leis and McCormick, 2002). Almany (2003) has shown experimentally that the presence and absence of residents and predators influences the species composition and abundance of settlers on patch reefs. Therefore, it is clear that the distribution and abundance of reef resident fishes can influence the distribution and abundance of settlement. Similarly, the distribution of habitat has an overriding influence on the distribution of settlement both at mesoscales and finer scales, but there are still poorly understood factors that lead larvae to accept or reject potential settlement sites, and these require more attention.

The settlement habitat and location for many demersal fishes is not known at all, and if it is known, understanding of variation in settlement location is limited. If larvae settle into more than one habitat, there is little information about the relative proportion of individuals that settle into each or possible differences among habitats in survival to adulthood (Gillanders and Kingsford, 1996; Lecchini and Galzin, 2005). Some species settle into intermediate habitats ranging from surf zones to mangroves that differ greatly from the adult habitat in location and biotic and physical characteristics (Finn and Kingsford, 1996; Lecchini and Galzin, 2005). Basic life-history information like this is a requirement for construction of realistic dispersal models, but it is lacking for the majority of species in the tropics and for many temperate species.

The mesoscale and smaller scale selectivity of settlement location, and particularly the very common tendency for larvae to reject settlement habitat and swim back into the pelagic environment, are clear violations of the simplifying assumption.

14. DIRECTLY TESTING THE SIMPLIFYING ASSUMPTION

The papers reviewed thus far provide ample evidence that larvae of at least some marine demersal fishes have the behavioural potential to influence their dispersal, a point made repeatedly in literature. It is important to go further and ask if there is evidence that larvae actually use this potential and, as a result, disperse differently than the simplifying assumption would predict (Mora and Sale, 2002; Leis and Carson-Ewart, 2003).

One approach is to predict dispersal by applying the simplifying assumption and then compare the predictions to observed distributions of larvae. Attempts to predict larval fish dispersal are based on hydrodynamic models that range in sophistication from surface currents averaged over 1-degree squares of latitude and longitude, to satellite depictions of surface temperature, to complex, ground-truthed numerical hydrodynamic models. Where these predictions are tested with data on distribution of larvae in the sea, there are frequent discrepancies between prediction and observation, for either all species, particular taxa or particular ontogenetic stages. In addition, field studies of larval distribution frequently identify features of the distributions that seem to violate the simplifying assumption. These discrepancies and features are often attributed to active behaviour by the larvae, sometimes with variably plausible behaviours nominated (e.g., Limouzy-Paris *et al.*, 1997; Hare *et al.*, 1999, 2001, 2002; Carleton *et al.*, 2001; Schmitt and Holbrook, 2002; Bradbury *et al.*, 2003). Seldom, however, are these largely *ad hoc* attributions based on empirical data on behavioural capabilities of the species under study. Where this does happen, vertical distribution data are most often applied. Usually, however, it is assumed that behaviour of the larva caused the discrepancy, and a faulty physical dispersal model is rarely suspected. This highlights the weakness of this approach; the accuracy and precision of the physical model is rarely known. Similarly, unless models include realistic variance in the invoked behaviour (e.g., vertical distribution), it is difficult to interpret any discrepancies between mean prediction and mean observation.

Some numerical models that have applied the simplifying assumption have successfully duplicated larval distributions or have predicted self-recruitment in a region where it had been demonstrated directly, providing evidence that passive drift, or an approximation of it, may be realistic in some cases. On the Scotian Shelf, "simple particle tracking" of gadid and clupeid larvae, combined with hydrographic data, led to a conclusion that retention of larvae on or near an offshore bank "requires neither convergence nor larval behaviour" (Reiss *et al.*, 2000). A two-dimensional hydrodynamic numerical model of a portion of the Great Barrier Reef that applied the simplifying assumption predicted that self-recruitment would vary among reefs (James *et al.*, 2002), but that Lizard Island would have consistently high levels of self-recruitment. The latter prediction seemingly corroborated an earlier field study at Lizard Island of a pomacentrid species that was one of the first to demonstrate self-recruitment in a reef fish (Jones *et al.*, 1999). The model, however, applied the simplifying assumption only to the first 14 days of the pelagic larval duration and included a 1-km wide "sensory zone" around each reef, where larvae older than 14 days could actively remain and settle once they entered it. Using a numerical hydrodynamic model, Jenkins *et al.* (1999) concluded that dispersal of larvae of a temperate Australian sillaginid was not influenced by observed vertical migration, and that assuming passive behaviour provided "excellent" predictions. However, because post-larvae were found closer to shore than predicted, Jenkins *et al.* stated that transport may have been influenced by other behaviours, such as horizontal swimming. Although this species is a relatively weak swimmer, capable of sustained speeds of only 6 cm s^{-1}, such performance was more than sufficient to bridge the gap between predicted and observed distribution.

In a limited number of cases, *in situ* observations of larvae have been combined with measurements of current speed or direction to determine the extent to which dispersal trajectories of larvae differ from the simplifying assumption of passive drift. Three such studies involving 10 species of five families are summarised in Table 2. In each, *in situ* measurements of speed or direction of settlement-stage larvae over the bottom (i.e., net speed or direction) were directly compared to measurements of the current: a direct test of the simplifying assumption. Larvae may either increase or retard displacement relative to the current alone, or they may have a net displacement that differs in direction from the current alone. It is only when net speed or direction does not differ from current speed or direction that the simplifying assumption is met.

In two blenniid species, there was no clear relationship between the direction the larvae swam and the direction of the current, but in 9 of 10 cases, the larvae had a net speed faster than the current alone, indicating that they used their swimming abilities to increase displacement (Ninos, 1984), in this case, toward shore where settlement habitat was located. The net movement of sillaginid larvae varied among locations, but the larvae consistently swam into the current (Hindell *et al.*, 2003). As a result, at all locations, displacement was less than predicted by current alone. At shallow locations (<3 m), current displacement and net displacement of the larvae were not related, whereas at a deeper location (~7 m), the two displacement measures were strongly correlated, but the net displacement of the larvae was roughly half of that of the current. In a tropical study, net movement of six of seven species studied differed from that of currents in either speed or direction (Leis and Carson-Ewart, 2003). Specifically, (1) net movement direction in three of four pomacentrids and two chaetodontids differed from the current direction, (2) one pomacentrid and a chaetodontid decreased dispersal along the current axis by about half the average current speed, and (3) another pomacentrid and a lutjanid may have increased dispersal along the current axis by lesser amounts, but this last result was equivocal. There were differences among locations in some of these results.

Each study, therefore, demonstrated that settlement-stage larvae were not behaving passively, which constitutes a direct falsification of the simplifying assumption. Notably, there was no consistency in the way the simplifying assumption was violated. Some species augmented dispersal, whereas some retarded dispersal. There were also differences among locations in the manner of the violation, for example, offshore movement relative to current in one location, but onshore movement relative to current in another; lack of correlation in one location, but strong correlation in another. The work in these three studies (Table 2) took place in relatively shallow water (3–35 m) where a view of the bottom might have informed the larvae of their movement and provided a reference external to the moving water. This could be

Table 2 Characteristics of *in situ* studies that examined whether movement of settlement-stage fish larvae differed from that of passive drift with current

Family (No. of species)	Habitat	Water depth (m)	Mean ambient current speed (measured by)	Mean fish speed through water	Mean fish speed over bottom	Source
Blenniidae (2)	Warm temperate island	5–6	7–12 (drifting beads)	No data	15–17 (8–10 SL s^{-1})	Ninos, 1984
Sillaginidae (1)	Temperate bay	2–7	7–12 (drogue, current meter)	No data	2–13 (3–6 SL s^{-1})	Hindell *et al.*, 2003
Pomacentridae (4)	Tropical island	8–35	10–16 (drogue)	12–27 (12–36 SL s^{-1})	15–20 (16–25 SL s^{-1})	Leis and Carson-Ewart, 2003
Chaetodontidae (2)	Tropical island	8–35	10–16 (drogue)	18–25 (12–23 SL s^{-1})	16–21 (15–20 SL s^{-1})	Leis and Carson-Ewart, 2003
Lutjanidae (1)	Tropical island	8–22	10–16 (drogue)	34 (17 SL s^{-1})	35 (18 SL s^{-1})	Leis and Carson-Ewart, 2003

Note: All speeds are cm s^{-1} except as noted.

relevant to how the larvae managed to avoid passive drift. A major limitation of this approach is that the observations of the larvae take place over periods of minutes, whereas the pelagic larval duration is days to weeks.

Using an approach less direct than *in situ* observations, it was possible to show that larvae of a pomacentrid species remained in the vicinity of their natal island by undertaking an ontogenetic vertical migration of about 60 m as notochord flexion approached (Paris and Cowen, 2004). Highest concentrations of young (preflexion) larvae were found in the upper 20 m, whereas those of older larvae were deeper. Simulations using a hydrographic model showed that larvae remaining in the upper 20 m for the 30-day pelagic larval duration of the species would be advected from the vicinity of the island. In contrast, larvae located deeper in the water column experienced shoreward flow and were retained near the island. The ontogenetic vertical descent of the larvae enabled them to avoid the offshore surface flow that would have advected passive larvae away from the island over the pelagic duration of the species. This was detected by intensive, vertically stratified sampling with plankton nets combined with physical oceanographic measurements that contributed to the development of a numerical model of the region—a labour-intensive, but exemplary, approach. This work provides another clear falsification of the simplifying assumption as retention of the larvae "cannot be explained without invoking behavior" (Paris and Cowen, 2004), in this case ontogenetic vertical migration. In addition, it seems likely that orientated horizontal swimming by the settlement-stage larvae would be required to reach settlement habitat on the island's reefs from their retained location several kilometres offshore.

15. THE IMPORTANCE OF HYDROGRAPHY

The impressive swimming and other behavioural capabilities of fish larvae, as well as their ability in some cases to avoid water-column flow, do not mean that hydrography is an unimportant component of dispersal (Cowen, 2002). Swimming speeds of larvae are of the same order of magnitude as current speeds of the water in which they swim (Leis and McCormick, 2002; Fisher, 2005). Therefore, currents must be a major factor in dispersal of fish larvae. Good hydrography—verified with field measurements—is required to discover the scale and nature of the influence of coastal and underwater topography on flow (Cowen, 2002). For example, the phenomenon of "sticky water" in the vicinity of coral reefs (whereby flow through a reef matrix is impeded by an unexpected increase in drag when incoming velocities are high) has only recently been discovered (Wolanski and Spagnol,

2000; Spagnol *et al.*, 2001), and it has large implications for dispersal in such systems. Such features are not only capable of retaining larvae (or making it easier for the larvae to effect retention through behaviour) but also represent a "hydrographic buffer" through which the propagules of benthic fishes much pass not only when departing from adult habitat, but also when returning to it from the far field circulation. Larvae are not generally released directly into the far field circulation, but into or near the adult habitat on a reef or soft bottom, bathed in these hydrographic buffers (Leis, 2002).

Numerical hydrographic models used in prediction of dispersal need to be three dimensional, because neither hydrography nor abundance and species composition of larvae are uniform with depth and because individual larvae actively change their vertical distributions. Two-dimensional models may, at times, provide predictions that agree with observed distributions of larvae of some species, possibly because either the larvae or the currents were not vertically stratified, but such predictions are unlikely to be generally valid and should be viewed with caution. Ground-truthing of all aspects of predictions of numerical models is also important. Recent work on surface flow near shore shows that some common assumptions about the nature of such flow are incorrect (Tapia *et al.*, 2004) and serves as a warning about uncritical acceptance of numerical models of circulation. Further, it may be necessary for the hydrographic component of dispersal models to use a much finer calculation cell size in the hydrodynamic buffer near topographic features (including the shoreline) than in the far field in order to accurately portray the movement of water at scales relevant to dispersal.

Weather predictions, which have a mixed record of reliability, are derived from numerical models that are the meteorological (atmospheric) equivalent of the numerical hydrographic models that are the basis for most dispersal models. Such meteorological models are continually being updated and improved and are the result of multiple iterations and constant ground truthing. Few numerical hydrographic models are ground truthed, and fewer still are subject to the types of corrections and improvements that have led meteorological models to their current state of accuracy and precision. Yet, marine biologists almost uniformly accept at face value the, usually first-generation, hydrographic models produced by physical oceanographers. Physical oceanographers know better, and there is at least one celebrated case in which conflict among physical oceanographers over the relative merits of the numerical models produced by different researchers for the same areas spilled over into the biological literature (Black, 1988, 1995; Black and Moran, 1991; Wolanski, 1993; Wolanski and Sarenski, 1997; Wolanski *et al.*, 1997, 1999). In this case, one group argued that the mismatch between prediction and observation found by the other was due to use of an inappropriate hydrodynamic model.

Armsworth *et al.* (Armsworth, 2000; Armsworth *et al.*, 2001) have, from a theoretical perspective, concluded that, at least in coral-reef systems, details of hydrographic circulation are less important to settlement at the end of the pelagic larval period than the sensory and swimming abilities of larvae. Using scenarios of larvae of different behavioural abilities attempting to reach a distant reef for settlement, Armsworth (2000) performed a scaling analysis including current structure, swimming ability and sensory ability and concluded that "hydrodynamically based considerations . . . appear less important than considerations of larval swimming, and the interaction of swimming with these physical transport processes. The extent of sensory capabilities of larvae proves to be a critical parameter, and the rate of larval supply depends sensitively upon it . . . ." This does not decrease the relevance of hydrography but emphasises the need to include both behaviour and hydrology into considerations of dispersal during the pelagic larval stage.

Upon hatching, larvae of most species of benthic fishes will have poor swimming abilities and, for a period, may well come close to conforming to the simplifying assumption (Fisher *et al.*, 2000; Fisher, 2005; Clark *et al.*, 2005; Leis *et al.*, 2006a,b), as may pelagic fish eggs before hatching. That period may be brief, however. More work on behavioural ontogeny in larval fishes is required to determine when, in terms of both time and morphological development, this period of quasi-acceptance of the simplifying assumption and dominance of hydrographic factors should end. Most likely, the transition from plankton to nekton will be gradual, rather than an abrupt end point. But, end it will, and after that, both hydrography and behaviour must be taken into account. During the expected gradual transition period, developing behavioural capabilities should also be taken into account. As the heuristic modelling has shown, even limited abilities can have an important influence on dispersal outcomes.

16. ARE FISH LARVAE PLANKTON OR NEKTON?

The simple answer is that larvae of demersal perciform fishes are both plankton and nekton. At the start of their pelagic period, and for a time following, larvae are both morphologically and behaviourally poorly developed; they are small, they lack fins other than fin folds, they have incompletely developed sensory organs and they swim slowly, probably in a viscous environment. During this time, they approach the planktonic condition implied by the simplifying assumption: passive drift with the currents. Yet, even at this stage of development, larvae have some locomotor abilities and the ability to regulate their vertical distribution, thus indirectly influencing their dispersal. At the end of the pelagic period, and for a time prior to this, larvae are both morphologically and behaviourally well developed; they are one or more centimetres in length, they have all or nearly all their fins, they have functional sensory organs that can detect predators and settlement habitat, and they swim well (faster than ambient currents in many cases), certainly in an inertial environment. During this time, larvae constitute gross violations of the simplifying assumption and are clearly nektonic. Perciform species have been the focus of this chapter, and it is not clear whether larvae from orders other than Perciformes (including Scorpaeniformes) are behaviourally as capable. Those who work on larvae of pleuronectiform fishes have seemingly concluded that flatfish larvae are closer to planktonic than nektonic throughout their pelagic phase (Bailey *et al.*, 2005). But knowledge of swimming abilities of pleuronectiform larvae is based almost exclusively on routine swimming, which returns lower speeds than other methods, and there are indications that behaviours other than vertical distribution may be important for pleuronectiform young (Dando, 1984), so it may be appropriate to reassess this view. Historic (i.e., phylogenetic), physical and habitat differences confound attempts to compare the behaviour and performance of these different orders (Leis and McCormick, 2002), and it is certainly inappropriate to assume that what applies to a temperate gadiform fish will apply to a tropical perciform fish (or, even a temperate perciform). There is, therefore, a clear need to conduct parallel studies of larval behaviour across different taxonomic groups and across taxa from different environments.

The real challenge is to describe how the gradual transition from plankton to nekton takes place during the pelagic larval phase and quantify the development of behavioural abilities. We need to know when behaviour can directly influence dispersal, but the answer to this will depend on context. It will differ among species, among locations and among scales, because species differ in their behavioural capabilities and ontogenies, because locations differ in their hydrography and because the relevance of behavioural inputs can depend on the scale over which dispersal takes place. Minimum states of morphological, sensory and behavioural development must each be attained before the transition from plankton to nekton can be considered complete. The simplifying assumption will largely apply to a larva that can swim well but lacks the sensory abilities to detect the cues needed to orientate its trajectory. Much remains to be learned about the ontogeny of behaviour in larvae of marine fishes, but at present, it appears that behaviour can have a large indirect influence on dispersal throughout the pelagic larval period and a large direct influence from about the time the caudal fin forms, or shortly thereafter.

Interspecific differences in abilities at any point in development are to be expected. For example, larvae that hatch from demersal eggs are typically larger and morphologically (and presumably behaviourally) better developed when they enter the pelagic environment than are larvae that hatch from pelagic eggs (Leis and McCormick, 2002; Leis and Carson-Ewart, 2004). Further, an acanthurid larva that settles at 25 mm SL will most likely have swimming and sensory abilities that are superior to those of a sparid that settles at 8 mm. Because growth rates differ both among and within species, the time that it takes to reach various developmental milestones will vary both among and within species. Although size, and not age, has typically been shown to be the best predictor of performance, the variation in performance at any size is typically large (both in wild and in reared larvae). All this means that, even in a given hydrological situation, the extent to which larvae are able to behaviourally influence dispersal trajectories will vary both among and within taxa, both among and within size groups and both among and within ages. Variation around mean performance (including consideration of the best performers) will be an important factor to include in dispersal models. And, of course, the interaction of behaviour with the hydrology in the area of interest is an essential component of any realistic consideration of dispersal.

It does not necessarily follow, however, that high variation in performance leads to high variation in dispersal outcomes. Larvae with poor performance might have poor survival, thus providing a filter that limits variation in dispersal outcomes. Further, many of the available estimates of performance are best thought of as measures of potential performance. Potential performance, either in whole or in part, may not be applied to influencing dispersal, or indeed to events in the pelagic period at all. It is conceivable that the behavioural capabilities of pelagic larvae may have more to do with preparation for survival in the postsettlement stage than with anything during the presettlement stage. Alternatively, larvae may attain the ability to directly influence dispersal much sooner than it is used; for example, swimming abilities might develop at about the time of notochord flexion, but not actually come into play until competency to settle is attained, or particular cues from settlement habitat are detected. These are important areas for future research.

Larvae may use their behavioural capabilities either to enhance or to retard dispersal, as shown in the section on direct tests. Evidence for recruitment of demersal fishes at local scales is increasing, and researchers have tended to attribute this to the behavioural abilities of larvae, but although there is some indirect evidence that behaviour, rather than hydrology alone, is involved, there is little direct evidence. Where studies find a discrepancy between distribution of larvae predicted by hydrodynamics and that actually observed, most instances seem to involve retarded dispersal (e.g., Cowen *et al.*, 2003), although others involve enhanced dispersal (e.g., Hare *et al.*, 2001, 2002), even if the enhancement is simply one in which the larvae actually reach settlement habitat on the continental shelf rather than remaining in deep water off the shelf.

The research reviewed here clearly shows that the simplifying assumption is invalid for at least a large proportion of the pelagic larval phase of demersal perciform fishes. The simplifying assumption can be shown to be wrong in each of its parts:

- Larvae are strong swimmers and can swim at speeds of the same magnitude as ambient currents for long periods.
- Biological variables other than duration of the pelagic larval stage are important in determining dispersal, including swimming abilities, vertical distribution, sensory abilities and orientation.
- Behavioural abilities of larvae that are of relevance to dispersal differ both ontogenetically and among species, and behaviour of larvae can differ both temporally and spatially.
- Larvae are highly selective about when and where they will settle at spatial scales from centimetres to 100s of kilometres.

Major questions remain about variation among taxa (particularly non-perciform fishes), the ontogeny of the change from plankton to nekton and the extent to which larvae actually use their behavioural potential to influence dispersal. It is perhaps not surprising that some aspects of the simplifying assumption are at odds with the data, but the degree to which behaviour in larval fishes violates this set of assumptions is remarkable. This violation is central, not marginal.

The research reviewed here shows that behavioural capabilities of larval fishes are so well developed that they cannot be ignored in dispersal models for which realism is a goal. The simplifying assumption really is dead (even if some refuse to attend the funeral), but how best to incorporate behaviour into considerations of dispersal is unclear, particularly when understanding of behaviour in fish larvae is so incomplete (Armsworth *et al.*, 2001). Attempts to date have focused on vertical distribution (see above), probably because vertical movements require only modest swimming abilities of which larvae were assumed to be capable, and because it has long been recognised that larvae do control their vertical distributions, and that vertical distribution does influence dispersal. The few inclusions of other behaviours, such as horizontal swimming and orientation to sensory cues, into predictive models have been based largely on speculation of what larvae might be capable of rather than empirical data (e.g., Wolanski *et al.*, 1997). Heuristic use of models to estimate the type and scale of larval behaviour that might be influential in dispersal can provide valuable information (e.g., Porch, 1998; Armsworth, 2000; Armsworth *et al.*, 2001; Cowen *et al.*, 2003). Scaling analysis can also be valuable in this regard, although it is revealing that one early scaling analysis of dispersal of fish larvae stated that "a growing body of evidence indicates that this [passive drift] is often not the case," but still applied the simplifying assumption (Hatcher *et al.*, 1987). A more recent variation on scaling analysis combined laboratory observations of diel behaviour, measures of swimming performance by two laboratory methods and current measurements (Fisher, 2005), to examine the potential for behaviour to influence dispersal. In this analysis, a conservative empirical relationship between routine speed and U_{crit} was used, with U_{crit} measurements taken throughout larval development to estimate the cumulative potential distance that larvae could swim during their pelagic phase. When compared with current transport distance in a particular area, this showed that, based on mean values, none of the tested taxa could swim as far as the potential transport distance, but that the 95% confidence interval of mean predicted distance swum broadly overlapped that of mean transport distance in most taxa (Fisher, 2005). In short, an average larva of any taxon might not be able to overcome by horizontal swimming alone the average advection by currents throughout its pelagic period, but some individual larvae could. Further, all larvae could accomplish this when minimal transport distances were considered.

17. FUTURE DIRECTIONS

As dispersal during the larval stage is seen as determining the geographic scale of population connectivity for demersal fishes, an understanding of connectivity depends on an understanding of dispersal. An understanding of dispersal during the pelagic larval stage of marine demersal fishes—both how hydrography and behaviour interact to produce dispersal trajectories, and how to accurately predict these trajectories—is increasingly recognised as essential to viable management of the populations of these fishes (Cowen, 2002; Kinlan and Gaines, 2003; Palumbi, 2003; Sale, 2004). Therefore, an important task ahead is providing the empirical base that will allow realistic incorporation of larval behaviour into predictive dispersal models. This will be labour-intensive and far from easy, but it can no longer be considered irrelevant or placed into the "too hard" basket. Possibly equally challenging will be the inclusion of this empirical element, and its inherent variability, into numerical dispersal models. It is increasingly evident that predictive dispersal models that strive for realism must also grapple with the influence of both larval mortality and physiological condition (including their temporal and spatial variation), and that these must also be based on empirical input from both the field and the laboratory (Cowen *et al.*, 2000; Cowen, 2002; Leis and McCormick, 2002).

Behavioural work on fish larvae must have a broader taxonomic coverage. Just as work in temperate areas has concentrated on a few gadids and a few clupeiform species, tropical work has concentrated on pomacentrids. Given the diversity of fishes (~24,000 teleost species of which about 60% are marine [Nelson, 1994]), the diversity of their larval morphologies and the diversity of the marine habitats they occupy, the taxonomic base upon which our understanding of larval fish behaviour is based is embarrassingly thin. Differences among taxa in behaviour of larvae are already clear. Further, we need to more carefully take into account the phylogeny of the species we attempt to compare. At the same time, it is important to seek morphological predictors of swimming performance in larvae, such as body shape, muscle mass or muscle histology so that performance of species whose behaviour is unstudied can be predicted.

We need to place more emphasis on the ontogeny of behaviour so the gradual transition from plankton to nekton can be identified. We must also place more emphasis on studies of behaviour in the sea. Laboratory studies of the behaviour of pelagic animals need to be verified in the pelagic environment. Perhaps use of mesocosims to study behaviour in fish larvae could be a useful approach for testing whether laboratory observations "scale up" accurately. Currently, most *in situ* observations of larval behaviour have been carried out by SCUBA divers, and we have no direct evaluations of how this might bias the observations. Remote sensing by the use of sonar (Jaffe *et al.*, 1995) or video (Holbrook and Schmitt, 1999) might enable such evaluations and offer a better means to study behaviour in the sea, including at night. The divers who study larval behaviour should consider the use of rebreathers to lessen disturbance by eliminating bubbles.

The study of sensory abilities of fish larvae requires more attention, especially the relatively "speculative senses" such as magnetoreception and electroreception. The possibility that larvae can sense the movement of the water column in which they swim should be investigated, and if demonstrated, the means by which it is achieved would be important to ascertain. It will be very challenging, indeed, to determine the spatial scales over which the demonstrated abilities of larvae to hear and smell settlement site cues actually do operate in the sea.

At a procedural level, evaluation of the relative behavioural performance of reared and wild larvae (Smith and Fuiman, 2004) is important whenever reared larvae are used in behavioural studies. Similarly, contrasting average versus exceptional individuals, not only in relation to performance itself but also in relation to the implications of that performance, is important (Fuiman and Cowan, 2003). Increasing evidence points to the possibility that it is the exceptional performers, rather than average individuals, that survive in the sea, and predictive and heuristic dispersal models that consider both may offer considerable insight.

We have a growing body of knowledge of the behavioural capabilities of larval fishes and of the potential influence these might have on dispersal. Much slower, however, is the growth of knowledge about the actual (as opposed to potential) influence these have (Hindell *et al.*, 2003; Leis and Carson-Ewart, 2003; Leis and Fisher, 2006). Growth of the latter must accelerate if our understanding of dispersal is to avoid another potentially misleading path. Direct tests of the simplifying assumption can tell us the extent to which it might be violated. More important and difficult to determine will be demonstrations of the factors that led to any violation. As Peter Sale (2004) puts it, "we ... know that larval reef fish are well-adapted, behaviourally complex, pelagic organisms that do not disperse passively on water currents, but inadequate information concerning the extent of their abilities to control their larval dispersal, and whether these abilities are used to stay close to home or to travel widely." The same applies to larvae of demersal fishes that live in habitats other than reefs. Increasing use of otolith microchemistry and cutting edge genetic tools to identify the source of settling larvae will be useful in determining "where the babies come from" and identifying the difference between actual dispersal and the null hypothesis of accepting the simplifying assumption. It may be possible to use genetic engineering to "tag" propagules in this effort. As nanotechnology develops, it may also become possible to produce nanotags to achieve even finer scale identification of the source of settling larvae. "Smart tagging" of adults of many marine organisms has proved enormously useful in understanding migrations in the sea, and if this were possible in larval stages, it would revolutionise our understanding of dispersal trajectories.

The larvae of perciform fishes tend to reach morphological developmental milestones at smaller sizes than larvae of other orders, such as clupeiform, gadiform and pleuronectiform (CGP) fishes (Moser *et al.*, 1984; Okiyama, 1988; Moser, 1996; Leis and Carson-Ewart, 2004; Richards, 2005). The same may be true of behavioural milestones, based on the limited information available on behavioural ontogeny in larvae of these orders (Leis and McCormick, 2002). This latter conclusion must, however, be viewed with caution. Comparisons of swimming ability among orders are at least partially confounded by differences in methodology. Vertical distribution behaviour is well developed in larval CGP fishes from an early stage, but there has been relatively little behavioural work on larvae of CGP fishes that is relevant to the issues of dispersal addressed here, specifically, orientation and the sensory abilities of hearing and olfaction. Given the large differences amongst orders in larval morphology and development, comparisons of behaviour must be made carefully if they are not to be misleading.

Perciform fishes dominate marine fish communities in warmer waters, whereas in cooler waters, other orders dominate. Further, there are sound physical and physiological reasons to expect that swimming by fish larvae might be more efficient in warmer water (Fuiman and Batty, 1997; Hunt von Herbing, 2002). Finally, in warmer water, larvae probably take longer to reach a given size than in cooler water, which, combined with the size-related factors mentioned in the previous paragraph, means that particular ontogenetic milestones, both morphological and behavioural, will be reached much sooner in warm waters. Therefore, it might be anticipated that in warmer waters, fish larvae may have more behavioural influence over dispersal than occurs in cooler waters, not only because they become behaviourally competent sooner, but also because they may swim more efficiently. This is based, then, on differences in the species mix between areas, and on the physical and physiological effects of differences in temperature. This will be an interesting idea to test, but we are a long way from being able to conclude that larvae of tropical fishes are more behaviourally competent than their temperate-water counterparts.

The larvae of perciform fishes have proved to be remarkably proficient at swimming, at orientating in the sea and at sensing cues. They are capable in ways that would have been unimaginable 15 yr ago, and they will probably continue to surprise us with their capabilities. The challenge is to derive ways to clearly measure these capabilities and their ontogeny and then to determine the extent to which they actually do apply (or not) to the questions at hand, particularly dispersal. The simplifying assumption is indeed dead. The imperative is to ensure that the replacement paradigm is more realistic and less resistant to change when the facts contradict the assumptions.

ACKNOWLEDGEMENTS

Preparation of this review (completed in mid 2005) was supported by an ARC Discovery Grant (DP0345876), and a DST International Science Linkages Grant (IAP-IST-CG03-0043) to the author, and by the Australian Museum. I thank Lee Fuiman for inviting me to write this review; the many colleagues who commented on oral presentations of the ideas contained herein or who provided me with copies of their work prior to publication; Rebecca Fisher, Barry Goldman, Kelly Wright and two reviewers for comments on the manuscript; Sue Bullock for editorial assistance; and the fish larvae for being so interesting.

REFERENCES

Allen, G. R. (1991). "Damselfishes of the World". Mergus Publishers, Melle, Germany.

Almany, G. R. (2003). Priority effects in coral reef fish communities. *Ecology* **84**, 1920–1935.

Anderson, T. W., Bartels, C. T., Hixon, M. A., Bartels, E., Carr, M. H. and Shenker, J. M. (2002). Current velocity and catch efficiency in sampling settlement-stage larvae of coral-reef fishes. *Fishery Bulletin, U.S.* **100**, 404–413.

Armsworth, P. R. (2000). Modelling the swimming response of late stage larval reef fish to different stimuli. *Marine Ecology Progress Series* **195**, 231–247.

Armsworth, P. R. (2001). Directed motion in the sea: efficient swimming by reef fish larvae. *Journal of Theoretical Biology* **210**, 81–91.

Armsworth, P. R., James, M. K. and Bode, L. (2001). When to press on, wait or turn back: Dispersal strategies for reef fish larvae. *American Naturalist* **157**, 434–450.

Arvedlund, M., McCormick, M. I., Fautin, D. G. and Bildsoe, M. (1999). Host recognition and possible imprinting in the anemonefish *Amphiprion melanopus*. *Marine Ecology Progress Series* **188**, 207–218.

Atema, J., Kingsford, M. J. and Gerlach, G. (2002). Larval fish could use odour for detection, retention and orientation to reefs. *Marine Ecology Progress Series* **241**, 151–160.

Bailey, K. M., Nakata, H. and van der Veer, H. (2005). The planktonic stages of flatfishes: Physical and biological interactions in transport processes. *In* "The Biology of the Flatfishes (Pleuronectiformes)" (R. Gibson, ed.), pp. 94–119. Blackwell Science, Oxford.

Bainbridge, R. (1952). Underwater observations on the swimming of marine zooplankton. *Journal of the Marine Biological Association of the United Kingdom* **31**, 107–112.

Barnett, A. M., Jahn, A. E., Sertic, P. D. and Watson, W. (1984). Distribution of ichthyoplankton off San Onofre, California, & methods for sampling very shallow coastal waters. *Fishery Bulletin, U.S.* **82**, 97–111.

Bartsch, J. (1988). Numerical simulation of the advection of vertically migrating herring larvae in the North Sea. *Meeresforschung* **32**, 30–45.

Bartsch, J. and Knust, R. (1994). Simulating the dispersion of vertically migrating sprat larvae (*Sprattus sprattus* (L.)) in the German Bight with a circulation and transport model system. *Fisheries Oceanography* **3**, 92–105.

Bellwood, D. R. and Fisher, R. (2001). Relative swimming speeds in reef fish larvae. *Marine Ecology Progress Series* **211**, 299–303.

Beyst, B., Mees, J. and Cattrijsse, A. (1999). Early postlarval fish in the hyperbenthos of the Sutch Delta (south-west Netherlands). *Journal of the Marine Biological Association of the United Kingdom* **79**, 709–724.

Billman, E. J. and Pyron, M. (2005). Evolution of form and function: Morphology and swimming performance in North American minnows. *Journal of Freshwater Ecology* **20**, 221–232.

Black, K. P. (1988). The relationship of reef hydrodynamics to variations in numbers of planktonic larvae on and around coral reefs. *Proceedings of the 6th International Coral Reef Symposium* **2**, 125–130.

Black, K. P. (1995). Response to (E. Wolanski) 'Facts and numerical artefacts in modelling the dispersal of crown-of-thorns starfish larvae in the Great Barrier Reef'. *Marine and Freshwater Research* **46**, 883–887.

Black, K. P. and Moran, P. J. (1991). Influence of hydrodynamics on the passive dispersal and initial recruitment of larvae of *Acanthaster planci* (Echinodermata: Asteroidea) on the Great Barrier Reef. *Marine Ecology Progress Series* **69**, 55–65.

Black, K. P., Moran, P. J. and Hammond, L. S. (1991). Numerical models show coral reefs can be self-seeding. *Marine Ecology Progress Series* **74**, 1–11.

Blaxter, J. H. S. (1976). Reared and wild fish – how do they compare? *In* "Proceedings of the 10th European Symposium on Marine Biology" (G. G. Persoone and E. Jaspers, eds), pp. 11–26. Universa Press, Wettern.

Blaxter, J. H. S. (1986). Development of sense organs and behavior of teleost larvae with special reference to feeding and predator avoidance. *Transactions of the American Fisheries Society* **115**, 98–114.

Boehlert, G. W. (1996). Larval dispersal and survival in tropical reef fishes. *In* "Reef Fisheries" (N. V. C. Polunin and C. M. Roberts, eds), pp. 61–84. Chapman & Hall, London.

Booth, D. J. and Wellington, G. (1998). Settlement preferences in coral-reef fishes: Effects on patterns of adult and juvenile distributions, individual fitness and population structure. *Australian Journal of Ecology* **23**, 274–279.

Bradbury, I. R. and Snelgrove, P. V. R. (2001). Contrasting larval transport in demersal fish and benthic invertebrates: The roles of behaviour and advective processes in determining spatial pattern. *Canadian Journal of Fisheries and Aquatic Sciences* **58**, 811–823.

Bradbury, I. R., Snelgrove, R. V. R. and Pepin, P. (2003). Passive and active behavioural contributions to patchiness and spatial pattern during the early life history of marine fishes. *Marine Ecology Progress Series* **257**, 233–245.

Breitburg, D. L. (1989). Demersal schooling prior to settlement by larvae of the naked goby. *Environmental Biology of Fishes* **26**, 97–103.

Breitburg, D. L. (1991). Settlement patterns and presettlement behavior of the naked goby, *Gobiosoma bosci*, a temperate oyster reef fish. *Marine Biology* **109**, 213–221.

Brett, J. R. (1964). The respiratory metabolism and swimming performance of young sockeye salmon. *Journal of the Fisheries Research Board of Canada* **21**, 1183–1226.

Brown, C. and Laland, K. (2001). Social learning and life skills training for hatchery reared fish. *Journal of Fish Biology* **59**, 471–493.

Caley, M. J., Carr, M. H., Hixon, M. A., Hughes, T. P., Jones, G. P. and Menge, B. A. (1996). Recruitment and the local dynamics of open marine populations. *Annual Review of Ecology, Evolution and Systematics* **27**, 477–500.

Carleton, J. H., Brinkman, R. H. and Doherty, P. J. (2001). The effects of water flow around coral reefs on the distribution of pre-settlement fish (Great Barrier Reef, Australia). *In* "Oceanographic Processes of Coral Reefs" (E. Wolanksi, ed.), pp. 209–230. CRC Press, Boca Raton.

Carroll, R. L. (1988). "Vertebrate Paleontology and Evolution". Freeman, New York.

Choat, J. H., Doherty, P. J., Kerrigan, B. A. and Leis, J. M. (1993). A comparison of towed nets, purse seine, and light-aggregation devices for sampling larvae and pelagic juveniles of coral reef fishes. *Fishery Bulletin, U.S.* **91**, 195–209.

Clark, D. L., Leis, J. M., Hay, A. C. and Trnski, T. (2005). Swimming ontogeny of larvae of four temperate marine fishes. *Marine Ecology Progress Series* **292**, 287–300.

Codling, E. A., Hill, N. A., Pitchford, J. W. and Simpson, S. D. (2004). Random walk models for the movement and recruitment of reef fish larvae. *Marine Ecology Progress Series* **279**, 215–224.

Copp, G. H., Faulkner, H., Doherty, S., Watkins, M. S. and Majecki, J. (2002). Diel drift behaviour of fish eggs and larvae, in particular barbel, *Barbus barbus* (L.), in an English chalk stream. *Fisheries Management and Ecology* **9**, 95–103.

Cowen, R. K. (2002). Larval dispersal and retention and consequences for population connectivity. *In* "Coral Reef Fishes: Dynamics and Diversity in a Complex Ecosystem" (P. F. Sale, ed.), pp. 149–170. Academic Press, San Diego.

Cowen, R. K. and Sponaugle, S. (1997). Relationships between early life history traits and recruitment among coral reef fishes. *In* "Early Life History and Recruitment in Fish Populations" (R. C. Chambers and E. A. Trippel, eds), pp. x–xx. Chapman & Hall, London.
Cowen, R. K., Lwiza, K. M. M., Sponaugle, S., Paris, C. B. and Olson, D. B. (2000). Connectivity of marine populations: Open or closed? *Science* **287**, 857–859.
Cowen, R. K., Paris, C. B., Olson, D. B. and Fortuna, J. L. (2003). The role of long distance dispersal versus local retention in replenishing marine populations. *Gulf and Caribbean Research* **14**, 129–138.
Cushing, D. H. (1990). Plankton production and year class strength in fish populations: An update of the match/mismatch hypothesis. *Advances in Marine Biology* **26**, 249–293.
Dando, P. R. (1984). Reproduction in estuarine fish. *In* "Fish Reproduction, Strategies and Tactics" (G. W. Potts and R. J. Wooton, eds), pp. 155–170. Academic Press, London.
Danilowicz, B. S. (1996). Choice of coral species by naive and field caught damselfish. *Copeia* **1996**, 735–739.
Danilowicz, B. S. (1997). The effects of age and size on habitat selection during settlement of a damselfish. *Environmental Biology of Fishes* **50**, 257–265.
Davis, M. W. (2001). Behavioural responses of walleye pollock, *Theragra chalcogramma*, larvae to experimental gradients of seawater flow: Implications for vertical distribution. *Environmental Biology of Fishes* **61**, 253–260.
Davis, M. W. and Olla, B. L. (1995). Formation and maintenance of aggregations in walleye pollock, *Theragra chalcogramma*, larvae under laboratory conditions: Role of visual and chemical stimuli. *Environmental Biology of Fishes* **44**, 385–392.
Ditty, J. G., Fuiman, L. A. and Shaw, R. F. (2003). Characterizing natural intervals of development in the early life of fishes: An example using blennies. *In* "The Big Fish Bang: Proceedings of the 26th annual Larval Fish Conference" (H. I. Browman and A. B. Skiftesvik, eds), pp. 405–418. Institute of Zoology, University of Bergen, Bergen.
Doherty, P., Kingsford, M., Booth, D. and Carleton, J. (1996). Habitat selection before settlement by *Pomacentrus coelestis*. *Marine and Freshwater Research* **47**, 391–399.
Doherty, P. and McIlwain, J. (1996). Monitoring larval fluxes through the surf zones of Australian coral reefs. *Marine and Freshwater Research* **47**, 383–390.
Doherty, P. J. (1987). Light traps: Selective but useful devices for quantifying the distributions and abundances of larval fishes. *Bulletin of Marine Science* **41**, 423–431.
Doherty, P. J. and Carleton, J. H. (1997). The distribution and abundance of pelagic juvenile fish near Grub Reef, central Great Barrier Reef. *Proceedings of the 8th International Coral Reef Symposium* **2**, 1155–1160.
Doi, M., Kohno, H., Taki, Y. and Ohno, A. (1998). Development of swimming and feeding functions in larvae and juveniles of the red snapper, *Lutjanus argentimaculatus*. *Journal of Tokyo University of Fisheries* **85**, 81–95.
Dudley, B., Tolimieri, N. and Montgomery, J. (2000). Swimming ability of the larvae of some reef fishes from New Zealand waters. *Marine and Freshwater Research* **51**, 783–787.
Dufour, V. (1991). Variations of fish larvae abundance in reefs: Effects of light on the colonization of the reefs by fish larvae. *Comptes rendus hebdomadaires de l'Académie des Sciences, Paris* **313**(sèr III), 187–194.

Dufour, V. (1994). Colonization of fish larvae in lagoons of Rangiroa (Tuamotu Archipelago) and Moorea (Society Archipelago). *Atoll Research Bulletin* **416**, 1–12.

Dufour, V. and Galzin, R. (1993). Colonization patterns of reef fish larvae to the lagoon at Moorea Island, French Polynesia. *Marine Ecology Progress Series* **102**, 143–152.

Døving, K. B., Mårstøl, M., Andersen, J. R. and Knutsen, J. A. (1994). Experimental evidence of chemokinesis in newly hatched cod larvae (*Gadus morhua* L.). *Marine Biology* **120**, 351–358.

Eckert, G. L. (2003). Effects of the planktonic period on marine population fluctuations. *Ecology* **84**, 372–383.

Egner, S. A. and Mann, D. A. (2005). Auditory sensitivity of sergeant major damselfish *Abudefduf saxatilis* from post-settlement juvenile to adult. *Marine Ecology Progress Series* **285**, 213–222.

Elliott, J. K., Elliott, J. M. and Mariscal, R. N. (1995). Host selection, location, and association behaviours of anemonefishes in field selection experiments. *Marine Biology* **122**, 377–389.

Finn, M. D. and Kingsford, M. J. (1996). Two-phase recruitment of apogonids (Pisces) on the Great Barrier Reef. *Marine and Freshwater Research* **47**, 423–432.

Fisher, R. (2004). Nocturnal vertical distribution of late-stage larval coral reef fishes off the leeward side of Lizard Island, Great Barrier Reef, Australia. *Bulletin of Marine Science* **75**, 439–451.

Fisher, R. (2005). Swimming speeds of larval coral reef fishes: Impacts on self-recruitment and dispersal. *Marine Ecology Progress Series* **285**, 223–232.

Fisher, R. and Bellwood, D. R. (2001). Effects of feeding on the sustained swimming abilities of late-stage larval *Amphiprion melanopus*. *Coral Reefs* **20**, 151–154.

Fisher, R. and Bellwood, D. R. (2002a). The influence of swimming speed on sustained swimming performance of late-stage reef fish larvae. *Marine Biology* **140**, 801–807.

Fisher, R. and Bellwood, D. R. (2002b). A light trap design for stratum-specific sampling of reef fish larvae. *Journal of Experimental Marine Biology and Ecology* **269**, 27–37.

Fisher, R. and Bellwood, D. R. (2003). Undisturbed swimming behaviour and nocturnal activity of coral reef fish larvae. *Marine Ecology Progress Series* **263**, 177–188.

Fisher, R. and Wilson, S. K. (2004). Maximum sustainable swimming speeds of late-stage larvae of nine species of reef fishes. *Journal of Experimental Marine Biology and Ecology* **312**, 171–186.

Fisher, R., Bellwood, D. R. and Job, S. D. (2000). Development of swimming abilities in reef fish larvae. *Marine Ecology Progress Series* **202**, 163–173.

Fisher, R., Leis, J. M., Clark, D. L. and Wilson, S. K. (2005). Critical swimming speeds of late-stage coral reef fish larvae: Variation within species, among species and between locations. *Marine Biology* **147**, 1201–1212.

Folkard, A. M. (2005). Hydrodynamics of model *Posidonia oceanica* patches in shallow water. *Limnology and Oceanography* **50**, 1592–1600.

Forward, R. B. and Tankersley, R. A. (2001). Selective tidal-stream transport of marine animals. *Oceanography and Marine Biology, an Annual Review* **39**, 305–353.

Frank, K. T., Carscadden, J. E. and Leggett, W. C. (1993). Causes of spatio-temporal variation in the patchiness of larval fish distributions: Differential mortality or behaviour? *Fisheries Oceanography* **2**, 114–123.

Frith, C. A., Leis, J. M. and Goldman, B. (1986). Currents in the Lizard Island region of the Great Barrier Reef Lagoon and their relevance to potential movements of larvae. *Coral Reefs* **5**, 81–92.

Fuiman, L. A. and Batty, R. S. (1997). What a drag it is getting cold: Partitioning the physical and physiological effects of temperature on fish swimming. *Journal of Experimental Biology* **200**, 1745–1755.

Fuiman, L. A. and Cowan, J. H. (2003). Behavior and recruitment success in fish larvae: Repeatability and covariation of survival skills. *Ecology* **84**, 53–67.

Fuiman, L. A. and Higgs, D. M. (1997). Ontogeny, growth and the recruitment process. *In* "Early Life History and Recruitment in Fish Populations" (R. C. Chamber and E. Trippel, eds), pp. 225–249. Chapman and Hall, London.

Fuiman, L. A. and Magurran, A. E. (1994). Development of predator defenses in fishes. *Reviews in Fish Biology and Fisheries* **4**, 145–185.

Fuiman, L. A., Higgs, D. M. and Poling, K. R. (2004). Changing structure and function of the ear and lateral line system of fishes during development. *American Fisheries Society Symposium* **40**, 117–144.

Fuiman, L. A., Smith, M. E. and Malley, V. N. (1999). Ontogeny of routine swimming speed and startle responses in red drum, with a comparison of responses to acoustic and visual stimuli. *Journal of Fish Biology* **55**(suppl. A), 215–226.

Fukuhara, O. (1985). Functional morphology and behavior of early life stages of Red Sea bream. *Bulletin of the Japanese Society of Scientific Fisheries* **51**, 731–743.

Fukuhara, O. (1987). Larval development and behavior in early life stages of black sea bream reared in the laboratory. *Nippon Suisan Gakkaishi* **53**, 371–379.

Gebauer, P., Paschke, K. and Anger, K. (1999). Costs of delayed metamorphosis: Reduced growth and survival in early juveniles of an estuarine grapsid crab, *Chasmagnathus granulata*. *Journal of Experimental Marine Biology and Ecology* **238**, 271–281.

Gillanders, B. M. and Kingsford, M. J. (1996). Elements in otoliths may elucidate the contribution of estuarine recruitment to sustaining coastal reef populations of a temperate reef fish. *Marine Ecology Progress Series* **141**, 13–20.

Grout, G. G. (1984). Quantitative distribution of Black Sea flounder larvae, *Platichthys flesus luscus*, in Molochniy Lagoon. *Journal of Ichthyology* **24**, 149–155.

Hadfield, M. G. (1998). The DP Wilson Lecture. Research on settlement and metamorphosis of marine invertebrate larvae: Past, present and future. *Biofouling* **12**, 9–29.

Hare, J. A. and Govoni, J. J. (2005). Comparison of average larval fish vertical distributions among species exhibiting different transport pathways on the southeast United States continental shelf. *Fishery Bulletin, U.S.* **103**, 728–736.

Hare, J. A., Quinlan, J. A., Werner, F. E., Blanton, B. O., Govoni, J. J., Forward, R. B., Settle, L. R. and Hoss, D. E. (1999). Larval transport during winter in the SABRE study area: Results of a coupled vertical larval behaviour-three-dimensional circulation model. *Fisheries Oceanography* **8**(suppl. 2), 57–76.

Hare, J. A., Fahay, M. P. and Cowen, R. K. (2001). Springtime ichthyoplankton of the slope region off the north-eastern United States of America: Larval assemblages, relation to hydrography and implications for larval transport. *Fisheries Oceanography* **10**, 164–192.

Hare, J. A., Churchill, J. H., Cowen, R. K., Berger, T. J., Cornillon, P. C., Dragos, P., Glenn, S. M., Govoni, J. J. and Lee, T. N. (2002). Routes and rates of larval fish transport from the southeast to the northeast United States continental shelf. *Limnology and Oceanography* **47**, 1774–1789.

Harmelin-Vivien, M. L., Harmelin, J. G. and Leboulleux, V. (1995). Microhabitat requirements for settlement of juvenile sparid fishes on Mediterranean rocky shores. *Hydrobiologia* **300/301**, 309–320.

Hatcher, B. G., Imberger, J. and Smith, S. V. (1987). Scaling analysis of coral reef systems: An approach to problems of scale. *Coral Reefs* **5**, 171–181.

Hecht, T., Battaglene, S. and Talbot, B. (1996). Effect of larval density and food availability on the behaviour of pre-metamorphosis snapper, *Pagrus auratus* (Sparidae). *Marine and Freshwater Research* **47**, 223–231.

Hedgpeth, J. W. (1957). Classification of marine environments. *In* "Treatise on Marine Ecology and Paleoecology" (J. W. Hedgpeth, ed.), pp. 17–27. The Geological Society of America, Washington, DC.

Hendriks, I. E., Wilson, D. T. and Meekan, M. I. (2001). Vertical distributions of late-stage larval fishes in the nearshore waters of the San Blas Archipelago, Caribbean Panama. *Coral Reefs* **20**, 77–84.

Hindell, J. S., Jenkins, G. P., Moran, S. M. and Keough, M. J. (2003). Swimming ability and behaviour of post-larvae of a temperate marine fish re-entrained in the pelagic environment. *Oecologia* **135**, 158–166.

Hogan, D. J. and Mora, C. (2005). Experimental analysis of the contribution of swimming and drifting to the displacement of reef fish larvae. *Marine Biology* **147**, 1213–1220.

Holbrook, S. J. and Schmitt, R. J. (1999). In situ nocturnal observations of reef fishes using infared video. *In* "Proceedings of the 5th Indo-Pacific Fish Conference, Noumea" (B. Seret and J.-Y. Sire, eds), pp. 805–812. Societie Francaise Ichtyologie, Paris.

Hunt von Herbing, I. (2002). Effects of temperature on larval fish swimming performance: The importance of physics to physiology. *Journal of Fish Biology* **61**, 865–876.

Hunt von Herbing, I. and Boutilier, R. G. (1996). Activity and metabolism of larval Atlantic cod (*Gadus morhua*) from Scotian Shelf and Newfoundland source populations. *Marine Biology* **124**, 607–617.

Hunt von Herbing, I., Gallagher, S. M. and Halteman, W. (2001). Metabolic costs of pursuit and attack in early larval Atlantic cod. *Marine Ecology Progress Series* **216**, 201–212.

Iles, T. D. and Sinclair, M. (1982). Atlantic herring: Stock discreteness and abundance. *Science* **215**, 627–633.

Jaffe, J. S., Ruess, E., McGehee, D. and Chandran, G. (1995). FTV: A sonar for tracking macrozooplankton in three dimensions. *Deep Sea Research* **42**, 1495–1512.

Jahn, A. E. and Lavenberg, R. J. (1986). Fine-scale distribution of nearshore, suprabenthic fish larvae. *Marine Ecology Progress Series* **31**, 223–231.

James, M. K., Armsworth, P. R., Mason, L. B. and Bode, L. (2002). The structure of reef fish metapopulations: Modelling larval dispersal and retention patterns. *Proceedings of the Royal Society, London, B* **269**, 2079–2086.

Jenkins, G. P., Black, K. P. and Keough, M. P. (1999). The role of passive transport and the influence of vertical migration on the pre-settlement distribution of a temperate, demersal fish: Numerical model predictions compared with field sampling. *Marine Ecology Progress Series* **184**, 259–271.

Jenkins, G. P. and Welsford, D. C. (2002). The swimming abilities of recently settled post-larvae of *Sillaginodes punctata*. *Journal of Fish Biology* **60**, 1043–1050.

Job, S. D. and Bellwood, D. R. (1996). Visual acuity and feeding in larval *Premnas biaculeatus*. *Journal of Fish Biology* **48**, 952–963.

Johnson, G. D. and Patterson, C. L. (1993). Percomorph phylogeny: A survey of acanthomorphs and a new proposal. *Bulletin of Marine Science* **52**, 554–626.

Johnson, M. P. (2005). Is there confusion over what is meant by 'open population'? *Hydrobiologia* **544**, 333–338.

Johnson, M. W. (1957). Plankton. *In* "Treatise on Marine Ecology and Paleoecology" (J. W. Hedgepeth, ed.), pp. 443–460. The Geological Society of America Memoir, Washington, DC.

Jones, G. P., Milicich, M. J., Emslie, M. J. and Lunow, C. (1999). Self-recruitment in a coral reef fish population. *Nature* **402**, 802–804.

Jones, G. P., Planes, S. and Thorrold, S. R. (2005). Coral reef fish larvae settle close to home. *Current Biology* **15**, 1314–1318.

Kaufman, L., Ebersole, J., Beets, J. and McIvor, C. C. (1992). A key phase in the recruitment dynamics of coral reef fishes: Post-settlement transition. *Environmental Biology of Fishes* **34**, 109–118.

Kellison, G. T., Eggleston, D. B. and Burke, J. S. (2000). Comparative behaviour and survival of hatchery-reared versus wild summer flounder (*Paralichthys dentatus*). *Canadian Journal of Fisheries and Aquatic Sciences* **57**, 1870–1877.

Kendall, A. W., Ahlstrom, E. H. and Moser, H. G. (1984). Early life history stages of fishes and their characters. *In* "Ontogeny and Systematics of Fishes" (H. G. Moser, W. J. Richards, D. M. Cohen, M. P. Fahay, A. W. Kendall and S. L. Richardson, eds), pp. 11–22. American Society of Ichthyologists and Herpetologists, Lawrence, Kansas.

Kingsford, M. J. (1988). The early life history of fish in coastal waters of northern New Zealand. *New Zealand Journal of Marine and Freshwater Research* **22**, 463–479.

Kingsford, M. J. (2001). Diel patterns of abundance of presettlement reef fishes and pelagic larvae on a coral reef. *Marine Biology* **138**, 853–867.

Kingsford, M. J. and Choat, J. H. (1989). Horizontal distribution patterns of presettlement reef fish: Are they influenced by the proximity of reefs? *Marine Biology* **101**, 285–297.

Kingsford, M. J., Leis, J. M., Shanks, A., Lindeman, K., Morgan, S. and Pineda, J. (2002). Sensory environments, larval abilities and local self-recruitment. *Bulletin of Marine Science* **70**, 309–340.

Kingsford, M. J. and Milicich, M. J. (1987). The presettlement phase of *Parika scaber* (Monacanthidae): A temperate reef fish. *Marine Ecology Progress Series* **36**, 65–79.

Kinlan, B. P. and Gaines, S. D. (2003). Propagule dispersal in marine and terrestrial environments: A community perspective. *Ecology* **84**, 2007–2020.

Kobayashi, D. R. (1989). Fine-scale distribution of larval fishes: Patterns and processes adjacent to coral reefs in Kaneohe Bay, Hawaii. *Marine Biology* **100**, 285–294.

Lecchini, D. and Galzin, R. (2005). Spatial repartition and ontogenetic shifts in habitat use by coral reef fishes (Moorea, French Polynesia). *Marine Biology* **147**, 47–58.

Leis, J. M. (1982). Nearshore distributional gradients of larval fish (15 taxa) and planktonic crustaceans (6 taxa) in Hawaii. *Marine Biology* **72**, 89–97.

Leis, J. M. (1991). The pelagic phase of coral reef fishes: Larval biology of coral reef fishes. *In* "The Ecology of Fishes on Coral Reefs" (P. F. Sale, ed.), pp. 183–230. Academic Press, San Diego.

Leis, J. M. (2002). Pacific coral-reef fishes: The implications of behaviour and ecology of larvae for biodiversity and conservation, and a reassessment of the open population paradigm. *Environmental Biology of Fishes* **65**, 199–208.

Leis, J. M. (2004). Vertical distribution behaviour and its spatial variation in late-stage larvae of coral-reef fishes during the day. *Marine and Freshwater Behaviour and Physiology* **37**, 65–88.
Leis, J. M. and Carson-Ewart, B. M. (1997). Swimming speeds of the late larvae of some coral reef fishes. *Marine Ecology Progress Series* **159**, 165–174.
Leis, J. M. and Carson-Ewart, B. M. (1998). Complex behaviour by coral-reef fish larvae in open-water and near-reef pelagic environments. *Environmental Biology of Fishes* **53**, 259–266.
Leis, J. M. and Carson-Ewart, B. M. (1999). *In situ* swimming and settlement behaviour of larvae of an Indo-Pacific coral-reef fish, the Coral Trout (Pisces, Serranidae, *Plectropomus leopardus*). *Marine Biology* **134**, 51–64.
Leis, J. M. and Carson-Ewart, B. M. (2001). Behavioural differences in pelagic larvae of four species of coral-reef fishes between two environments: Ocean and atoll lagoon. *Coral Reefs* **19**, 247–257.
Leis, J. M. and Carson-Ewart, B. M. (2002). *In situ* settlement behaviour of damselfish larvae (Pisces: Pomacentridae). *Journal of Fish Biology* **61**, 325–346.
Leis, J. M. and Carson-Ewart, B. M. (2003). Orientation of pelagic larvae of coral-reef fishes in the ocean. *Marine Ecology Progress Series* **252**, 239–253.
Leis, J. M. and Carson-Ewart, B. M. (2004). "The Larvae of Indo-Pacific Coastal Fishes: A Guide to Identification" Leiden, Brill.
Leis, J. M. and Clark, D. L. (2005). Feeding greatly enhances endurance swimming of settlement-stage reef-fish larvae (Pomacentridae). *Ichthyological Research* **52**, 185–188.
Leis, J. M. and Fisher, R. (2006). Swimming speed of settlement-stage reef-fish larvae measured in the laboratory and in the field: A comparison of critical speed and *in situ* speed. *In* "Proceedings of the 10th International Coral Reef Symposium, Okinawa".
Leis, J. M. and Lockett, M. M. (2005). Localization of reef sounds by settlement-stage larvae of coral-reef fishes (Pomacentridae). *Bulletin of Marine Science* **76**, 715–724.
Leis, J. M. and McCormick, M. I. (2002). The biology, behaviour and ecology of the pelagic, larval stage of coral-reef fishes. *In* "Coral Reef Fishes: New Insights into Their Ecology" (P. F. Sale, ed.), pp. 171–199. Academic Press, San Diego.
Leis, J. M. and Miller, J. M. (1976). Offshore distributional patterns of Hawaiian fish larvae. *Marine Biology* **36**, 359–367.
Leis, J. M. and Stobutzki, I. C. (1999). Swimming performance of late pelagic larvae of coral-reef fishes: *In situ* and laboratory-based measurements. *In* "Proceedings of the 5th Indo-Pacific Fish Conference, Noumea, 1997" (B. Seret and J.-Y. Sire, eds), pp. 575–583. Societe Francaise d'Ichtyologie & Institut de Recherche pour le Developpment, Paris.
Leis, J. M., Goldman, B. and Reader, S. E. (1989). Epibenthic fish larvae in the Great Barrier Reef Lagoon near Lizard Island, Australia. *Japanese Journal of Ichthyology* **35**, 428–433.
Leis, J. M., Sweatman, H. P. A. and Reader, S. E. (1996). What the pelagic stages of coral reef fishes are doing out in blue water: Daytime field observations of larval behaviour. *Marine and Freshwater Research* **47**, 401–411.
Leis, J. M., Carson-Ewart, B. M. and Cato, D. H. (2002a). Sound detection in situ by the larvae of a coral-reef damselfish (Pomacentridae). *Marine Ecology Progress Series* **232**, 259–268.

Leis, J. M., Carson-Ewart, B. M. and Webley, J. (2002b). Settlement behaviour of coral-reef fish larvae at subsurface artificial-reef moorings. *Marine and Freshwater Research* **53**, 319–328.

Leis, J. M., Carson-Ewart, B. M., Hay, A. C. and Cato, D. H. (2003). Coral-reef sounds enable nocturnal navigation by some reef-fish larvae in some places and at some times. *Journal of Fish Biology* **63**, 724–737.

Leis, J. M., Hay, A. C. and Trnski, T. (2006a). *In situ* behavioural ontogeny in larvae of three temperate, marine fishes. *Marine Biology* **148**, 655–669.

Leis, J. M., Hay, A. C., Clark, D. A., Chen, I.-S. and Shao, K.-T. (2006b). Behavioral ontogeny in larvae and early juveniles of the giant trevally, *Caranx ignobilis* (Pisces: Carangidae). *Fishery Bulletin, U.S.* **104**, 401–444.

Limouzy-Paris, C. B., Graber, H. C., Jones, D. L., Röpke, A. W. and Richards, W. J. (1997). Translocation of larval coral reef fishes via sub-mesoscale spin-off eddies from the Florida Current. *Bulletin of Marine Science* **60**, 966–983.

Lindeman, K. C. (1986). Development of larvae of the French grunt, *Haemulon flavolineatum* and comparative development of twelve western Atlantic species of *Haemulon*. *Bulletin of Marine Science* **39**, 673–716.

Marliave, J. B. (1977a). Development of behavior in marine fish. *Memorial University Marine Science Research Laboratory Technical Reports* **20**, 240–267.

Marliave, J. B. (1977b). Substratum preferences of settling larvae of marine fishes reared in the laboratory. *Journal of Experimental Marine Biology and Ecology* **27**, 47–60.

Marliave, J. B. (1986). Lack of planktonic dispersal of rocky intertidal fish larvae. *Transactions of the American Fisheries Society* **115**, 149–154.

Marliave, J. B. (1989). Epibenthic associations of *Merluccius productus* larvae with fjord walls. *Rapports et Procès-verbaux des Réunions, Conseil international pour l'Exploration de la Mer* **191**, 146–152.

Masuda, R. and Tsukamoto, K. (1998). The ontogeny of schooling behaviour in the striped jack. *Journal of Fish Biology* **52**, 483–493.

Masuda, R. and Tsukamoto, K. (1999). School formation and concurrent developmental changes in carangid fish with reference to dietary conditions. *Environmental Biology of Fishes* **56**, 243–252.

McCormick, M. I. (1998). Condition and growth of reef fish at settlement: Is it important? *Australian Journal of Ecology* **23**, 258–264.

McCormick, M. I. (1999). Delayed metamorphosis of a tropical reef fish (*Acanthurus triostegus*): A field experiment. *Marine Ecology Progress Series* **176**, 25–38.

McHenry, M. J. and Lauder, G. V. (2005). The mechanical scaling of coasting in zebrafish (*Danio rerio*). *Journal of Experimental Biology* **208**, 2289–2301.

McKinnon, A. D., Meekan, M. G., Carleton, J. H., Furnas, M. J., Duggan, S. and Skirving, W. (2003). Rapid changes in shelf waters and pelagic communities on the southern Northwest Shelf, Australia, following a tropical cyclone. *Continental Shelf Research* **23**, 93–111.

Meng, L. (1993). Sustainable swimming speeds of striped bass larvae. *Transactions of the American Fisheries Society* **122**, 702–708.

Miller, T. J., Crowder, L. B., Rice, J. A. and Marschall, E. A. (1988). Larval size and recruitment mechanisms in fishes: Toward a conceptual framework. *Canadian Journal of Fisheries and Aquatic Science* **45**, 1657–1670.

Montgomery, J., Carton, G., Voight, R., Baker, C. and Diebel, C. (2000). Sensory processing of water currents by fishes. *Philosophical Transactions of the Royal Society of London,* **B 355**, 1325–1327.

Montgomery, J. C., Baker, C. F. and Carton, A. G. (1997). The lateral line can mediate rheotaxis in fish. *Nature* **389**, 960–963.

Montgomery, J. C., Tolimieri, N. and Haine, O. S. (2001). Active habitat selection by pre-settlement reef fishes. *Fish and Fisheries* **2**, 261–277.

Mooi, R. D. and Gill, A. C. (1995). Association of epiaxial musculature with dorsal-fin pterygiophores in acanthomorph fishes, and its phylogenetic significance. *Bulletin of the Natural History Museum, London (Zoology)* **61**, 121–137.

Mora, C. and Sale, P. F. (2002). Are populations of coral reef fish open or closed? *Trends in Ecology and Evolution* **17**, 411–428.

Morgan, M. J., Anderson, J. T. and Brown, J. A. (1995). Early development of shoaling behaviour in larval capelin (*Mallotus villosus*). *Marine Behaviour and Physiology* **24**, 197–204.

Morgan, S. G. (2001). The larval ecology of marine communities. *In* "Marine Community Ecology" (M. D. Bertness, S. D. Gaines and M. E. Hay, eds), pp. 159–181. Sinauer Press, Sunderland, MD.

Moser, H. G. (1996). The early stages of fishes in the California Current region. *California Cooperative Oceanic Fisheries Investigations Atlas* **33**, 1–1505.

Moser, H. G. Richards, W. J. Cohen, D. M. Fahay, M. P. Kendall, A. W. and Richardson, S. L. (eds), (1984). Ontogeny and Systematics of Fishes. *American Society of Ichthyologists and Herpetologists Special Publication* **1**, 1–760.

Myrberg, A. A. and Fuiman, L. A. (2002). The sensory world of coral reef fishes. *In* "Coral Reef Fishes: Dynamics and Diversity in a Complex Ecosystem" (P. F. Sale, ed.), pp. 123–148. Academic Press, San Diego.

Neilson, J. D. and Perry, R. I. (1990). Diel vertical migration of marine fishes: An obligate or facultative process. *Advances in Marine Biology* **26**, 115–168.

Nelson, J. S. (1994). "Fishes of the World". Wiley, New York.

Ninos, M. (1984). Settlement and metamorphosis in *Hypsoblennius* (Pisces, Blenniidae) Unpublished Ph.D. dissertation, University of Southern California, Los Angeles.

Nowell, A. R. M. and Jumars, P. A. (1984). Flow environments of aquatic benthos. *Annual Review of Ecology and Systematics* **15**, 303–328.

Öhman, M. C., Munday, P. L., Jones, G. P. and Caley, M. J. (1998). Settlement strategies and distribution patterns of coral-reef fishes. *Journal of Experimental Marine Biology and Ecology* **225**, 219–238.

Okiyama, M. (1988). "An Atlas of the Early Stage Fishes in Japan". Tokai University Press, Tokyo.

Olla, B. L. and Davis, M. W. (1993). The influence of light on egg buoyancy and hatching rate of the walleye pollock, *Theragra chalcogramma*. *Journal of Fish Biology* **42**, 693–698.

Olla, B. L., Davis, M. W. and Ryer, C. H. (1998). Understanding how the hatchery environment represses or promotes the development of behavioral skills. *Bulletin of Marine Science* **62**, 531–550.

Olla, B. L., Davis, M. W., Ryer, C. H. and Sogard, S. (1996). Behavioural determinants of distribution and survival in early stages of walleye pollock, *Theragra chalcogramma*: A synthesis of experimental studies. *Fisheries Oceanography* **5** (suppl 1), 167–178.

Olney, J. E. and Boehlert, G. W. (1988). Nearshore ichthyoplankton associated with seagrass beds in the lower Chesapeake Bay. *Marine Ecology Progress Series* **45**, 33–43.

Osse, J. W. M. and Drost, M. P. (1989). Hydrodynamics and mechanics of fish larvae. *Polish Archive of Hydrobiology* **36**, 455–465.

Osse, J. W. M. and van den Boogaart, J. G. M. (2000). Body size and swimming types in carp larvae; effects of being small. *Netherlands Journal of Zoology* **50**, 233–244.

Palumbi, S. R. (2001). The ecology of marine protected areas. *In* "Marine Community Ecology" (M. D. Bertness, S. D. Gaines and M. E. Hay, eds), pp. 509–530. Sinauer Press, Sunderland, MD.

Palumbi, S. R. (2003). Population genetics, demographic connectivity, and the design of marine reserves. *Ecological Applications* **13**, S146–S158.

Pankhurst, P. M., Pankhurst, N. W. and Montgomery, J. C. (1993). Comparison of behavioural and morphological measures of visual acuity during ontogeny in a teleost fish, *Forsterygion varium*, Tripterygiidae (Forster, 1801). *Brain, Behaviour and Evolution* **42**, 178–188.

Paris, C. B. and Cowen, R. K. (2004). Direct evidence of a biophysical retention mechanism for coral reef fish larvae. *Limnology and Oceanography* **49**, 1964–1979.

Pearre, S. (1979). Problems of detection and interpretation of vertical migration. *Journal of Plankton Research* **1**, 29–44.

Pearre, S. (2003). Eat and run? The hunger/satiation hypothesis in vertical migration: History, evidence and consequences. *Biological Reviews* **78**, 1–79.

Pepin, P. and Helbig, J. A. (1997). Distribution and drift of Atlantic cod (*Gadus morhua*) eggs and larvae on the northeast Newfoundland Shelf. *Canadian Journal of Fisheries and Aquatic Sciences* **54**, 670–685.

Peterson, R. H. and Harmon, P. (2001). Swimming ability of pre-feeding striped bass larvae. *Aquaculture International* **9**, 361–366.

Pitcher, T. J. and Parrish, J. K. (1993). Functions of shoaling behaviour in teleosts. *In* "The Behaviour of Teleost Fishes" (T. J. Pitcher, ed.), pp. 363–439. Chapman and Hall, London.

Plaut, I. (2001). Critical swimming speed: Its ecological relevance. *Comparative Biochemistry and Physiology A* **131**, 41–50.

Poling, K. R. and Fuiman, L. A. (1997). Sensory development and concurrent behavioural changes in Atlantic croaker larvae. *Journal of Fish Biology* **51**, 402–421.

Poling, K. R. and Fuiman, L. A. (1998). Sensory development and its relation to habitat change in three species of sciaenids. *Brain, Behavior and Evolution* **52**, 270–284.

Popper, A. N. and Carlson, T. J. (1998). Application of sound and other stimuli to control fish behavior. *Transactions of the American Fisheries Society* **127**, 673–707.

Porch, C. E. (1998). A numerical study of larval fish retention along the southeast Florida coast. *Ecological Modelling* **109**, 35–59.

Potts, G. W. and McGuigan, K. M. (1986). Preliminary survey of the distribution of postlarval fish associated with inshore reefs and with special reference to *Gobiusculus flavescens* (Fabricius). *Progress in Underwater Science* **11**, 15–25.

Queiroga, H. and Blanton, J. (2005). Interactions between behaviour and physical forcing in the control of horizontal transport of decapod crustacean larvae. *Advances in Marine Biology* **47**, 109–214.

Reiss, C. S., Panteleev, G., Taggart, C. T., Sheng, J. and deYoung, B. (2000). Observations on larval transport and retention on the Scotian Shelf in relation to geostrophic circulation. *Fisheries Oceanography* **9**, 195–213.

Richards, W. J. (2005). "Early Stages of Atlantic Fishes: An Identification Guide for the Western Central North Atlantic". Taylor & Francis, Boca Raton, FL.

Richardson, S. L. and Pearcy, W. G. (1977). Coastal and oceanic fish larvae in an area of upwelling of Yaquina Bay, Oregon. *Fishery Bulletin, U.S.* **75**, 125–145.

Roberts, C. M. (1997). Connectivity and management of Caribbean coral reefs. *Science* **278**, 1454–1456.
Sakakura, Y. and Tsukamoto, K. (1996). Onset and development of cannibalistic behaviour in early life stages of yellowtail. *Journal of Fish Biology* **48**, 16–29.
Sale, P. F. (1991a). "The Ecology of Fishes on Coral Reefs" Academic Press, San Diego.
Sale, P. F. (1991b). Reef fish communities: Open nonequilibrial systems. *In* "The Ecology of Fishes on Coral Reefs" (P. F. Sale, ed.), pp. 564–598. Academic Press, San Diego.
Sale, P. F. (2002). "Coral Reef Fishes: Dynamics and Diversity in a Complex Ecosystem" Academic Press, San Diego.
Sale, P. F. (2004). Connectivity, recruitment variation, and the structure of reef fish communities. *Integrative and Comparative Biology* **44**, 390–399.
Sale, P. F., Douglas, W. A. and Doherty, P. J. (1984). Choice of microhabitats by coral reef fishes at settlement. *Coral Reefs* **3**, 91–99.
Sale, P. F. and Kritzer, J. P. (2003). Determining the extent and spatial scale of population connectivity: Decapods and coral reef fishes compared. *Fisheries Research* **65**, 153–172.
Sand, O. and Karlsen, H. E. (2000). Detection of infrasound and linear acceleration in fishes. *Philosophical Transactions of the Royal Society of London,* **B 355**, 1295–1298.
Sarkisian, B. L. (2005). From viscous to inertial forces: Defining the limits of hydrodynamic regimes for larval fishes. Unpublished M.Sc thesis, University of Texas, Austin.
Schmitt, R. J. and Holbrook, S. J. (2002). Spatial variation in concurrent settlement of three damselfishes: Relationships with near-field current flow. *Oecologia* **131**, 391–401.
Searcy, S. P. and Sponaugle, S. (2001). Selective mortality during the larval-juvenile transition in two coral reef fishes. *Ecology* **82**, 2452–2470.
Shanks, A. L. (1995). Orientated swimming by megalopae of several eastern North Pacific crab species and its potential role in their onshore migration. *Journal of Experimental Marine Biology and Ecology* **186**, 1–16.
Shanks, A. L. and Graham, W. M. (1987). Orientated swimming in the jellyfish *Stomolopus meleagris* L. Agassiz (Scyphozoan: Rhizostomida). *Journal of Experimental Marine Biology and Ecology* **108**, 159–169.
Shulman, M. J. (1998). What can population genetics tell us about dispersal and biogeographic history of coral-reef fishes? *Australian Journal of Ecology* **23**, 216–225.
Simons, A. M. (2004). Many wrongs: The advantage of group navigation. *Trends in Ecology and Evolution* **19**, 453–455.
Simpson, S. D., Meekan, M. G., McCauley, R. D. and Jeffs, A. (2004). Attraction of settlement-stage coral reef fishes to reef noise. *Marine Ecology Progress Series* **276**, 263–268.
Simpson, S. D., Meekan, M., Montgomery, J., McCauley, R. and Jeffs, A. (2005a). Homeward sound. *Science* **308**, 221.
Simpson, S. D., Yan, H. Y., Wittenrich, M. L. and Meekan, M. G. (2005b). Response of embryonic coral reef fishes (Pomacentridae: *Amphiprion* spp.) to noise. *Marine Ecology Progress Series* **287**, 201–208.
Sinclair, M. (1988). "Marine Populations: An Essay on Population Regulation and Speciation" University of Washington Press, Seattle.

Smith, M. E. and Fuiman, L. A. (2004). Behavioral performance of wild-caught and laboratory-reared red drum *Sciaenops ocellatus* (Linnaeus) larvae. *Journal of Experimental Marine Biology and Ecology* **302**, 17–33.

Smith, N. P. and Stoner, A. W. (1993). Computer simulation of larval transport through tidal channels: Role of vertical migration. *Estuarine, Coastal and Shelf Science* **37**, 43–58.

Spagnol, S., Wolanski, E. and Deleersnjider, E. (2001). Steering by coral reef assemblages. *In* "Oceanographic Processes of Coral Reefs" (E. Wolanski, ed.), pp. 231–236. CRC Press, Boca Raton.

Sponaugle, S., Cowen, R. K., Shanks, A., Morgan, S. G., Leis, J. M., Pineda, J., Boehlert, G. W., Kingsford, M. J., Lindeman, K., Grimes, C. and Munro, J. L. (2002). Predicting self-recruitment in marine populations: Biophysical correlates. *Bulletin of Marine Science* **70**, 341–376.

Srinivasan, M. (2003). Depth distributions of coral reef fishes: The influence of microhabitat structure, settlement, and post-settlement processes. *Oecologia* **137**, 76–84.

Steffe, A. S. (1990). Epibenthic schooling by larvae of the athrinind fish *Leptatherina presbyteroides*: An effective mechanism for position maintenance. *Japanese Journal of Ichthyology* **36**, 488–491.

Stobutzki, I. C. (1997). Energetic cost of sustained swimming in the late pelagic stages of reef fishes. *Marine Ecology Progress Series* **152**, 249–259.

Stobutzki, I. C. (1998). Interspecific variation in sustained swimming ability of late pelagic stage reef fish from two families (Pomacentridae and Chaetodontidae). *Coral Reefs* **17**, 111–119.

Stobutzki, I. C. and Bellwood, D. R. (1994). An analysis of the sustained swimming abilities of pre- and post-settlement coral reef fishes. *Journal of Experimental Marine Biology and Ecology* **175**, 275–286.

Stobutzki, I. C. and Bellwood, D. R. (1997). Sustained swimming abilities of the late pelagic stages of coral reef fishes. *Marine Ecology Progress Series* **149**, 35–41.

Stobutzki, I. C. and Bellwood, D. R. (1998). Nocturnal orientation to reefs by late pelagic stage coral reef fishes. *Coral Reefs* **17**, 103–110.

Sundby, S. (1991). Factors affecting the vertical distribution of eggs. *ICES Marine Science Symposium* **192**, 33–38.

Suthers, I. M. (1998). Bigger? Fatter? Or is faster growth better? Considerations on condition in larval and juvenile coral-reef fish. *Australian Journal of Ecology* **23**, 265–273.

Suthers, I. M. and Frank, K. T. (1991). Comparative persistence of marine fish larvae from pelagic versus demersal eggs of southwestern Nova Scotia, Canada. *Marine Biology* **108**, 175–184.

Swearer, S. E., Caselle, J. E., Lea, D. W. and Warner, R. R. (1999). Larval retention and recruitment in an island population of a coral-reef fish. *Nature* **402**, 799–802.

Swearer, S. E., Shima, J. S., Hellberg, M. E., Thorrold, S. R., Jones, G. P., Robertson, D. R., Morgan, S. G., Selkoe, K. A., Ruiz, G. M. and Warner, R. R. (2002). Evidence of self-recruitment in demersal marine populations. *Bulletin of Marine Science* **70**, 251–272.

Sweatman, H. P. A. (1988). Field evidence that settling coral reef fish larvae detect resident fishes using dissolved chemical cues. *Journal of Experimental Marine Biology and Ecology* **124**, 163–174.

Tapia, F. J., Pineda, J., Ocampo-Torres, F. J., Fuchs, H. L., Parnell, P. E., Montero, P. and Ramos, S. (2004). High-frequency observations of wind-forced onshore transport at a coastal site in Baja California. *Continental Shelf Research* **24**, 1573–1585.

Taylor, C. A., Watson, W., Chereskin, T., Hyde, J. and Vetter, R. (2005). Retention of larval rockfishes, *Sebastes*, near natal habitat in the southern California Bight, as indicated by molecular identification methods. *CalCOFI Reports* **45**, 152–165.
Theilacker, G. and Dorsey, K. (1980). Larval fish diversity, a summary of laboratory and field research. *UNESCO Intergovernmental Oceanography Committee Workshop Report* **28**, 105–142.
Thorrold, S. R. and Williams, D. M. (1996). Meso-scale distribution patterns of larval and pelagic juvenile fishes in the central Great Barrier Reef lagoon. *Marine Ecology Progress Series* **145**, 17–31.
Thresher, R. E., Colin, P. L. and Bell, L. J. (1989). Planktonic duration, distribution and population structure of western and central Pacific damselfishes (Pomacentridae). *Copeia* **1989**, 420–434.
Tolimieri, N., Jeffs, A. and Montgomery, J. C. (2000). Ambient sound as a cue for navigation by the pelagic larvae of reef fishes. *Marine Ecology Progress Series* **207**, 219–224.
Tolimieri, N., Haine, O., Montgomery, J. C. and Jeffs, A. (2002). Ambient sound as a navigational cue for larval reef fish. *Bioacoustics* **12**, 214–217.
Tolimieri, N., Haine, O., Jeffs, A., McCauley, R. and Montgomery, J. (2004). Directional orientation of pomacentrid larvae to ambient reef sound. *Coral Reefs* **23**, 184–194.
Trnski, T. (2002). Behaviour of settlement-stage larvae of fishes with an estuarine juvenile phase: *in situ* observations in a warm-temperate estuary. *Marine Ecology Progress Series* **242**, 205–214.
Utne-Palm, A. C. (2004). Effects of larvae ontogeny, turbidity, and turbulence on prey attack rate and swimming activity of Atlantic herring larvae. *Journal of Experimental Marine Biology and Ecology* **310**, 147–161.
Vickers, N. J. (2000). Mechanisms of animal navigation in odor plumes. *Biological Bulletin* **198**, 203–212.
Victor, B. C. (1986a). Delayed metamorphosis with reduced larval growth in a coral reef fish (*Thalassoma bifasciatum*). *Canadian Journal of Fisheries and Aquatic Sciences* **43**, 1208–1213.
Victor, B. C. (1986b). Duration of the planktonic larval stage of one hundred species of Pacific and Atlantic wrasses (family Labridae). *Marine Biology* **90**, 317–326.
Vigliola, L. and Meekan, M. G. (2002). Size at hatching and planktonic growth determine post-settlement survivorship of a coral reef fish. *Oecologia* **131**, 89–93.
Warner, R. R. and Cowen, R. K. (2002). Local retention of production in marine populations: Evidence, mechanisms and consequences. *Bulletin of Marine Science* **70**, 245–249.
Webb, P. W. and Weihs, D. (1986). Functional locomotor morphology of early life history stages of fishes. *Transactions of the American Fisheries Society* **115**, 115–127.
Wellington, G. M. and Victor, B. C. (1989). Planktonic larval duration of one hundred species of Pacific and Atlantic damselfishes (Pomacentridae). *Marine Biology* **101**, 557–567.
Werner, F. E., Page, F. H., Lynch, D. R., Loder, J. W., Lough, R. G., Perry, R. I., Greenberg, D. A. and Sinclair, M. M. (1993). Influences of mean advection and simple behavior on the distribution of cod and haddock early life stages on Georges Bank. *Fisheries Oceanography* **2**, 43–64.
von Westernhagen, H. and Rosenthal, H. (1979). Laboratory and in-situ studies on larval development and swimming performance of Pacific herring *Clupea harengus pallasi*. *Helgoländer Wissenschaftliche Meeresuntersuchungen* **32**, 539–549.

Wiltschko, W. and Wiltschko, R. (2005). Magnetic orientation and magentoreception in birds and other animals. *Journal of Comparative Physiology A* **191**, 675–693.
Wolanski, E. (1993). Facts and numerical artefacts in modelling the dispersal of Crown-of-thorns starfish larvae in the Great Barrier Reef. *Australian Journal of Marine and Freshwater Research* **44**, 427–436.
Wolanski, E., Doherty, P. J. and Carelton, J. (1997). Directional swimming of fish larvae determines connectivity of fish populations on the Great Barrier Reef. *Naturwissenschaften* **84**, 262–268.
Wolanski, E., King, B. and Spagnol, S. (1999). The implications of oceanographic chaos for coastal management. *In* "Environmental Science: Perspectives on Integrated Coastal Zone Management" (W. Salomons, R. K. Turner, L. Drude de Lacerda and S. Ramachandran, eds), pp. 129–141. Springer-Verlag, Berlin.
Wolanski, E. and Sarenski, J. (1997). Larvae dispersion in coral reefs and mangroves. *American Scientist* **85**, 236–243.
Wolanski, E. and Spagnol, S. (2000). Sticky waters in the Great Barrier Reef. *Estuarine, Coastal and Shelf Science* **50**, 27–32.
Wright, K. J., Higgs, D. M., Belanger, A. J. and Leis, J. M. (2005). Auditory and olfactory abilities of pre-settlement larvae and post-settlement juveniles of a coral reef damselfish (Pisces: Pomacentridae). *Marine Biology* **147**, 1425–1434.

Sound as an Orientation Cue for the Pelagic Larvae of Reef Fishes and Decapod Crustaceans

John C. Montgomery,* Andrew Jeffs,† Stephen D. Simpson,‡ Mark Meekan§ and Chris Tindle¶

*Leigh Marine Laboratory and School of Biological Sciences, University of Auckland, Auckland, New Zealand
†National Institute of Water and Atmospheric Research, Newmarket, Auckland, New Zealand
‡School of Biological Sciences, University of Edinburgh, United Kingdom
§Australian Institute of Marine Science, Northern Territory, Australia
¶Department of Physics, University of Auckland, Auckland, New Zealand

ADVANCES IN MARINE BIOLOGY VOL 51

0065-2881/06 $35.00
DOI: 10.1016/S0065-2881(06)51003-X

The pelagic life history phase of reef fishes and decapod crustaceans is complex, and the evolutionary drivers and ecological consequences of this life history strategy remain largely speculative. There is no doubt, however, that this life history phase is very significant in the demographics of reef populations. Here, we initially discuss the ecology and evolution of the pelagic life histories as a context to our review of the role of acoustics in the latter part of the pelagic phase as the larvae transit back onto a reef. Evidence is reviewed showing that larvae are actively involved in this transition. They are capable swimmers and can locate reefs from hundreds of metres if not kilometres away. Evidence also shows that sound is available as an orientation cue, and that fishes and crustaceans hear sound and orient to sound in a manner that is consistent with their use of sound to guide settlement onto reefs. Comparing particle motion sound strengths in the field (8×10^{-11} m at 5 km from a reef) with the measured behavioural and electrophysiological threshold of fishes of (3×10^{-11} m and 10×10^{-11}, respectively) provides evidence that sound may be a useful orientation cue at a range of kilometres rather than hundreds of metres. These threshold levels are for adult fishes and we conclude that better data are needed for larval fishes and crustaceans at the time of settlement. Measurements of field strengths in the region of reefs and threshold levels are suitable for showing that sound could be used; however, field experiments are the only effective tool to demonstrate the actual use of underwater sound for orientation purposes. A diverse series of field experiments including light-trap catches enhanced by replayed reef sound, in situ observations of behaviour and sound-enhanced settlement rate on patch reefs collectively provide a compelling case that sound is used as an orientation and settlement cue for these late larval stages.

1. INTRODUCTION

The spatial mosaic of hard substrate that makes up reef habitat supports distinctive communities of reef organisms. Many reef species tend to be site associated as adults but have a pelagic phase at an early part of their life history cycle. This pelagic period in the water column is typically considered a "dispersal phase," although the evolutionary forces driving this common life history strategy are still controversial and largely speculative. Whatever the evolutionary drivers, the pelagic larvae must return to the reef habitat of adults, and this habitat selection occurs at the time of settlement. As a consequence, each of these early life history processes of survival and dispersal in the pelagic environment, habitat selection and successful settlement back onto the reef play a major role in the demographics of reef populations and generate a correspondingly high level of research interest.

This review deals with the later part of the pelagic phase: the processes underlying habitat selection and settlement in reef fishes and crustaceans. The focus is on those processes mediating settlement and in particular the use of underwater sound in active habitat selection. Different terms are used to describe the settlement stages of different taxa, but here, we most often use the term "larvae" in a generic sense to refer to the entire pelagic phase that precedes settlement and "metamorphosis" to the juvenile stage. Despite the obvious adaptive advantages of long-distance active habitat selection, until fairly recently presettlement larvae were thought to lack that behavioural capability. It was recognised that settlement could be enhanced by behaviours that took advantage of onshore transport by oceanographic features such as slicks and tidal fronts (Kingsford and Choat, 1986; Kingsford and Finn, 1997) and the tides themselves (Forward and Tankersley, 2001; Queiroga and Blanton, 2004). However, the general picture was that larvae are largely distributed by physical processes. This reasonable null hypothesis has been challenged by evidence that presettlement larvae are more behaviourally competent than previously thought. Close to settlement, larvae have been shown to have quite remarkable swimming capabilities (Dudley *et al.*, 2000; Leis, 2006; Stobutzki and Bellwood, 1994). An ability for long-distance movement that is energetically costly for these small pelagic organisms will be most advantageous only if it can reliably be directed toward desirable settlement sites. There is now clear evidence for such directed swimming capabilities of both reef fishes and decapod crustacean larvae and the sensory cues that mediate this ability are now being discovered. For longer distances, underwater sound is emerging as a leading candidate to guide active habitat selection.

While we might normally think of animal migration as occurring over tens to hundreds of kilometres, the directed swimming of a larval fish or crustacean, typically $\leq$20 mm in length, over a distance of kilometres or tens of kilometres, can be thought of within the same framework. Able (1996) recognises three predominating themes of contemporary research in animal orientation and navigation. First, animal orientation systems are replete with interacting mechanisms and are highly flexible. Second, there is a compelling need to take studies of animal orientation and navigation back into the field. Third, it is necessary to uncover, and tease apart, the "rules of thumb" that migrators use. Although this review targets the use of underwater sound as an orientation cue, the first of Able's themes provides a useful reminder that where available, other cues also are likely to be used. It will also become apparent that the second of the aforementioned themes is particularly relevant to our consideration of the use of sound as an orientation cue. The physical nature of underwater sound means that many studies can simply not be done in the confines of laboratory tanks and we need to study acoustic orientation of presettlement animals in a field setting. The primary goal of this review, though, addresses the third of Able's themes, and we explore the evidence that one of the key "rules of thumb" that presettlement individuals use for active habitat selection is to orient to the underwater soundscape. This evidence is addressed on three levels to show that (1) underwater sound has the appropriate characteristics to provide a useful orientation cue; (2) the animals have the capacity to detect the relevant sensory information; and (3) that there is direct behavioural evidence that the animals do use sound in their orientation and settlement choice.

After our detailed consideration of sound as an orientation cue, we briefly compare sound with other candidate cues. Finally, we draw out some of the implications of these findings and highlight some knowledge gaps and prospects for further work.

2. BACKGROUND

2.1. Evolutionary and ecological context

The purpose of this section is to further define the scope of this review and to place the issue of acoustic orientation in the settlement of reef species into a wider context. There is extensive literature on dispersal, its evolutionary origins and its population and community consequences (e.g., Bradbury and Snelgrove, 2001; Sponaugle *et al.*, 2002; Strathmann *et al.*, 2002; Pittman and McAlpine, 2003; Sale and Kritzer, 2003; Cook and Crisp, 2005). It is not the intention here to provide an extensive review of this literature, but to indicate some of the current issues and thereby provide a context for our detailed review of the evidence for sound as an orientation cue. This context is also important for deriving the conclusions that might be made from this evidence and to appreciate the potential significance of acoustic orientation and active habitat selection in the life histories of reef species.

For this review, we adopt a broad definition of reef communities as those inhabiting shallow-water hard-substrate environments, including coral reefs and rocky reefs (cf. Bellwood and Wainwright's, 2002 definition of reef fishes as "those taxa that are found on, and are characteristic of, coral reefs"). These reef habitats do support distinctive communities of organisms different from pelagic or soft-bottom communities. The current review considers both fish and decapod crustacean components of reef communities and their dispersal from, but particularly their return to, reef habitat. The choice of including both fishes and decapod crustaceans arises from the fact that both groups are important members of reef communities, they share similarities in overall life history and there is evidence for both groups that sound may be used to guide their return to reefs. In addition, as Bradbury and Snelgrove (2001) remark, there are relatively few studies that take a taxonomically broad approach to explore the processes involved in dispersal and settlement, yet the comparative perspective has much to offer.

Although a general feature of biology is the difficulty of making general statements about widely separated taxonomic groups such as fishes or decapod crustaceans, there does seem to be a general rule that many reef species tend to be site associated as adults but have a pelagic phase at an early part of their life history cycle. For example, coral reef fish species, of which there are thousands (~800 species on the Great Barrier Reef [GBR], Australia alone [Bellwood and Wainwright, 2002]), provide only a handful of examples that bypass the pelagic phase. In their review of coral reef larval biology, Leis and McCormick (2002) talk of the "near ubiquity of a pelagic stage" and list only about half a dozen species that have effectively eliminated the pelagic phase. The length of this phase varies from as little as a week in anemone fishes (Pomacentridae) to >64 wk in some porcupine fishes (Diodontidae). More typical coral reef fish families have larval durations of ~30 d, which may extend to 60 d in some families such as the Chaetodontidae and the Lutjanidae (Jones *et al.*, 2002). Like the coral reef fishes, the majority of temperate reef fish species also show a pelagic larval phase, although there are a few exceptions such as *Sebastes* spp. on the Californian coast that are live bearers (Boehlert and Yoklavich, 1984).

Like reef fishes, decapod crustaceans also commonly have a pelagic larval phase. We do not have summary statistics of the relative numbers of reef crustacean species within this group with and without a pelagic phase. However, Bradbury and Snelgrove (2001) provide data showing that 90% of the benthic invertebrates included in their review had a pelagic life history phase of a week or more. Perhaps more importantly for our discussion, spiny lobsters of the family Palinuridae provide extreme examples of extended pelagic development (up to 18 mo) that conclude with long-distance migrations of the post-larvae back to reef habitat from offshore oceanic waters (Phillips and Sastry, 1980).

2.2. Adaptive value of dispersal

The predominance of a pelagic larval phase among reef fishes and decapod crustaceans underlines the importance of understanding this life history phase for understanding the demographics of reef populations. This "near ubiquity" of the pelagic larval phase also implies an evolutionarily stable life history strategy with adaptive value in its own right. Typically, the adaptive value is considered the ability to disperse so that the spatial mosaic of reef inhabitants is connected and replenished by the pelagic larval supply and reef populations are "open" both in ecological and in evolutionary terms. A contrasting view that has gained currency is that the larval period is not driven by dispersal at all, but that reef systems self-recruit, that larvae are retained within the reef system and that larvae may even "home" to natal reefs, resulting in populations that are more "closed" than "open." But as Mora and Sale (2002) argue, the status of reef fish populations as "open" or "closed" is yet to be determined. Different points of view along the continuum from dispersal to retention to homing imply different views on the evolutionary drivers of life history patterns (e.g., Johannes, 1978; Strathmann *et al.*, 2002). If the pelagic phase is dispersive and dispersal provides a selective advantage, then this alone may be sufficient to explain the great predominance of this life history trait. But other factors such as high larval predation and limited opportunities for larval feeding on reefs might also be important. Understanding the evolutionary drivers of the pelagic phase clearly has implications for understanding the proximal mechanisms of settlement. However, the converse also applies, that is, that an improved understanding of proximal mechanisms may provide insight into evolutionary drivers. So before addressing the proximal processes directly, it is appropriate to explore in a little more detail the status of the arguments for the adaptive value of dispersal.

2.3. Dispersal and dispersal kernels

Is there direct evidence for the adaptive value of dispersal? To address this question, it is instructive to broaden our perspective to the parallels between larval dispersal in reef fishes and seed dispersal in terrestrial plants. The production of large numbers of wind-dispersed seed is a common strategy. Given that seeds are simpler than fish larvae in that they do not feed or behave in ways that directly influence their own dispersal, the plant literature does provide some insight into the evolution of dispersal and its community ecology ramifications. The most cited unifying concept in seed dispersal is the Janzen-Connell hypothesis, in which dispersal away from the parent plant confers greatly reduced density-dependent mortality (Levine and Murrell, 2003). Combined with genetic and environmental considerations, the major forces selecting for dispersal are recognised as kin competition, inbreeding depression and spatiotemporal variability in environmental conditions (Levine *et al.*, 2003). Both modelling and empirical studies show a positive relationship between dispersal and abundance, implying an attendant adaptive value of dispersal. However, even with seed dispersal, the situation can be complex. Some modelling studies show that long-range dispersal can be disadvantageous if the landscape is variable in quality. Depending on the spatial scale of favourable and unfavourable patches, short-range dispersal may lead to higher abundance than long-range dispersal. In these cases, the importance of remaining in a good patch outweighs the increased intraspecific competition that often results from short-range dispersal (Levine and Murrell, 2003). Patch dynamics and dispersal can also be important in maintaining metapopulations. Given stochastic and asynchronous variation in conditions among patches, dispersal can be an effective bet-hedging strategy, potentially allowing metapopulation persistence through dispersal from one transiently favourable site to another, even when the expected growth rate in all local populations is negative (Metz *et al.*, 1983). Thus in comparison to the situation for pelagic fish larvae, the plant literature supports the idea that dispersal is adaptive in its own right. It is also instructive that even within the simpler seed dispersal system, the plant literature is still short of a satisfactory synthesis that links dispersal, community structure and evolution. The impediment turns out to be the practical difficulties of determining the distribution of wind-dispersed seed.

To quantify dispersal, ideally we need to know the dispersal kernel. "By dispersal kernel, we mean the probability density function describing the probability of seed transport to various distances from the parent plant" (Levine and Murrell, 2003). For ecological considerations, the dispersal kernel needs to be assembled over the appropriate ecological time scale, and connectivity needs to be at a reasonably high level to be ecologically meaningful (Cowen *et al.*, 2000). For evolutionary considerations, much longer time scales are appropriate and the connectivity can be considerably less (Palumbi, 2003). It is obvious that even for plants, where the physics is arguably more tractable than that in the ocean, the conditions for tracking seeds are better and the seeds are totally passive propagules, the task of determining the dispersal kernel is still a difficult challenge. Given the added difficulties of determining the dispersal kernel for reef larvae, it is not surprising that the task is daunting. For reef fishes, the empirical evidence is sparse and hard won. Jones *et al.* (1999) tagged ~10 million eggs of the small damselfish *Pomacentrus amboinensis* with tetracycline at Lizard Island, GBR. Recapture of tagged presettlement larvae indicated that somewhere between 15 and 60% of recruits to Lizard Island over that spawning period originated from the Lizard Island spawning. Swearer *et al.* (1999) used elemental composition between larvae developing in coastal waters (locally retained) and larvae developing in open ocean waters (produced in distant locations) to evaluate the source of recruits. They, too, found that recruitment to an island population of a widely distributed coral reef fish also had a strong component of local retention even on leeward reefs. Paris and Cowen (2004) sampled patches of stage-specific larvae and found evidence for retention based on vertical migration behaviour in bicolour damselfish. Miller and Shanks (2004) estimated along-shore larval dispersal in the black rockfish as $<$120 km. Such results are still considerably short of what is required to define a probability density function over the appropriate time scales or to generalise across reef fish species. As others have argued, the balance across the continuum from dispersal to retention and homing is essentially unknown and the case for reef fish populations as being open or closed is far from being resolved (Mora and Sale, 2002; Sale, 2004). The current lack of evidence favours a shift in perspective from one of contrasting dispersal versus retention to a more neutral consideration of a probability density function that encompasses both. Hopefully, future work will better define the actual dynamic range of kernels both within and among species.

Given the state of the art and the difficulties of dealing with marine systems, is it an attainable goal to understand the linkage between dispersal, population dynamics, community structure and evolution? Are we dealing with a system that is just complex or hopelessly chaotic? And is there a useful distinction between the two? To move from the chaotic to the merely complex, we need better empirical data that define the dispersal kernel across species, across habitats and across time. This critical knowledge gap impedes both our fundamental understanding and our management of reef species (Sale, 2004). The tagging study of Jones *et al.* (1999) mentioned above indicates a way forward. Other advances in using both "natural" and artificial tags evident in otolith microchemistry also offer potential (Campana and Thorrold, 2001). Indirect methods that infer dispersal through population and parental genetics will also be critical in defining population connectivity (Kinlan and Gaines, 2003; Palumbi, 2003). Population models and physical models can also provide insight and need to be informed by the underlying evolutionary and proximal processes driving the pelagic larval phase (James *et al.*, 2002; Codling *et al.*, 2004). Evolutionary processes can be debated from theoretical and comparative considerations of the commonalities and differences among species and among higher order taxa. But one assertion of this review is that there is also a contribution to be made in nudging chaos towards complexity by understanding the proximal processes that govern the pelagic phase and the process of settlement.

The proximal processes underlying the pelagic phase are: (1) the physics of the environment; (2) the physical attributes of the larvae; and (3) their sensory and locomotor behavioural capabilities. In the early pelagic phase, physical attributes, such as buoyancy of eggs and newly hatched larval fishes, will be important and interact with the physics of the ocean environment to determine distribution. A behavioural contribution to distribution may be significant but will be limited by behavioural competence and further limited where other behavioural imperatives, such as feeding, take precedence. The focus of this review is the late pelagic phase where behavioural competence has increased and return to settlement on a suitable reef has become the clear survival priority. The main question addressed in here is what are the behavioural competencies that develop prior to settlement, and how are these used to modify distribution and settlement success? An answer to this question is a step towards addressing the more general questions of the relative importance of active habitat selection and the contribution of larval behaviour, specifically acoustic orientation, to the final distribution kernel?

2.4. Behavioural competence

Behavioural competence, particularly the swimming ability, of late-stage reef fish larvae is examined in detail in the accompanying review (Leis, 2006). In order for fishes and decapod larvae to actively seek out suitable settlement habitats, they not only need appropriate swimming capabilities but also must have the ability to locate reefs from a distance. Before examining the evidence for sound as the orientation cue, it is appropriate to briefly summarise the evidence relating to reef orientation by presettlement larvae, to describe how this orientation differs between day and night and to show how this relates to our understanding of the timing of settlement.

Fish larvae show oriented behaviour to reefs and are capable of orienting to reefs from distances of at least 1 km. For example, apogonid, chaetodontid and pomacentrid larvae released during the day at distances >1 km from reefs were shown to swim offshore regardless of the position around the small island where they were released (Leis *et al.*, 1996). Stobutzki and Bellwood (1998) showed that apogonid and pomacentrid larvae move onshore at night. It has been proposed that larvae move offshore during the day to avoid reef-based predators and to feed, and they move onshore at night to settle (Leis and Carson-Ewart, 1998; Stobutzki and Bellwood, 1998; Kingsford *et al.*, 2002). During daytime observations, Leis and Carson-Ewart (1998) noted that 8.5% of larvae released near the reef were eaten by wrasses and lizard fishes, while none of those released offshore were eaten. Approximately 10% of larvae fed when released in open water, but <1% fed when released near reef waters, again providing circumstantial evidence that the larvae know when they are close to a reef. Resident reef fish adults were aggressive towards larvae attempting to settle. Larvae may, therefore, avoid reefs during the day when many of these residents are active. In line with this view, nocturnal settlement is generally assumed, but few hard data exist that show this to be the case. Kingsford (2001) found that most reef fishes on One Tree Island, GBR, settled at night and suggested that studies on settlement cues should focus on nocturnal phenomena. However, while nocturnal settlement may be common, it is not ubiquitous, and Leis and Carson-Ewart (1999) noted that some larvae settle during the day. Likewise, when Kingsford (2001) converted data to an hourly rate, some pomacentrids showed similar rates of settlement during the night and day. As cues may vary between night and day, the timing of settlement is a crucial issue and needs to be addressed in more detail. In the meantime, it is a reasonable assumption that most settlement occurs at night. Thus, nocturnal cues are likely to be most significant, but from the evidence cited, it is apparent that reef fish larvae can determine the location of reefs both during the day and at night.

Rock lobster larvae also show evidence of shoreward movement from considerable distances offshore. Distributional data of spiny lobsters is consistent with the final stage larvae and post-larvae actively swimming toward their coastal settlement grounds from distances of >100 km offshore (Chiswell and Booth, 1999; Jeffs *et al.*, 2001). The larvae and post-larvae of *Jasus edwardsii* were estimated to be achieving net shoreward-directed swimming speeds of 4–6 cm s^{-1} and 8–10 cm s^{-1}, respectively. These swimming speeds are sufficient to break out of a large permanent oceanic eddy system into which the earlier larval stages are thought to be entrained (Chiswell and Booth, 1999).

So the combined evidence reviewed above is that presettlement fish and crustacea are active swimmers and capable of locating reefs from a considerable distance. But how good is the evidence that one of the "rules of thumb" that presettlement individuals use for this behaviour is to orient to the underwater soundscape? We now address this question by reviewing this evidence at three levels to show that (1) underwater sound has the appropriate characteristics to provide a useful orientation cue; (2) the animals can detect the relevant information; and (3) that there is direct behavioural evidence that the animals do use sound in their orientation and settlement choice.

3. SOUND PRODUCTION, SIGNAL-TO-NOISE CONSIDERATIONS AND THE MARINE SOUNDSCAPE

This section addresses the first of the three levels of evidence that make up the case for establishing sound as an orientation cue, namely that underwater sound has the appropriate characteristics to provide a useful orientation cue. In general terms, underwater sound has long been recognised as one of the strongest candidates for onshore orientation by pelagic organisms because it is conducted over long distances and can also carry biologically significant information about distant coastal locations, such as reefs (Myrberg, 1978; Hawkins and Myrberg, 1983; Stobutzki and Bellwood, 1998; Montgomery *et al.*, 2001; Kingsford *et al.*, 2002). The source of such sound can be either abiotic, such as wave break, or biotic in origin, such as snapping shrimp, urchins or fishes (Tait, 1962; McCauley and Cato, 2000). To provide a more detailed analysis of the potential for underwater sound to provide an orientation cue, we consider the physics of underwater sound, how it is produced and how it propagates. We address some of the relevant complexities of sound propagation in reef environments and how noise sources, such as sea state and rain, could mask useful directional information. The ideal would be to be able to predict the acoustic "footprint" of a reef from such considerations, but the real world complexities of underwater sound make this difficult. So we selectively review underwater sound recordings, which provide the information required to assess the potential use of underwater sound in distance orientation to reefs.

The physical properties of underwater sound as a biological stimulus have been well reviewed by Rodgers and Cox (1988) and the general characteristics of underwater sound and ambient noise in the sea are well known (Albers, 1965; Urick, 1983; Medwin and Clay, 1998). Consideration of the physics starts with the idealised situation of a water body with no boundaries or obstructions before it can be extended to the much more complex acoustic environment of a shallow water reef.

In the idealised situation where a specified sound is produced by a source in a homogenous infinite environment, the sound at any location can be precisely defined. A number of biological sources can be described as "small pulsating sources." Under these conditions, the sound field consists of a pressure wave that propagates radially from the source, and a corresponding radial water particle motion. If R is the distance from the source, the amplitude of the pressure waveform is proportional to R^{-1}, because of spherical spreading. The particle motion, or particle velocity, is composed of two components. The first is due to the compression of the fluid by the pressure wave and is considered "true sound." The second is a "flow" component, which for a pulsating source decreases with the square of the distance (i.e., is proportional to R^{-2}). Close to the source, the flow component will dominate, and this is the region termed the *acoustic nearfield*. The region beyond that is termed the *acoustic farfield*, where the particle motion is directly related to the propagating pressure wave ($p = \rho c v$ where: p = pressure, ρ = density, c = speed of sound in water, v = particle velocity). For the purposes of sound as an orientation cue, we are only concerned with the acoustic farfield. As a rough guide, the nearfield is confined to an area within one or two wavelengths of the source. At 30 Hz, the wavelength of sound is 50 m, at higher frequencies the wavelengths are correspondingly less (10 m at 150 Hz and 1 m at 1500 Hz) ($\lambda = c/f$, where λ = wavelength, c = speed of sound in water, f = frequency in Hz).

In order to describe sound intensity and compare sounds, the convention is to use decibels (dB). The level in decibels of a sound with intensity I is defined with respect to a reference intensity, I_0, as $10 \log_{10}(I/I_0)$. Sound intensity is proportional to the square of the acoustic pressure. Because most hydrophones measure the acoustic pressure rather than the intensity, it is more convenient in underwater acoustics to measure the sound pressure level as $20 \log_{10}(p/p_0)$, where p_0 is the reference pressure level of 1 μPa. The sound pressure used in the decibel determination is the root mean square (rms), and the standard reference level is chosen so that for all practical purposes the decibel levels for underwater sound will have positive values. Source level is always quoted at a standard distance of 1 m from the source but can be measured at any convenient greater distance that is less than half the distance to the nearest reflector such as the ocean surface. The level at 1 m can then be calculated. We note in passing that it is not appropriate to compare decibel levels in water and air because the reference levels are quite different.

Most of the loss of sound intensity as you move away from a source is due to spreading rather than absorption. For example, at 500 Hz, which is within the hearing range of fishes, sound suffers only 1 dB of attenuation due to absorption in 100 km of propagation (Rogers and Cox, 1988). Low attenuation can result in sounds propagating over large distance, which leads to high ambient noise backgrounds. In addition to low attenuation, other distinctive properties of underwater sound compared with airborne sound include a much higher speed (1500 m s^{-1}); a greater tendency to refract due to density and temperature gradients; a greater tendency to be scattered by objects, particularly objects containing a gas inclusion; and a smaller particle velocity for a given pressure. All of these properties have implications for the way in which reef noise may propagate outward from the reef and the mechanisms that pelagic larvae may use to detect reefs and orient towards them.

Whether the sound sources on a reef are physical or biological in origin, the most important sound sources are likely to occur in shallow water, in the top 10–20 m. It is obvious that noise generated by waves breaking on the reef is generated in the shallows. The top 20 m is also the most productive part of the reef, with the greatest concentration of fishes and invertebrates, so the most significant sources of biotic sounds are also likely to be found in relatively shallow water. For example, in the temperate reef systems of northern New Zealand, the rasping feeding activity of the sea urchin *Evechinus chloroticus* is thought to be one of the principal sound producers and part of the evening chorus. *Evechinus* has its greatest density in the region from the surface down to ~7 m, with few occurring below 10 m (Choat and Schiel, 1982). The significance of the shallow distribution of noise sources is that this immediately takes us away from the idealised situation of sound production in a boundless medium. Furthermore, these shallow regions are often strongly affected by temperature and density gradients that can influence sound transmission. For example, in tropical regions, strong temperature stratification leads to an increase in sound velocity close to the surface. This velocity gradient leads to downward refraction of sound such that at distances of ~500 m and beyond, there can be a sound shadow close to the water surface. Such a sound shadow may be of particular importance on outer reef slopes, where it is not filled in by bottom reflection. In shallower reef systems, such as inside the barrier reef, or in many temperate reef systems, the bottom forms a good sound reflector at low angles of incidence. Sound also reflects off the sea surface, where it undergoes a phase reversal on reflection. So, the combination of a distributed series of shallow sound sources, refraction, scattering, and multipath reflections between these sources and the receiver will give rise to constructive and destructive interference, resulting in a complex sound field. Nevertheless, under most conditions, this sound field will still retain the critical properties of a distinctive "reef sound" with an intrinsic directionality.

Sound propagation in shallow water has a number of other attributes, some of which may increase the opportunity for its use as an orientation cue and others that may decrease its potential. For example, where the range between the source and the receiver is much greater than the water depth, surface and bottom reflection retain the sound within the water column, giving rise to cylindrical spreading. With cylindrical spreading, the amplitude of the sound decreases as $1/(\sqrt{R})$ and this can increase the effective range of sound in comparison with the spherical spreading of sound in a free-field situation. However, shallow water also limits the propagation of low-frequency sound. For example, it has been estimated that sounds at frequencies of less than ~70 Hz will be lost to a fine sand seafloor in water depth of 10 m (Hamilton and Bachman, 1982; Medwin and Clay, 1998). However, this does not necessarily mean that low frequencies generated in the nearshore environment do not propagate offshore. For example, intense low-frequency sound generated by wave break may generate waves that propagate through the seafloor substrate, which subsequently generate underwater sound in deeper water offshore through acoustic coupling. So in shallow water reef settings, it may be legitimate to concentrate our attention on frequencies that would be considered to be in the "normal" acoustic range for fishes, that is, in the tens to hundreds of Hz (see Section 4.2). However, that is not to say that infrasound is not a potential or important navigational cue under some circumstances.

The main conclusion to draw from our consideration of the properties of underwater sound is that, given the complexities of sound in the near-reef environment, there is no real substitute for direct measurement of sound in a manner appropriate to determining its suitability as an orientation cue.

There are relatively few studies that provide us with the information required to assess the potential use of underwater sound in distance orientation to reefs. In many studies, "reef sound" is background "noise" and usually considered a nuisance. From the perspective of this review, "reef sound" is the focus of our attention and is the "signal" that we would like to characterise. Noise sources that will interfere with the ability of larvae seeking out reef habitats will be sounds that contain no useful directional information but mask the detection of directionally useful sound. Noise generated by sea state or rain on the surface of the sea would be obvious examples. Anthropogenic sources may or may not contain useful directional information and may represent either signal or noise depending on the circumstances. Standard deepwater noise spectra (Knudsen curves [see Urick, 1983]) show that in the range 50–1000 Hz, noise due to shipping and wind-generated surface waves can significantly raise the "noise floor" and potentially mask the detection of the reef signals. Concerns have also been raised that anthropogenic sound levels have risen significantly in recent decades (Ross, 1993) and may have a negative impact on fish (Popper, 2003) and other marine animals (Foote *et al.*, 2004).

Underwater sound recordings near coastal reefs of New Zealand and Australia have a component of ambient noise with high amplitude at frequencies of 1200–1600 Hz (Tait, 1962; Cato, 1978). This noise originates from rocky coastlines and includes snapping shrimp, sea urchins and other biological sources (Figures 1 and 2). Figure 1 shows data with many snapping shrimp and background noise at ~118 dB with a broad peak centred at 5 kHz. Figure 2 shows quieter conditions with snapping shrimp sounds on a background level of 100 dB. The snapping shrimp give a broad peak centred at 11 kHz. The sound levels increase markedly for about 3 h after sunset and these studies allow us to calculate the particle displacement amplitude due to these sources at a known distance from the reef. Tait (1962) measured a 7- to 10-dB directional increase in ambient noise at a point 5 km off shore from a reef and a 6-dB fall-off with distance doubled. At this frequency, attenuation is 1 dB per 10 km, so the directional noise would be detectable at least 10 km off shore. In absolute terms, sound pressure levels were reported as 16–20 dB re 1 microbar (note the use of the microbar as a reference standard as this study predates the adoption of current standard of reference to μPa). Taking 18 dB as representative, this equates to 118 dB re 1 μPa, 0.8 Pa rms, or a p_{max} of 1.1 Pa. The energy peak is at 1400 Hz. Using $p = \rho c v$ and $v = \omega A$ gives a calculated particle displacement, A, of 8×10^{-11} m. This particle displacement amplitude calculation will be useful for comparison

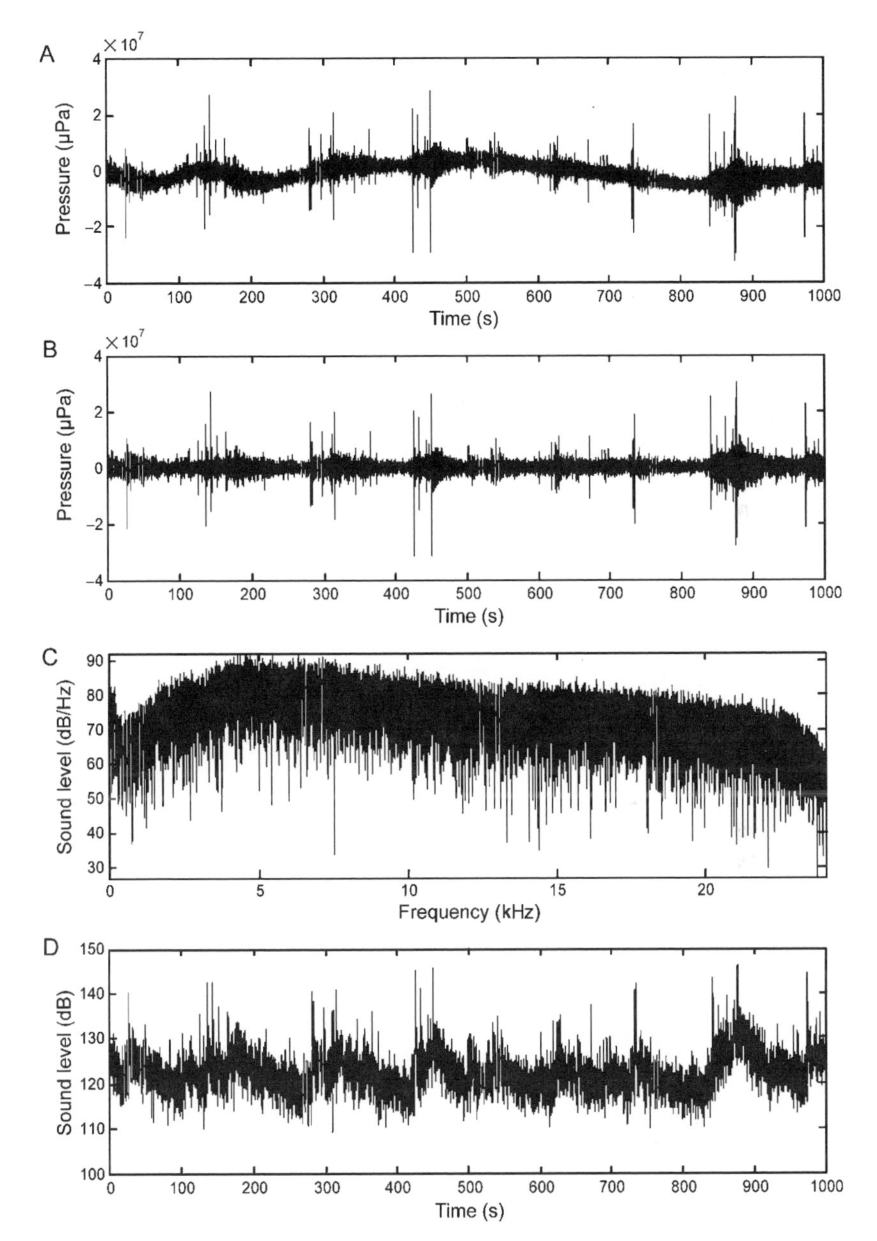

Figure 1 Sound recordings 250 m offshore from rocky coastline in northern New Zealand. (A) 1000 s of raw data at 250 m showing many individual snapping shrimp on a background of biological noise plus some low-frequency waves due to surface motion. (B) As for (A) but with low-frequency (<20-Hz) waves removed by filtering. (C) Relative sound level as a function of frequency in the conventional units of dB/Hz. (D) Filtered sound level versus time. Background noise level is 118 ± 2 dB with individual shrimp snaps up to 145 dB.

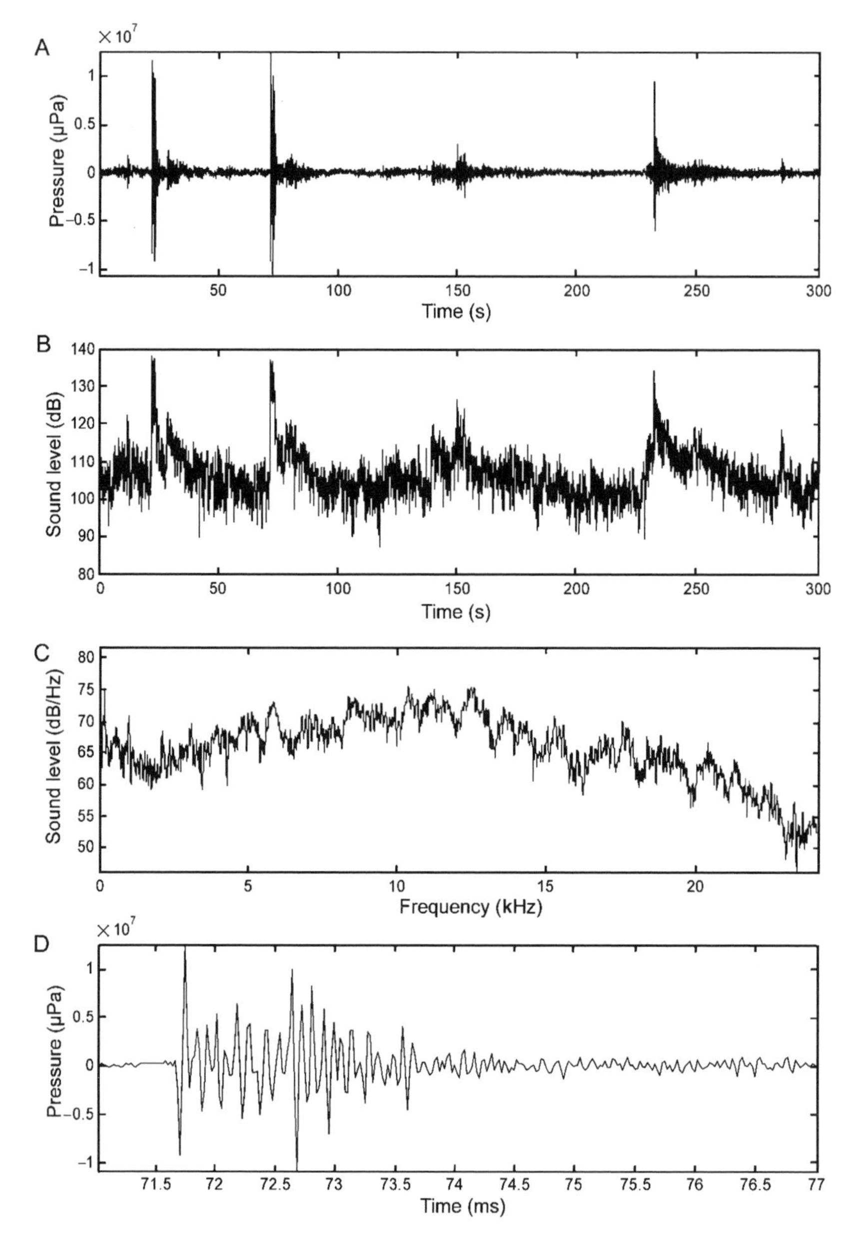

Figure 2 Snapping shrimp sounds recorded from northern New Zealand. (A) 300 ms of raw underwater sound data recorded near a wharf in 3 m of water with the hydrophone 0.75 m below the surface. The sampling rate was 48 kHz. There are some nearby snapping shrimp and three loud snaps are shown in the figure. (B) Sound level in dB re 1 microPascal. Calibration was achieved using a standard 10-kHz pinger.

with auditory sensitivity measurements (see Section 4). McCauley and Cato (2000) reported that nocturnal noise from reefs can be detected at levels above background noise for distances of >10 km and Cato (1978) reported detecting reef noise that would be louder than surface wind noise at up to 25 km in Australian waters.

4. SOUND RECEPTION IN MARINE LARVAE

4.1. Auditory sense organs

Sound reception in the farfield is based on detecting either the radial particle velocity or the pressure field. In some cases, we can identify specific auditory receptor systems that respond to one or other of these components of the acoustic field. These anatomical specialisations are particularly evident in pressure reception mechanisms. The identification of such specialisations is good evidence for hearing. However, the converse is not true; the absence of such anatomically identified hearing end-organs does not imply an inability to detect sound.

We are used to the idea of sound reception as it occurs in mammals, consisting of an array of sensory transducers, or hair cells, complete with the elaborate supporting anatomy of the cochlea and a specialised sound-transduction pathway, including the externally obvious pinna or outer ear and the impedance matching ear bones of the middle ear. Sound reception in aquatic animals is based on quite different physical principles with, in most cases, no obvious external anatomy that is indicative of hearing ability. In many cases there may be no obvious anatomical sensory specialisation at all for the detection of sound, but we know from behavioural and physiological evidence that a good sense of hearing is present. This lack of a 1-to-1 mapping of specific receptor systems onto "hearing" is due to the obvious

---

The background noise level is ~100 dB, with individual snapping shrimp events at ~135 dB at the hydrophone. Assuming the shrimp was on the bottom ~2 m below the hydrophone, there is a factor of 4 increase of intensity at the standard distance of 1 m, giving a source level for the shrimp of 141 dB. (C) Power spectrum of the recorded signal. It is a broad peak centered at ~11 kHz. (D) Loudest snapping shrimp pulse on an expanded scale. The initial pulse at 71.7 ms is the single oscillation that goes first negative and then positive. The surface reflection arrives at ~72.7 ms, indicating that the hydrophone was ~0.75 m below the surface. The surface reflection is clearly reversed in phase, as expected. The oscillations between 71.8 and 72.6 ms are probably due to reflections from rocks near the shrimp and the corresponding surface reflections arrive between 72.8 and 73.7 ms. After 73.8 ms, there is just background noise again.

fact that sound is but one biologically important source of water motion and pressure change. Receptors responding to whole-body motion of the animal or to pressure change will provide a variety of useful information and respond to a very wide "sensory space," only part of which constitutes "hearing" as it is normally defined.

For the motional component of sound, the lack of an identity between the acoustic stimulus and a specific hearing end-organ arises from the fact that equilibrium sensors based on the principle of differential density accelerometers are widespread in the animal kingdom. Linear accelerometers convey sensitivity to the motional component of the underwater sound field. But they also respond to movement generated actively by the animal itself or by other external sources, such as current turbulence and wave action. So it is somewhat arbitrary as to where self-movement stimulation of the equilibrium sensors grades into movements of the animal induced by movements of the surrounding medium, such as those induced by wave action or those produced by an acoustic source. Low-frequency sources in particular produce spatially extensive, nearfield, oscillating flows in addition to the radial particle motion of the farfield. So low-frequency sound, or infrasound, with its extensive and sometimes complex local flow fields, is intermediate between equilibrium stimuli and sound itself.

It follows from these considerations that the most common acoustic receptors in marine animals are these motion detectors based on the differential density accelerometers of the vertebrate otolithic inner ear and the invertebrate statolith organs. In many cases, particularly in fishes, we can recognise anatomical and physiological specialisations of components of these organs for sound reception *per se*. But in many other cases, particularly for invertebrates, behavioural and physiological data are absent, and the anatomy alone provides little or no insight into the relative equilibrium or acoustic function of these receptor systems.

Similar considerations apply to the detection of the pressure component of the acoustic field. In a limited number of cases, such as in clupeid (Gray and Denton, 1979) or, notably for this review, chaetodontid fishes (Webb and Smith, 2000), there are extensive anatomical specialisations that use pressure for sound reception. These detector mechanisms are based around the presence of a gas bubble, either the swim bladder itself or a gas bubble derived from it. However, there are indications that even without such anatomical specialisations, the swim bladder may be important for pressure detection and a source of information for sound source localisation. Some of the larval fishes and all the crustaceans of interest to this review do not have gas inclusions in their body. Does this imply that they are insensitive to acoustic pressure? Evidence in crustaceans (detailed below) shows that pressure receptors based on other physical principles are possible, but the extent to which these may confer sound reception is unknown.

In summary, absence of anatomical evidence for sound reception is not evidence of absence of hearing. The sections that follow detail our understanding of water motion and pressure detectors that confer sound reception in reef fishes and decapod crustaceans. However, we are still a long way from a satisfactory account of sound reception across the full range of these two groups, particularly the crustaceans, but also the presettlement larval forms of both. Given the unreliability of inferring acoustic function from anatomy, progress depends on the development of behavioural (psychophysical) and physiological methods to characterise the acoustic capabilities of presettlement fishes and crustaceans.

4.2. Sound reception based on motion detection

The pelagic larvae of fishes and crustaceans are approximately the same density as the surrounding water and are small in comparison with the wavelength of sound. For example, presettlement fishes and crustaceans are typically of the order of tens of mm in length, and the wavelength of sound at 1.5 kHz is 1 m. As a result, these fishes and crustaceans move in sympathy with the sound field. Their otoliths and statoliths are about three times denser than water (Fay and Megela Simmons, 1999) and so have more inertia and move less. This forms the physical basis of a differential density accelerometer. The sensory receptors are typically mechanosensory cells that detect the differential motion between the otoliths or statoliths and the surrounding tissue. The acoustic function of the inner ear of fishes is much better understood than the equivalent invertebrate organ, so fish otolithic hearing is described first.

A comprehensive description of the auditory periphery of bony fishes is provided by Popper and Fay (1999). The fish inner ear consists of three semicircular canals and their associated sensory epithelium or cristae, and three otolith organs (Figure 3). The otolith organs are the saccule, the lagena and the utricle, each with its patch of sensory epithelium called a *macula* and an overlying dense calcareous otolith. In addition to these six commonly described sensory end-organs of the inner ear, there is an often overlooked additional macula in some fish species, fittingly called the *macula neglecta*, which has no otolith. The sensory epithelium in all these organs is composed of mechanosensory hair cells and support cells. The hair cells have a characteristic apical ciliary bundle with a single kinocilium and a graded series of stereocilia located to one side of the kinocilium. In the otolithic organs, the otolith is suspended alongside the sensory epithelium and the cilia of the hair cells are presumed to be coupled to the otolith via an otolithic membrane (Popper, 1971; Dunkelberger *et al.*, 1980). Hair cells are sensitive

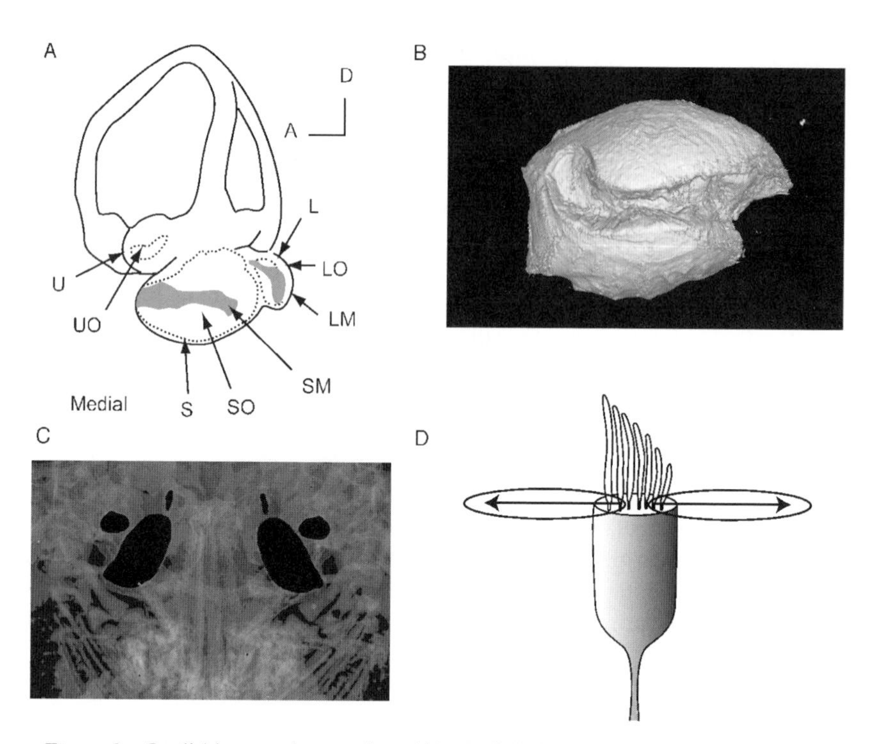

Figure 3 Otolithic sound reception. (A) Medial view of the right inner ear of the reef fish *Sebrasoma veliferum*, an acanthurid from Hawaii, showing the layout of the two vertical semicircular canals and the three otolith organs (A: anterior, D: dorsal, L: lagena, LM: lagena macular, LO: lagena otolith, S: sacculus, SM: saccular macular, SO: saccular otolith, U: utriculus, UO: utricular otolith). (B) Surgeon fish saccular otolith reconstructed from microCT. Note the sensory groove, which in life sits over the sensory macula. (C) The otoliths of the triplefin *Fosterygion lapillum* modelled *in situ* using x-ray microCT. The otoliths of the left and right otic capsule are shown *in situ* in their normal location and orientation as seen looking from the front of the fish. The large sagitta of the sacculus is flanked laterally by the lapullus of the utriculus and dorsally by the asteriscus of the lagena. (D) Representation of the cosine directional response of a hair cell, deflection of the cilia towards the kinocilium (to the left in this diagram) is excitatory; deflection in the opposite direction is inhibitory. (A, Redrawn after Fay and Popper, 1980.)

mechanoelectrical transducers. Displacement of the ciliary bundle in the direction of the kinocilium depolarises the hair cell and is excitatory. Orthogonal to this direction, there is no response. Displacement of the ciliary bundle in the direction from the kinocilium to the stereocilia results in hyperpolarisation and is inhibitory. So the response of the hair cell to displacement of the cilia is inherently directional, with the response following a cosine function (Roberts *et al.*, 1988).

Hair cells have proven to be highly useful detectors and form the basis of a wide range of sensory systems (Webster *et al.*, 1992). To a large degree, it is the associated anatomical structure around the hair cell epithelium that determines the effective stimulus, although heterogeneity in hair cell function also contributes. Hair cells are exquisitely sensitive, so much so that they can, and do, respond to all sorts of fluid motions and mechanical disturbances. Signal-to-noise ratios and signal specification have been issues throughout the evolution of hair cell–based sensory systems. It is possible in many cases to identify features of the anatomy and physiology of hair cells and surrounding tissues that improve signal-to-noise ratio. Although there is a tendency over evolution for the separation of composite, or ambiguous, stimulus dimensions into distinctive sensory channels, the separation of equilibrium and hearing modalities across otolith organs in fishes remains problematic.

The standard model of hearing in the vast majority of fish species is that "all otoliths in all species tend to respond to sound-induced motions of the fish's body" (Popper and Fay, 1999), but the principal hearing end-organ responsible for hearing is considered to be the sacculus. The compartmentalisation, or separation of equilibrium (i.e., vestibular) and acoustic function could occur along the dimensions of both frequency and amplitude. Vestibular stimuli tend to be low in frequency but can be large in amplitude. For example, the maximum frequency of self-induced head rotation is likely to be that generated by maximal swimming, and in a presettlement fish (20 mm in length) is of the order of 25 Hz (Batty and Blaxter, 1992), with an amplitude measured in millimetres. Passive displacement by other stimuli in the environment, such as wave motion or the turbulence generated by water currents or the motion of other animals, would also be low in frequency. Some of these sources may have low amplitudes, but others such as wave motion could have amplitudes measured in metres at least close to the surface. By comparison, behavioural and physiological measures give a frequency range for otolithic hearing in the range of tens to hundreds of hertz and at much lower amplitudes of movement (Popper and Fay, 1999). The tiny movement amplitudes of otolithic hearing can be calculated from behavioural thresholds that show fishes can detect an acoustical pressure of 0.01 Pa at 500 Hz. This represents an amplitude of particle motion of 2×10^{-12} m. Given that otolithic movement relative to the sensory epithelium is less than half that of the particle motion, this results in a stimulus to the receptors of $<10^{-12}$ m at threshold, or $\sim$1/100th of the diameter of a hydrogen atom (Rodgers and Cox, 1988). This analysis assumes that the behavioural response threshold quoted is mediated by motion detection. However, direct measurements of motion thresholds support these extremely high sensitivities. From these considerations, one might predict that the "acoustic channel," or specialisation of a particular otolith organ for sound reception, might be accompanied by reduced sensitivity at low frequencies and within the bandwidth of operation an increased sensitivity, perhaps at the expense of a reduced dynamic range. Direct physiological evidence presented below shows that saccular afferents do have these properties; however, there is considerable overlap between the response properties of the sacculus and the other otolith end-organs. As Popper and Fay (1999) note, we still lack an understanding of the relative roles of the saccule, lagena and utricle in vestibular and auditory function, and even whether these functions are mixed within an organ or parsed among organs. Given the rather smooth transition from self-induced accelerations and those induced by turbulence, wave motion, infrasound and true acoustic stimulation, it may be that vestibular, infrasound and auditory sensory categories are somewhat artificial constructs from the perspective of the fish ear.

Direct recordings from saccular afferents have been made in only a few bony fish species. Of most interest in this section of the review are the recordings made in "hearing generalists," that is, species that lack a specialised connection between the inner ear and the swim bladder. For example, in the toadfish *Opsanus tau* (Fay and Edds-Walton, 1997), the responses of saccular afferents to full-body accelerations do indeed show a low-frequency cutoff more suited to the detection of acoustic rather than vestibular stimuli (Figure 4). Frequency/response properties show two major categories of saccular afferents: One group responds best at frequencies of ~70 Hz, whereas the other responds best at ~140 Hz. The suggestion has been made (Rodgers and Cox, 1988) that the low-frequency component of the filter (high-pass filter) is generated by elastic coupling of the hair cells to the overlying otolith, but the mechanical and electrical tuning of hair cells themselves may also play a role (Popper and Fay, 1999). For example, hair cells with shorter cilia are associated with the transduction of higher frequencies (Popper and Platt, 1983; Platt and Popper, 1984; Sugihara and Furukawa, 1989).

The frequency/response characteristics of acoustic receptors are also presented as "audiograms." These are generated either behaviourally or using the acoustic brainstem response (ABR) technique. There is reasonable agreement between the results obtained using the two techniques (Kenyon *et al.*, 1998) at least for the shape of the audiogram. The absolute sensitivity can differ widely between the two techniques and among different investigators (Higgs, 2002). Results are typically presented as threshold tuning curves. For hearing generalists, the lowest threshold reported is ~100 dB re: 1 μPa at 100 Hz. However, the necessity of conducting behavioural and ABR experiments within the confines of a tank makes the measurement and interpretation of stimulus intensities problematic. Fay and Megela Simmons (1999) make the strong statement that "the only behavioural thresholds reported for fishes in the literature that could be interpreted are sound pressure levels for hearing specialists, and particle motion thresholds for several hearing

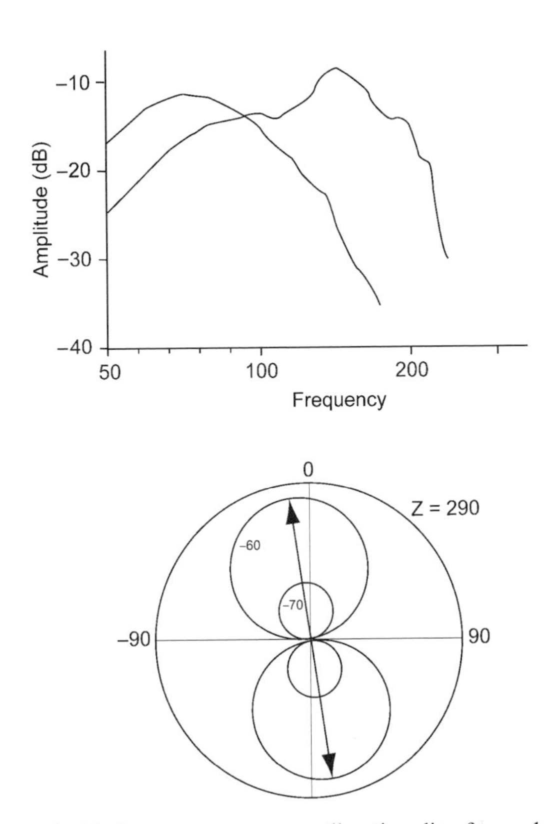

Figure 4 Threshold, frequency response, directionality from electrophysiology. (A) Averaged filter functions for the two major categories of saccular afferents of the toadfish (*Opsanus tau*). (B) Polar plot of the responsiveness (Z) of a saccular afferent as a function of the direction of a 100-Hz translatory motion in the horizontal plane showing the predicted cosine directional sensitivity, with a clear axis of maximal response. (Redrawn from Popper and Fay, 1999.)

generalists." Following this view, the best behavioural evidence to use to evaluate hearing thresholds in hearing generalists is the particle motion thresholds, which have been reported for cod and plaice (Fay and Megela Simmons, 1999) (Figure 5). These data show the greatest sensitivity to particle motion of −90 dB re: 1 μm, which equates to a displacement of 0.03 nm (i.e., 3×10^{-11}m) (rms).

Across a range of sensory modalities, the general finding is that electrophysiologically measured thresholds are higher than behaviourally measured thresholds. The reason for this is that electrophysiology typically relies on the response properties of single nerve fibers, whereas the central nervous

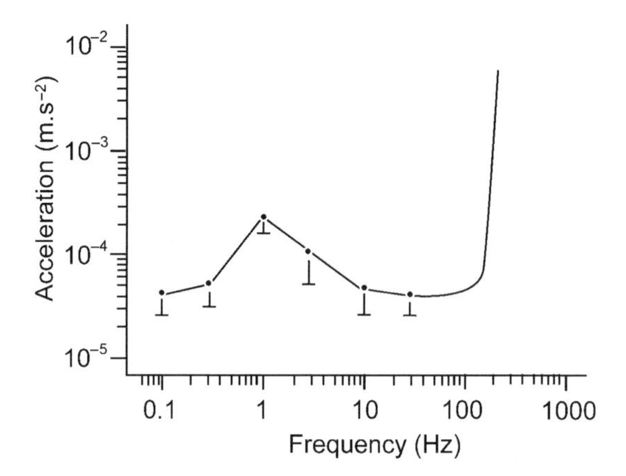

Figure 5 Audiogram for a hearing generalist. Behaviourally measured acceleration threshold for plaice combined from Karlson (1992) and Chapman and Sand (1974). (From Sand and Karlson, 2000.)

system (CNS) has access to a population of inputs and can improve signal detection through ensemble averaging. Taking account of these factors, the directly measured electrophysiological responses of saccular afferents in toadfish support the very low thresholds for motion detection. These recordings show a threshold response at stimuli of <0.1 nm (i.e., 10^{-10} m) and responses that saturate at displacements >1 μm. These recordings also very nicely demonstrate the directional sensitivity of individual afferents and the directional distribution pattern of the population of afferent fibres (Fay and Edds-Walton, 1997).

The inherent directionality of individual hair cell responses and the orientation pattern of the population of saccular hair cells provide a substrate for directional hearing. Theoretically, two populations of orthogonally oriented hair cells lying within a plane can determine sound direction within that plane. Taking the standard orientation of the saccular hair cells, motion of the otolith relative to the sensory epithelium along the rostrocaudal axis will excite the cells at the rostral end of the epithelium (i.e., those oriented rostrocaudally). The vertically oriented hair cells would be unaffected. Vertical stimulation would have the converse effect. The two populations thus give rise to a rostrocaudal and a vertical response vector. Intermediate sound directions can be determined by combining the components of the two vectors. Sound direction in three dimensions can be determined if the sensory epithelium is not planar but curved, by comparison across differently oriented epithelia from different otoliths or by binaural comparison in the CNS between the differently oriented saccular otoliths of the left and right inner ear. It should be noted that this mechanism provides information to the fish on the axis of the particle motion but does not determine the direction of wave propagation. Possible mechanisms to resolve this 180-degree ambiguity are discussed after the section on pressure reception mechanisms.

Given the focus of this review on the use of sound for orientation in presettlement larvae, it is important to know whether the hearing capabilities described earlier extend to this early life history phase. There have been relatively few studies on the development of hearing in fishes and those that have been conducted provide contradictory results. Using heart rate conditioning, Popper (1971) found no change in auditory sensitivity with growth in the juvenile and adult stages of goldfish (*Carassius auratus*). Using evoked brainstem responses, Higgs *et al.* (2002) found a similar result for zebrafish (*Danio rerio*). In other teleosts, there are either large increases in auditory sensitivity over the entire range of detectable frequencies (using behavioural conditioning in damselfishes, *Pomacentrus* spp. [Kenyon, 1996]) or small improvements in sensitivity over a much narrower range of audible frequencies (red sea bream, *Pagrus major*, with heart rate conditioning [Iwashita *et al.*, 1999]; gourami, *Trichopsis vittata*, with brainstem responses [Wysocki and Ladich, 2001]). Behavioural work has shown increases in responsiveness to a broadband auditory stimulus during the larval and juvenile periods of several species of fishes (Atlantic herring, *Clupea harengus* [Blaxter and Batty, 1985]; red drum, *Sciaenops ocellatus* [Fuiman *et al.*, 1999]). A study by Simpson *et al.* (2005) shows that even embryonic clownfishes (*Amphiprion ephippium* and *A. rubrocinctus*) can detect sound, and that the frequency range of detected sounds and the sensitivity of the response both increase through the embryonic period. Wright *et al.* (2005) used the ABR technique to investigate auditory abilities of presettlement and postsettlement stages of a damselfish, *Pomacentrus nagasakiensis.* Audiograms of the two ontogenetic stages were similar. Presettlement larvae, as well as their postsettlement counterparts, heard at all but two of the tested frequencies.

So far this discussion has concentrated on the "normal" frequency range of hearing, since the extent to which infrasound propagates away from shallow reefs is not known. However, even if reefs are not a source of infrasound, infrasound itself may still provide a potential orientation cue. Work (Sand and Karlsen, 2000) shows that the sensitivity of otolith organs can extend down to very low frequencies with sensitivities $<5 \times 10^{-5}$ m s^{-2}. Ambient noise in the sea increases towards lower frequencies, and turbulence along the edges of ocean currents and seismic motion of the ocean floor could be a source of the high level of ambient infrasound. The speculation is that the directional pattern of infrasound in the oceans may be an additional potential cue for onshore migration. However, the precise nature of the cue requires infrasound recordings to be made in the areas around reefs. The problem of resolving the 180-degree ambiguity also applies to infrasound. An accelerometer on its own is insufficient to use an onshore directional cue generated by an offshore infrasound source.

One additional point to consider is that the inner ear of some fishes has a macula neglecta, a patch of hair cells without an otolith. Rodgers and Cox (1988) argue that these hair cells could still respond to sound. In a "free-field" situation, the response might be created by the distortion of the cell generated by the fluid shear of the sound wave. The threshold of this response would be higher, and the nature of the directional response would be different from that provided by the otolithic end-organs. The suggestion has also been made that particle motion generated by sound at the water–air interface may also be an effective stimulus to the hair cells of the macula neglecta (Montgomery *et al.*, 2001). Recordings from the macula neglecta of sharks with the dorsal part of the head held at the water surface (Corwin, 1981) do indeed show an effective response to airborne sound. Corwin (1981) also suggests that the combination of otolithic and nonotolithic hearing inputs could perhaps be useful in determining the direction of the source.

From these considerations, it is apparent that otolithic hearing for fishes is well established and that other parallel hearing pathways are also possible. We have reasonable descriptions of the sensory mechanisms, frequency/threshold data and an understanding of the potential directional capabilities of otolithic hearing. However, these data are limited to very few species, only some of which belong to our group of interest, the reef fishes (which are mostly perciform teleosts). The data available for the auditory capabilities of presettlement reef fishes are sparser still.

Given the greater amount of work devoted to the sensory physiology of fishes as opposed to crustaceans, it is no surprise that our understanding of hearing in crustaceans is still rudimentary. Decapod crustaceans have a wide variety of sensory structures that have attracted considerable research interest, but their auditory sensory and behavioural functions are still relatively poorly understood (Budelmann, 1992; Popper *et al.*, 2001). Some receptors have been identified that may have the ability to respond to parameters of underwater sound such as hydrodynamic flows, particle motion and pressure changes, but their operation, sensory thresholds, range of sensitivity and especially their behavioural significance are not well defined (Popper *et al.*, 2001).

The obvious suggestion is that the crustacean statocyst is the analogue of the fish otolith and in adult crayfish (*Orconectes*) Breithaupt and Tautz (1988) reported vibration sensitivity of the statocyst with a peak-to-peak threshold of 0.1 μm over a range of frequencies from 150 to 2350 Hz. This earlier work has been backed up by further anatomical and physiological evidence. Lovella *et al.* (2005) describe the anatomy of the statocyst sensory structures of the prawn (*Palaemon serratus*) and provide direct electrophysiological evidence of sound reception by this organ. The basis of reception is an array of sensory hairs projecting from the floor of the statocyst into a mass of sand granules embedded in a gelatinous substance. Using an ABR technique, for the first time in invertebrates, their study shows that the statocyst of *P. serratus* is sensitive to the motion of water particles displaced by low-frequency sounds ranging from 100 to 3000 Hz, with a hearing acuity similar to that of a generalist fish. However, as with the fish, only limited information is available on these structures in the early life history stages of crustaceans (Budelmann, 1992; Popper *et al.*, 2001). A study by Sekiguchi and Terazawa (1997) of the statocyst of the puerulus (post-larva) of the spiny lobster *Jasus edwardsii* failed to find sensory hairs, secretory pores and fluid within the statocyst cavity, so statocyst function of these life history phases remains an open question.

Some studies have investigated the antennal sensory structures of the early life history stages of several species of spiny lobsters. These have found an almost continuous array of pinnate setae along the flagella of the antennae present in both pueruli and early juveniles, but absent from the late stage phyllosoma (larvae) (Phillips and Penrose, 1985; Phillips and Macmillan, 1987; Macmillan *et al.*, 1992; Nishida and Kittaka, 1992; Jeffs *et al.*, 1997). Similar arrays of sensory setae have been found in other decapods without a shoreward migrating life-cycle phase and it has been speculated that they may be used for detecting low-frequency water vibrations (Ball and Cowan, 1977; Denton and Gray, 1985, Wilkens *et al.*, 1996). Setae of this sort may represent an analogue of the lateral line of fishes, and if so, they would not respond to far-field sound, but to pressure gradients and to hydrodynamic flows close to a sound source.

Electrophysiological recordings from such setae have produced conflicting results. Pinnate setae of the antennae of both *Palinurus elephas* and *Panulirus japonicus* were found to have only limited sensitivity, and it was concluded that they could only detect gross water movement or act as proprioceptors (Tazaki and Ohnishi, 1974; Vedel, 1985). However, other workers have found that the pinnate setae on the antennae of some freshwater crayfishes are coupled with neighbouring sensory setae and consequently are highly sensitive to low-frequency vibrations (Tautz *et al.*, 1981; Masters *et al.*, 1982; Bender *et al.*, 1984).

Based on a boundary layer analysis and assumptions of the sensory modality of mechanoreceptors observed on the antennae of pueruli of *Panulirus cygnus*, Phillips and Penrose (1985) concluded that pueruli of this species would only be able to effectively detect directionality of reef noise above 1750 Hz, and that the attenuation of reef noise at this frequency would prevent it from being detected at distances >40 km away from the source. This analysis contravenes the traditional view that there is little or no differential movement of the animal with respect to the surrounding water because animals of this size move with the acoustic field.

Overall, these studies suggest that while aquatic crustaceans in general appear to have some capacity to detect waterborne sound and vibration, the sensory capabilities of the early stages of crustaceans remain unclear and require further work to identify the sensory mechanisms and sensitivities of hearing.

4.3. Sound reception based on pressure

Because of the relative incompressibility of water, it is the sound pressure wave that propagates and dominates the farfield (Rodgers and Cox, 1988). So it is somewhat paradoxical that acoustic senses in aquatic animals tend to be dominated by particle motion detectors. However, particular groups of fishes have developed acoustic pressure detectors and are collectively known as "hearing specialists" (Popper and Fay, 1999). Typically hearing specialists have lower threshold sensitivities and an extended upper frequency range of hearing (up to 2–5 kHz). These extended hearing capabilities are due to a close connection between a gas-filled cavity and the inner ear. In otophysan fishes, such as carp and goldfish, vibrations of the swim bladder are conducted to the inner ear by the mechanical linkage of Weberian ossicles. Anabantoids (labyrinth fishes) hold a gas bubble in a suprabranchial chamber close to the inner ear, and mormyrid fishes have specialised bilateral tympanic bladders (Saidel and Popper, 1987). All of these groups mentioned so far are freshwater species, and there is a view that evolution of high-frequency hearing, which is facilitated by these specialisations, has been driven by the shallow water cutoff for low frequencies. In the marine environment, clupeoid fishes have an elaborate acoustic bulla derived from anterior extensions of the swim bladder, which is functionally associated with both the inner ear and the lateral line and is thought to be associated with the evolution of ultrasonic hearing and the avoidance of echolocating marine mammal predators (Mann *et al.*, 2001). Among reef fishes, we have relatively few examples of specialisations for acoustic pressure detection; however, given the diversity of reef species, it is likely that additional examples remain to be discovered. The only putative hearing specialists described among the reef fishes are the chaetodontids (or butterflyfishes) and the holocentrids (squirrelfishes). One genus of butterflyfishes (Chaetodon) has a very interesting laterophysic connection, a linkage not between the swim bladder and the inner ear, but between the swim bladder and the lateral line canal system (Webb and Smith, 2000). The other group of interest are the holocentrids. They show an interesting range of morphologies, from the standard teleost arrangement where the swim bladder and the inner ear are quite separate, to some species showing a close proximity of the anterior end of the swim bladder to the inner ear (Coombs and Popper, 1979). The functional anatomy of the laterophysic connection will be described, after first considering the more general and potentially widespread role of the swim bladder in sound reception.

Hearing generalists, which lack specialised pressure detection mechanisms, may still be sensitive to acoustic pressure. The most obvious mechanism for this sensitivity is the way in which the swim bladder vibrates in the acoustic field, acting as a pressure-to-displacement transducer. This displacement will radiate away from the swim bladder and stimulate the inner ear, providing what has been called the *indirect stimulus* to the inner ear. The pressure sensitivity so generated is unquantified, but this mechanism has been proposed as providing the acoustic pressure reference to allow the fish to determine the sound direction (Figure 6). The sound pressure is a scalar property that, on its own, provides no information on direction to the source. However, when combined with particle movement information, the phase reference of the pressure can be used to resolve the 180-degree ambiguity inherent in otolithic hearing. One way of articulating this effect is that, on its own, otolithic hearing would provide a movement of the otolith along the direction of propagation of the sound wave. The indirect stimulus re-radiated from the swim bladder would generate an orbital motion of the otolith. Swapping the source from one side of the fish to the other would

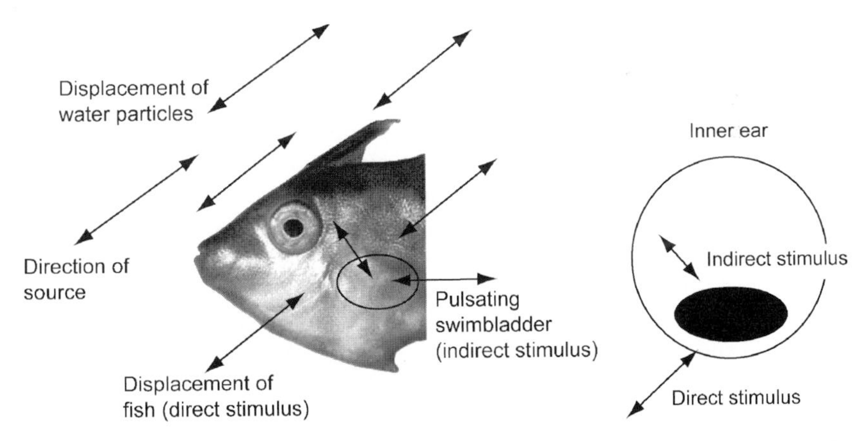

Figure 6 How fishes with swim bladders may resolve the 180-degree ambiguity according to the orbital model proposed by Wubbels and Schellart (1997). On its own, otolithic hearing would provide a movement of the otolith along the direction of propagation of the sound wave. The indirect stimulus re-radiated from the swim bladder would generate an orbital motion of the otolith. Swapping the source from one side of the fish to the other would reverse the apparent direction of orbital motion. In this description, the direction of orbital motion resolves the 180-degree ambiguity.

reverse the apparent direction of orbital motion. In this description, the direction of orbital motion resolves the 180-degree ambiguity (Wubbels and Schellart, 1997).

Given the potential utility of this indirect stimulus pathway, it is perhaps surprising that there are not more examples of swim bladder–inner ear connections that enhance pressure detection. The holocentrids vary in the extent to which the anterior end of the swim bladder connects to the inner ear (Coombs and Popper, 1979). The species with the closer connections have a wider auditory frequency range and greater sound pressure sensitivity. Pomacentrids do not have specialised hearing structures, but there is evidence for directional hearing in this family (Myrberg and Spires, 1980; Myrberg *et al.*, 1986). Female bicolour damselfish *Stegastes partitus* localise sound source direction of conspecific male courtship sounds in their natural environment (Myrberg and Spires, 1980; Myrberg *et al.*, 1986), and *S. dorsopunicans* is sensitive to both the pressure and the particle motion components of sound (Myrberg and Spires, 1980). Some families (e.g., pomacentrids, haemulids and sciaenids) have low thresholds, despite an apparent lack of auditory specialisations for enhanced pressure detection.

The best example we have of putative specialisation for pressure detection in reef fishes, with implications for settlement behaviour, is the laterophysic connection of chaetodontid fishes of the genus *Chaetodon* (Webb and Smith, 2000) (Figure 7). In this system (whose morphology varies among species [Smith *et al.*, 2003]), paired diverticula of the swim bladder ("horns") extend anteriorly and approach or directly contact a medial opening in the lateral line canal contained within the supracleithral bone. In its most developed form, only a thin membrane (the laterophysic tympanum) separates the fluid-filled system of the lateral line canal from the gas-filled swim bladder horns. The ability of the laterophysic tympanum to transmit a stimulus into the lateral line canal and the functional characteristics of the lateral line receptors receiving the stimulus via the otophysic connection have yet to be determined. The timing of the development of the connection would also be of interest, particularly so, given the long pelagic larval phase of chaetodontid fishes (60 d). The development of the laterophysic connection in a Caribbean species *Chaetodon ocellatus* shows that the medial opening in the supracleithrum is present in the smallest larvae examined (14.5 mm SL tholichthys) and that the cylindrical anterior swim bladder horns occur in all individuals $\geq$29 mm SL, right around the size at which they settle on the reef (J.F. Webb, personal communication).

It is worth noting that mechanisms of pressure reception other than those that depend on gas compression are possible. Fraser and Shelmerdine (2002) have demonstrated a direct response of hair cells in the dogfish to hydrostatic pressure. Both the spontaneous firing rate of afferent nerves and the gain of the rotation response are affected by hydrostatic pressure. This follows an

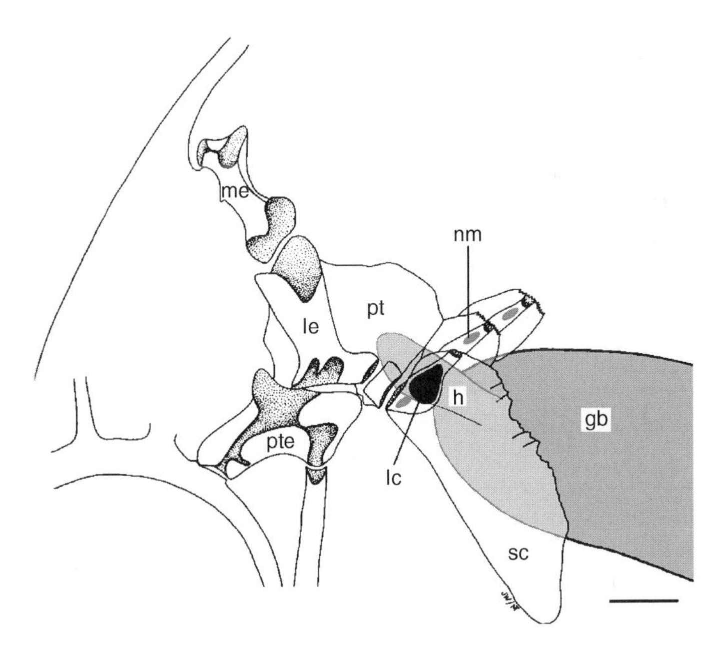

Figure 7 The swim bladder and laterophysic connection in *Chaetodon octofasciatus*. (A) Radiograph in lateral view shows the robust swim bladder, which approaches the posterior margin of the skull. A sphincter formed by the infolding of the tunica

earlier demonstration of a hydrostatic pressure–sensing mechanism in crabs (Fraser and Macdonald, 1994). A piston mechanism in crabs, generated by differential compression of the outer cuticle and the internal fluid of thread hairs in the balancing system, leads to nanometre-level displacements that are sensed by mechanoreceptors. The sensitivity and slow response time of both these receptor mechanisms seem more suited to the detection of hydrostatic pressure rather than acoustic pressure. But their recent discovery does suggest that novel pressure and/or acoustic sensors may yet be found.

5. EVIDENCE FOR THE USE OF SOUND IN ORIENTATION AND SETTLEMENT

The physics of sound propagation and ambient sound recordings provide evidence for the prospect of sound being used as an orientation cue. Psychophysics and electrophysiology provide evidence of the hearing competence of presettlement larvae and show that fishes and crustaceans may well have the capability to use sound to find reefs. However, only behavioural experiments can provide a direct demonstration that sound is used by these animals to guide their movements in relation to reefs.

A number of researchers have suggested that underwater sound, such as waves breaking on the coast, may provide a shoreward orientation cue for presettlement fish larvae (Stobutzki and Bellwood, 1998) and the pueruli of spiny lobsters (Phillips and Penrose, 1985; Phillips and Macmillan, 1987; Macmillan *et al.*, 1992). The "choice chamber" experiments of Stobutzki and Bellwood (1998) in particular showed that pomacentrid and apogonid larvae oriented onshore at night. The effect was observed at distances of a kilometre or so from the reef, and although the cue involved was not isolated, sound propagated from the reef was a plausible candidate.

Given the combined difficulties of working with sound and with presettlement larvae, it is perhaps not surprising that direct tests of the use of

---

interna divides the swim bladder into two internal chambers. (B) *Camera lucida* drawing of the bony elements just behind the orbit at the posterior margin of the skull. The anterior horns of the swim bladder (shaded) sit deep to the supracleithrum (sc). The laterophysic connection is shown in black, with the supracleithral neuromast just rostral to it, and the neuromasts of the first two lateral line scales (blue ovals) just caudal to it. Scale bar, ~1 mm. h: horn, gb: gas (swim) bladder, lc, laterophysic connection (medial opening in supracleithrum), le: lateral extrascapular, me: medial extrascapular, nm: neuromast, pt: post-temporal, pte: pterotic, sc: supracleithrum. (Modified from Webb and Smith, 2000; courtesy J. F. Webb.)

sound in larval orientation have taken time (Table 1). The difficulties of working with sound within the confines of the laboratory have already been alluded to in the section on fish psychophysics. The wavelengths of sound in water and the practical restrictions of the size of laboratory tanks make it essentially impossible to do meaningful behavioural studies involving the broadcast of sound in a tank. Hence, the need to take studies of orientation to sound into the field is compelling. The first successful direct demonstration of a behavioural effect of sound was published by Tolimieri *et al.* (2000). In this experiment, recorded reef sound was replayed in the vicinity of light traps. The sound level of the stimulus was 180 dB re: 1 μP at 1 m from the source, with the light traps several metres away. The traps were moored in a bay some distance from reefs, and light traps associated with replayed reef sound had a significantly higher catch of triplefin larvae (family: Tripterygiidae) than silent control traps. Interestingly, larvae of a pelagic clupeid were also caught, but their numbers did not differ between "sound" traps and the silent controls. Triplefins are temperate reef species, and in these areas, the catch rate of light traps is usually low, in terms of both numbers and species diversity. So it was of real interest to repeat these experiments in a coral reef location where the high numbers of settlement larvae increase the statistical power of the techniques and the high species diversity can address the generality of the effect. Two independent studies (Leis *et al.*, 2003; Simpson *et al.*, 2004) have repeated this experiment. Not only have they confirmed the original observation undertaken in temperate waters, but these experiments also showed that attraction to sound occurs over a very wide range of tropical reef fish species.

The study by Leis *et al.* (2003) used sound that was biologically "rich," mainly very short duration clicks and snaps caused by snapping shrimps with some contribution from other invertebrates and fishes. Hydrophone measurements of propagation loss showed that for frequencies $<$1.25 kHz broadcast sound was below the ambient sound at distances greater than $\sim$65 m from the speaker. At $\sim$8 m from the speaker, the measured broadcast sound level was comparable to that of a typical reef chorus and decreased with distance at approximately the rate of 6 dB per distance doubled, which is typical of free-field spreading loss. The level at the light trap, 3 m from the speaker, was $\sim$10 dB higher than typical chorus noise levels, but within the range of chorus noise levels measured close to reefs. Their main finding was that some taxa, particularly apogonids and pomacentrids, had catches up to 155% greater in noisy traps than silent traps, but this apparently varied with location and time. They calculated that sound-enhanced catches imply a radius of attraction of the sound 1.02–1.6 times that of the light.

Simpson *et al.* (2004) used methods similar to those of Leis *et al.* (2003) and compared the total catches for each reef fish family from light traps with sound treatments to total catches from silent light traps. In 37 samples,

Table 1 Representative selection of the direct evidence for orientation to reef sound

Study	Group	Experimental	Comparison
Light traps with and without sound (Tolimieri *et al.*, 2000)	Tripterygiidae	Median no. of fish larvae entering traps with sound: 350	Median no. of fish larvae entering traps without sound: 24
Binary choice chambers (Tolimieri *et al.*, 2002)	Tripterygiidae	No. of fish moving towards sound at night: 12	No. of fish moving away at night: 4
	Tripterygiidae	No. of fish moving away from sound during day: 15	No. of fish moving toward sound during day: 1
Light traps with and without sound at a near-reef site (Leis *et al.*, 2003)		No. of fish larvae entering traps with sound:	No. of fish larvae entering traps without sound:
	Apogonidae	156	61
	Pomacentridae	2639	1833
Binary choice chambers (Tolimieri *et al.*, 2004)	Pomacentridae	No. of trials where majority of fish moved towards sound at night: 24	No. of trials where majority of fish moved away from sound at night: 8
Light traps with and without sound, quarter moon phase (Jeffs *et al.*, 2003)	Brachyura:	No. entering traps with sound:	No. entering traps without sound:
	Post-larvae	97,713	52,630
	Larvae	10,591	3990
	Anomura:		
	Post-larvae	100,241	51,167

Light traps with and without sound (Simpson *et al.*, 2004)		Percentage of fish larvae entering traps with sound:	Percentage of fish larvae entering traps without sound:
	Apogonidae	60	40
	Blenniidae	64	36
	Holocentridae	70	30
	Lethirinidae	63	37
	Mullidae	76	24
	Nemipteridae	64	36
	Pomacentridae	68	32
	Pseudochromidae	61	39
	Syngnathidae	73	27
	Trichonotidae	75	25
Patch reef experiment (Simpson *et al.*, 2005)		Percentage of fish larvae settling on reefs with sound:	Percentage of fish larvae settling on reefs without sound:
	Apogonidae	62	38
	Pomacentridae	61	39

Note: In each case, the differences tabulated are statistically significant.

40,191 settlement-stage reef fish larvae were collected, and 10 families were represented by at least 20 individuals and deemed statistically viable. There was a strong consistency across families, with each of these families responding positively to reef noise with a mean ratio per family of 2.2:1 (sound:silent).

Jeffs *et al.* (2003) used the same technique to investigate the attraction of crustaceans to sound. Like the experiments on fishes, light traps with sound caught significantly more crab larvae and post-larvae than silent traps, although this effect was only evident at particular moon phases. No effect of sound was seen near full and new moon, when tidal currents would be strongest.

The use of light traps in these experiments requires some comment. It is somewhat ironic that the orientation to sound is demonstrated as a modulation of the catch rates of light traps. Clearly light is attractive to presettlement larval fishes and some crustaceans, and as a consequence, light traps have become a standard and convenient mode of catching them. However, we lack any satisfactory explanation for the natural function of this attraction. Like moths, larval fishes may maintain a constant direction of locomotion by holding a constant angle to celestial light sources, such as the moon. For a close light source, this tactic results in a spiral path inwards toward the source. Alternately, the attraction of larval fishes may be secondary to the attraction of the small planktonic crustaceans, and the light may simply provide a feeding opportunity by concentrating the plankton. This still begs the question as to why the plankton are attracted to the light! The increase in catch rate of light traps with replayed sound is presumably due to the sound attracting the larvae into the range of the light, thereby increasing the effective range of the trap. However, this proposed mechanism does not require the larvae to resolve the 180-degree ambiguity. Beyond the range of the light, even if half the larvae went towards the sound and half went away from the sound, the catch rate would still increase. The use of light traps in conjunction with sound to test orientation of larvae has been criticised on the grounds of a potential interactive effect between the light and sound behavioural cues. However, the most likely explanation for increased catch rates is the attractiveness of the sound, and several other field studies support this.

The second line of experiments that provide evidence for a behavioural effect of sound are the "*in situ*" observations made by Leis and colleagues (Leis *et al.*, 2002; Leis and Carson-Ewart, 2003; Leis and Lockett, 2005). By catching larvae in light traps and observing their behaviour on release at distances of 100–1000 m from the reef, these investigators were able to conclude that settlement-stage larvae of some butterflyfishes and damselfishes can determine the direction to the reef (Leis and Carson-Ewart, 2003).

Furthermore, differences in swimming behaviour (speed and direction) between ambient conditions and during broadcast sound (both natural and artificial) showed that larvae are able to hear the sounds and discriminate between them (Leis *et al.*, 2002). The behavioural differences were observed at distances of 25–100 m from the loudspeaker. Recordings of the sound played by the loudspeaker showed that the broadcast levels fell to near the background noise level beyond ~100 m. Leis *et al.* (2002) argue that under natural conditions, larvae will be able to hear reefs from a much greater distance primarily because the reef noise will attenuate less with distance than with a point source, such as a speaker. Like the light-trap studies, the *in situ* observations provided no clear indication that larval fishes can localise the sound.

The third line of evidence demonstrating a behavioural response to replayed reef sound was a development of the Stobutzki and Bellwood (1998) choice chamber. Instead of using the choice chambers to evaluate the orientation of the larvae towards and away from the reef, Tolimieri *et al.* (2004) observed the response to replayed reef sound. Larvae caught in light traps at night were held in the laboratory over the day. Choice chambers were deployed the following night parallel to the shore with an underwater loudspeaker broadcasting prerecorded nocturnal reef sound. The location of the speaker at one or the other end of the choice chamber was allocated at random. Larvae were released into the centre of the choice chamber via a holding compartment and a time-delay release mechanism. Funnel traps at either end of the choice chamber held the larvae in place once they had swum to one end or the other. These experiments were conducted on temperate triplefin (tripterygiid) species and repeated again on damselfish (pomacentrid) larvae. The triplefins showed a significant orientation towards the sound at night and away from the sound during the day, matching the original Stobutzki and Bellwood (1998) and Leis *et al.* (1996) observations that the presettlement fishes tended to orient towards the reef at night, but away from the reef in the day. Damselfish larvae also showed a significant orientation towards the sound at night. Fishes responding to the sound were, on average, larger than those that did not, suggesting that there may be a rather narrow window of response to sound that corresponds with readiness to settle. Leis and Lockett (2005) used a similar choice chamber experiment, which provided evidence that larvae of some pomocentrid species can localise and orient to a sound source, implying that sound emanating from reefs at night could be a usable sensory cue for locating reefs.

These choice chamber results clearly show that late-stage larvae are orienting to sound in a way that is appropriate to guide their movement towards reefs as suitable settlement habitat. In addition, these experiments provide evidence that the larvae are capable of determining not only the axis of sound propagation, but also the direction to the source. That is, they can resolve the 180-degree ambiguity inherent in otolithic hearing. This finding is particularly interesting for the triplefin larvae because they have no swim bladder and so cannot use a re-radiated stimulus from the swim bladder to determine sound direction. Only one other report indicates that fishes without swim bladders are capable of resolving the 180-degree ambiguity. The shark, *Chiloscyllim griseum*, is capable of detecting particle motion and sound pressure and can hear directionally (Van den Berg and Schuijf, 1983). If both these species can resolve the direction to the source without a swim bladder, it strongly suggests that another sensory mechanism is in operation. From first principles, an otolithic sensor could resolve the 180-degree ambiguity if it were possible to sample the spatial distribution of sound intensity or direction. In other words, a fish swimming towards the source would detect an increase in intensity level. However, within the confines of the choice chamber, such a possibility is unlikely. One other theoretical possibility is that the air–water interface at the surface of the sea could act in a manner similar to the swim bladder, providing an indirect stimulus path that imparts an orbital motion to the otolith. As with the indirect stimulus from the swim bladder, the direction to the source can be determined by the direction of the rotational motion. A final possibility remains that these fish possess some other pressure reference system not dependent on the swim bladder or any other gas inclusion.

The final line of evidence for the use of sound in habitat selection by coral reef fish larvae is the demonstration that replayed reef sound alone is effective at increasing settlement on small experimental patch reefs. In a randomised design, Simpson *et al.* (2005) were able to show that artificial patch reefs constructed from dead coral had higher levels of settlement when they were associated with replayed reef sound in a lagoon on the GBR. As most settlement is thought to occur at night, reefs were cleared of young fishes every morning. Of the 868 newly settled fishes collected, most were apogonids (79%) and pomacentrids (15%), two families that are key members of reef fish assemblages. Both families settled in greater numbers on noisy reefs than on silent reefs. This pattern was also evident in rarer fishes, with significantly more families and taxa on reefs with broadcast noise than on reefs without noise.

This experiment was extended by comparing settlement rates on silent reefs with settlement rates on reefs where only the high- (80% > 570 Hz) or low-frequency (80% < 570 Hz) portion of reef noise was broadcast. Apogonids settled on "high- and low-frequency" reefs in equivalent numbers, but pomacentrids were preferentially attracted to high-frequency noise (Table 1). Again, reefs without sound received much lower settlement and total numbers of families and taxa than reefs with broadcast sound.

6. SENSE AND SETTLEMENT: A COMPARISON OF SOUND AND OTHER SENSORY CUES

Able (1996) provided a useful reminder that animal orientation systems are replete with interacting mechanisms and are highly flexible. So although sound may be a prime candidate for long-distance orientation to reefs, if other cues are available, they are likely to be used. There is good evidence of vision and chemosensory information being used in habitat selection at close range, but what information is available at greater distances from the reef?

To locate a suitable reef habitat, the ideal is to have sensory information that indicates the direction, distance and quality of potential reef settlement sites. However, even in the absence of this information, other behavioural strategies may improve the odds of finding suitable settlement habitat. For example, innate directional preferences, location in the water column and orientation to physical cues, such as swell patterns, may bring the larvae closer to shore. In addition, maintaining directional orientation may also be important if the proximate clues indicating the direction to the reef are intermittent. Acoustic cues may, for example, be generated at particular times or may be masked by background noise associated with sea state. Under such conditions, the maintenance of directed headings triggered, or set previously, from underwater sound or other directional cues might make an important contribution to orientation behaviour. So the sensory mechanisms underlying the determination and maintenance of directional preferences make up part of the story of the potential sensory basis of orientation. To complete the picture, the sensory modalities in addition to sound that could allow for direct detection of a reef are also considered.

In some locations, innate directional preferences may be effective in increasing the odds of finding suitable settlement habitat. For example, east coast larvae swimming in a westerly direction will always be moving onshore. It is difficult to envisage how innate directional preferences may be programmed, and to our knowledge, there is no evidence for such a preference in fishes or crustacean larvae. However, the precedents that exist elsewhere in the animal kingdom make this a candidate mechanism for larval fishes. Perhaps it is more likely that a directional preference is triggered by some other directional information. For example, an intermittently available cue may trigger an orientation preference that is maintained in the absence of the cue. One of the clearest examples of this mechanism in operation in vertebrates is the compass orientation of turtle hatchlings (Lohmann and Lohmann, 1996; Avens and Lohmann, 2003). Initial movement in these hatchlings is directed towards the lowest brightest part of the horizon, and once this offshore orientation is initiated, it is maintained by a magnetic compass sense. There is also evidence for a magnetic compass sense in both fishes and crustaceans. Smith and Smith (1998) have shown that migrating larval gobies have a magnetic compass sense that is entrained by an upstream rheotaxis orientation. Boles and Lohmann (2003) and Lohmann *et al.* (1995) provide evidence for magnetism-based navigation in spiny lobsters and a magnetic compass sense. The maintenance of a particular directional preference is not necessarily dependent on a magnetic sense. Other directional orientation mechanisms have been proposed and/or demonstrated in fishes. These include inertial mechanisms (Steele, 1989), sun compass (Levin and Gonzalez, 1994), polarised light (Flamarique and Hawryshyn, 1997) and electric fields (Metcalfe *et al.*, 1993). The attraction of larval fishes and crustaceans to light traps has already been mentioned as a possible consequence of celestial orientation.

Beyond the potential use of sun and moon compass, star compass and polarised light for direction holding, it is difficult to imagine vision as being a useful long-distance orientation cue to find reefs under most conditions, particularly at night. However, given the compression of the above-water visual field into Snell's window, it is possible that elevated land masses could be seen from some distance offshore, particularly on calm days.

Reef fish and crustacean larvae appear to use chemosensory cues when settling. However, much of this information comes from experiments conducted at small spatial scales (e.g., coral heads on a reef, or small experimental setups) (Sweatman, 1988; Elliott *et al.*, 1995; Arvedlund and Nielson, 1996; Danilowicz, 1996; Lecchini *et al.*, 2005) or local nursery areas in estuaries (Forward and Tankersley, 2003), and its relevance to large-scale patterns is unknown. Cury (1994) has suggested that imprinting may be important for habitat selection for fishes and other marine vertebrates, and Arvedlund and Nielson (1996) have shown imprinting of chemical cues in anemone fishes. In the case of reef fishes, Cury's hypothesis would suggest that individuals imprint on reef habitat, and this imprinting affects their choice of habitat at settlement. Many marine and anadromous fishes do show homing behaviour (reviewed by Cury, 1994; Thorrold *et al.*, 2001). However, it is difficult to conceive how species with pelagic larvae, and especially those with pelagic eggs, could imprint on their natal reef habitat or on reef habitat in general.

Certainly, with or without imprinting, some fishes use chemoreception for large-scale navigation, the best known example being salmon returning to their natal streams (Cury, 1994). In addition, migrating whitebait (family: Galaxiidae) species preferentially select water that has been exposed to adults (Baker and Montgomery, 2001), and homing on pheromones released by adults would be an effective way for juveniles to locate suitable habitat. Islands can produce strong chemical signals that larvae might use to navigate shoreward (Kingsford, 1997). For example, during low tide, ponds develop within the lagoon at One Tree Reef, GBR. On flood tides, this warmer and chemically rich water exits the lagoon, providing a strong chemical signal up to 5 km from the island (Kingsford, 1997). So in some locations, an olfactory plume may provide useful information for orientation at considerable distances from the source, and behavioural experiments provide good evidence of presettlement larval preference for lagoon water (Atema *et al.*, 2002).

The use of olfactory cues has difficulties and limitations. In a moving fluid environment, water currents carry the olfactory stimuli. At the very least, this means that an olfactory approach to the reef system must be from downstream, which is energetically more demanding than detecting and swimming towards the reefs from any other direction (Armsworth, 2000; Armsworth *et al.*, 2001). In addition, in complex, turbulent and tidal current systems, olfactory search is a difficult task because search strategies based on olfactory cues alone (such as a gradient search) become more demanding. Olfactory stimuli break into patches within eddies and a gradient search strategy traps the searcher within the eddy system heading in the wrong direction. Under normal circumstances, a combined olfactory/rheotaxis (water current) strategy can provide a more reliable approach (Montgomery *et al.*, 2000) but is contingent on the larvae being able to detect water currents. This is problematic for fish moving in a body of water without a geostatic reference frame, because there is no basis on which to detect the water current. Typically the reference frame is visual or tactile (Montgomery *et al.*, 1997), but it is difficult to see how small pelagic larvae might establish such a reference frame in the water column. It is possible that a larva might change depth in response to chemical cues, and that in some circumstances, a change in depth will have predictable consequences for transport. But the extent to which fish larvae can solve these problems and utilise chemical senses for onshore transport is unknown. The best we can conclude at this stage is that olfactory cues certainly exist and they may exist over a range from several to many kilometres; however, the degree to which larvae use these cues for orienting to settlement sites on reefs is still an open question.

The direction of ocean swell presents one potentially useful orientation cue (Lewis, 1979). Large waves must move onshore because they require sufficient fetch to develop, and therefore, they cannot come from a shoreward direction. Waves also tend to refract towards the shore as water depth decreases near land (Tait, 1962; Kennish, 2001). The direction of these larger long period waves may provide a reasonably consistent cue. For example, in the Canterbury region of New Zealand, 66% of all waves, and all waves with heights of 3.0 m and above, move onshore (Kingsland and Macky, 1999). The vertebrate inner ear is capable of sensing wave direction in the ocean through the detection of the orbital motion of the wave. For example, sea turtles use wave direction to orient offshore after hatching (Lohmann and

Lohmann, 1992, 1996; Lohmann *et al.*, 1995). While there is no direct evidence that fishes use wave direction for orienting onshore, Cook (1984) found that juvenile salmon migrations in the north Atlantic followed the direction of surface swell. As noted earlier, within the coastal environment, long period swells will, in general, be directed onshore. Long period waves penetrate deeper in the ocean than shorter period waves (Kennish, 2001), potentially allowing larvae to detect wave motion at depths of tens of metres. One theoretical strategy would be to descend in the ocean until no wave motion is detected and then to ascend until an orbital motion is first detected. This will be the longest period wave detectable and could, in theory, provide clues that would tend to lead the larvae in an onshore direction. Sea turtle hatchlings again provide direct evidence of the use of wave motion for directional orientation (Lohmann and Lohmann, 1992; Lohmann *et al.*, 1995). Wave orientation could also, in theory, provide more specific clues as to the direction of nearby reefs. Polynesian navigators of ocean voyaging sailing vessels are known to use complicated observations of swell reflections and refractions to locate isolated reefs (Lewis, 1979). Unlike fishes, there is clear evidence that a number of decapod crustacean species can use swell information for long-distance orientation. For example, wave surge has been implicated in long-distance orientation behaviour of the spiny lobsters *Panulirus argus* and *Panulirus guttatus* moving around benthic reef habitats (Walton and Herrnkind, 1977; Lozano-Álvarez *et al.*, 2002).

7. CONCLUSIONS/PROSPECTS

The pelagic life history phase of reef fishes and decapod crustaceans is complex, and the evolutionary drivers and ecological consequences of this life history strategy remain largely speculative. However, this life history phase is very significant in the demographics of reef populations. One of the assertions of this review is that some progress can be made by studying the proximal processes driving the pelagic larval phase. This review concentrates on the latter part of the pelagic phase as the larvae transit back onto a reef. We reviewed evidence showing that larvae are actively involved in this transition. They are capable swimmers and can locate reefs from hundreds of metres if not kilometres away. There is good evidence that sound is available as an orientation cue, and that fishes and crustaceans hear sound and orient to sound in a manner consistent with their use of sound to guide settlement onto reefs. Comparing field sound strengths (8×10^{-11} m at 5 km from a reef) with the measured behavioural and electrophysiological threshold of fishes of (3×10^{-11} m and 10×10^{-11}, respectively) provides evidence that sound may be a useful orientation cue at a range of kilometres rather than hundreds of metres (Table 2). These threshold levels are for adult fishes and we clearly need better data for larval fishes and crustaceans at the time of settlement. Field experiments are the only effective tool to demonstrate the actual use of underwater sound for this orientation purpose. The diverse series of field experiments including light-trap catches enhanced by replayed reef sound, *in situ* observations of behaviour and sound-enhanced settlement rate on patch reefs collectively provide a compelling case that sound is used as an orientation and settlement cue for these late-larval stages.

This conclusion, on its own, does not imply any particular position on the debate as to the relative importance of dispersal, retention and homing. Comparative considerations argue that dispersal does appear to have adaptive value in its own right, but again such an argument does not rule out elements of retention and homing. Until we have better defined probability density functions to work with, it is reasonable to consider a default condition that includes both retention and dispersal. For the purposes of understanding proximal mechanisms of habitat selection at the end of the larval phase, dispersal and retention are equivalent—the settling larva in both instances is seeking suitable reef habitat. The demands of homing are significantly greater, with the requirement for the discrimination of the characteristics of the home reef or home site over all other possible settlement sites. Since homing is a more demanding task than habitat selection, establishing the potential for, and existence of, active habitat selection could be seen by those favouring settlement homing as a prelude to the more exacting task of the larvae finding their way to their natal reefs.

The basic phenomenon of a contribution of acoustic orientation to settlement may be established, but we still lack key elements to make up the full picture. What is the acoustic "footprint" of a reef? What is the "fingerprint" of a reef, that is, how much information on habitat quality does underwater sound from reefs convey? What is the spatial and temporal variability of the

Table 2 Comparison of sound stimulus at a known distance from the reef and measures of biological sensitivity

Condition	Sound pressure level (dB re: 1 Pa)	Equivalent particle motion	Reference
Field measurement at 5 km from reef	118	0.08 nm	Tait (1962)
Biological sensitivity (behavioural)	—	0.03 nm	Fay and Megela Simmons (1999)
Biological sensitivity (electrophysiological)	—	0.1 nm	Fay and Edds-Walton (1997)

directional cue? What are the true hearing capabilities of the larval stages of reef fishes and crustaceans? How well does this match the sound "footprint" produced by reefs? What elements of the sound are attractive? What is the masking impact of other sound such as rain, sea state or anthropogenic sources? How does sound interface with other potential cues? What is the contribution of acoustic orientation to the final distribution kernel? To answer these questions, we need better sound measurements in the field, better comparative anatomy and laboratory experiments of hearing capability and better behavioural assays.

With respect to field sound measurements, we need much greater spatial and temporal understanding of the soundscapes associated with, and emanating from, potential settlement habitats such as reefs. This should allow the distinctive attributes of reef noise to be determined and the strength of the orientation cue to be charted in the area around the reef. To determine the effective range of the cue requires better information on the hearing capabilities and corresponding behavioural responses of fishes and crustaceans, particularly of the presettlement life history stages. A comparative analysis of the sensory systems in all the larval species that show a significant response to the sound light traps versus those that do not might provide useful cues. Similarly, the timing of the ontogeny of these sensory systems could provide useful information. Many of the ABR and laboratory behaviour measures do not adequately control the stimulus or discriminate between particle motion and pressure components of the stimulus. It is still in the situation where it is possible to claim that the available evidence argues against larval fishes and crustaceans having the degree of sensitivity necessary to make use of sound as an orientation cue (Myrberg and Fuiman, 2002), or to argue as we have here that particle displacement thresholds and potential pressure-sensing mechanisms of the appropriate late-stage larvae are likely to have the appropriate characteristics. Field experiments argue strongly in favour of the latter proposition. In the short term, field experiments also probably offer the best hope for resolving some of the previous questions. However, to address these questions in the field, we need better behavioural assays that compress the time and effort required to obtain meaningful results. For example, if larvae could be demonstrated to show an unequivocal response to a short sequence of replayed reef sound, it would then become possible to play through an intensity response function over the course of tens of minutes. By comparison, at present, each data point for the choice chambers takes one night of collecting fishes and another night to generate the response to a single sound intensity. In addition, the larvae are subject to considerable disturbance through collection and handling. An improved behavioural assay would also be able to more efficiently identify the footprint of a reef, the nature of the attractive component of reef sound and the potential masking effect of natural and anthropogenic noise. These methods will also allow us to establish the importance of sound from natal reefs as a homing settlement cue. These kinds of findings will place us in a much better position to address the wider questions on the evolution and ecology of dispersal and active habitat selection and to evaluate the importance of active habitat selection in settlement. A better understanding of this phenomenon will also provide us with insight into the potential that sound has to structure marine communities and provide directional cues for other marine animals.

This review has concentrated on fish and decapod crustaceans as the taxa most behaviourally capable of influencing their settlement distribution. Some other taxa, such as stomatopod crustaceans, and cephalopods also are active swimmers as larvae (Villanueva *et al.*, 1997); however, nothing is known of their hearing capability or field orientation in relation to reefs. Even without a high level of swimming competence, larvae of other taxa may be able to enhance their settlement options by responding to sound. For example, altered vertical migration in response to reef sound could conceivably enhance transport towards the reef. The potential for sound to structure marine communities may also extend beyond the reef, but again little is known of the acoustic characteristics of other habitats. Finally, the demonstrated potential of artificial reef sound to enhance settlement may also provide the basis of management tools to manipulate and conserve reef species.

In summary, a case can be made at each of the required levels of evidence in favour of sound as an important orientation cue for some larval reef animals. Sound is generated and transmitted from reefs, some reef fishes and crustaceans have sensory structures suited to sound reception and behavioural studies show clear evidence of orientation to sound in the field. However, we still lack quantitative answers to properly gauge the significance and generality of sound-based active settlement. Echoing the predominating themes of contemporary research in orientation and navigation (Able, 1996), further work is needed to understand the interaction of different cues within and across species. In addition, Able's recognition of the compelling need to take studies of navigation back into the field is especially true for sound as an underwater orientation cue for settling larval stages. Finally, a broad spectrum of approaches will be necessary to uncover and tease apart the "rules of thumb" that guide settlement back onto the reef.

ACKNOWLEDGEMENTS

The authors work in this area has been supported by the Marsden Fund of the Royal Society of New Zealand, the British Ecological Society, the Fisheries Society of the British Isles and the Australian Institute of Marine

Science. We thank Dr. Jacqueline Webb for comments on the manuscript and for providing unpublished information and illustrations of the butterflyfish laterophysic system, and Craig Radford for providing the sound recordings for Figures 1 and 2.

REFERENCES

Able, K. P. (1996). Large scale navigation. *Journal of Experimental Biology* **199**, 1–2.

Albers, V. M. (1965). "Underwater Acoustics Handbook II." Pennsylvania State University Press, University Park, Pennsylvania.

Armsworth, P. R. (2000). Modeling the swimming response of late stage larval reef fish to different stimuli. *Marine Ecology Progress Series* **195**, 231–247.

Armsworth, P. R., James, M. K. and Bode, L. (2001). When to press on or turn back: dispersal strategies for reef fish larvae. *American Naturalist* **157**, 434–450.

Arvedlund, M. and Nielson, L. E. (1996). Do the anemone fish *Amphiprion ocelaris* (Pisces: Pomacentridae) imprint themselves to their host sea anemone *Heteractis magnifica* (Anthrozoa: Actinidae)? *Ethology* **102**, 1–15.

Atema, J., Kingsford, M. J. and Gerlach, G. (2002). Larval reef fish could use odour detection, retention and orientation to reefs. *Marine Ecology Progress Series* **241**, 151–160.

Avens, L. and Lohmann, K. J. (2003). Use of multiple orientation cues by juvenile loggerhead sea turtles *Caretta caretta. Journal of Experimental Biology* **206**, 4317–4325.

Baker, C. F. and Montgomery, J. C. (2001). Species-specific attraction of migratory banded kokopu juveniles to adult pheromones. *Journal of Fish Biology* **58**, 1221–1229.

Ball, E. E. and Cowan, A. N. (1977). Ultrastructure of the antennal sensilla of *Acetes* (Crustacea, Decapoda, Natania, Sergestidae). *Philosophical Transactions of the Royal Society of London* B **277**, 429–456.

Batty, R. S. and Blaxter, J. H. S. (1992). The effect of temperature on the burst swimming performance of fish larvae. *Journal of Experimental Biology* **170**, 187–201.

Bellwood, D. R. and Wainwright, P. C. (2002). The history and biogeography of fishes on coral reefs. *In* "Coral Reef Fishes: Dynamics and Diversity in a Complex Ecosystem" (P. F. Sale, ed.), pp. 5–32. Academic Press, San Diego.

Bender, M., Gnatzy, W. and Tautz, J. (1984). The antennal feathered hairs in the crayfish: a non-innervated stimulus transmitting system. *Journal of Comparative Physiology A* **154**, 45–47.

Blaxter, J. H. S. and Batty, R. S. (1985). The development of startle responses in herring larvae. *Journal of the Marine Biological Association of the United Kingdom* **65**, 737–750.

Boehlert, G. W. and Yoklavich, M. M. (1984). Reproduction embryonic energetics and the maternal–fetal relationship in the viviparous genus *Sebastes pisces* scorpaenidae. *Biological Bulletin* **167**, 354–370.

Boles, L. C. and Lohmann, K. J. (2003). True navigation and magnetic maps in spiny lobsters. *Nature* **421**, 60–63.

Bradbury, I. R. and Snelgrove, P. V. R. (2001). Contrasting larval transport in demersal fish and benthic invertebrates: The roles of behaviour and advective

processes in determining spatial pattern. *Canadian Journal of Fisheries and Aquatic Sciences* **58**, 811–823.
Breithaupt, T. and Tautz, J. (1988). Vibration sensitivity of the crayfish statocyst. *Naturwissenschaften* **75**, 310–312.
Budelmann, B. U. (1992). Hearing in Crustacea. *In* "The Evolutionary Biology of Hearing" (D. B. Webster, R. R. Fay and A. N. Popper, eds), pp. 131–139. Springer-Verlag, New York.
Campana, S. E. and Thorrold, S. R. (2001). Otoliths, increments, and elements: keys to a comprehensive understanding of fish populations? *Canadian Journal of Fisheries and Aquatic Sciences* **58**, 30–38.
Cato, D. H. (1978). "Features of Ambient Noise in Shallow Waters around Australia," Vol. 2. Sonobuoy Working Party, 26–29 September 1978, London.
Chapman, C. J. and Sand, O. (1974). Field studies of hearing in two species of flatfish *Pleuronectes platessa* (L.) and *Limanda limanda* (L.) (Family Pleuronectidae). *Comparative Biochemistry and Physiology A* **47**, 371–385.
Chiswell, S. M. and Booth, J. D. (1999). Rock lobster *Jasus edwardsii* larval retention by the Wairarapa Eddy off New Zealand. *Marine Ecology Progress Series* **183**, 227–240.
Choat, J. H. and Schiel, D. R. (1982). Patterns of distribution and abundance of large brown algae and invertebrate herbivores in subtidal regions of northern New-Zealand. *Journal of Experimental Marine Biology & Ecology* **60**, 129–162.
Codling, E. A., Hill, N. A., Pitchford, J. W. and Simpson, S. D. (2004). Random walk models for the movement and recruitment of reef fish larvae. *Marine Ecology-Progress Series* **279**, 215–224.
Cook, P. H. (1984). Direction information from surface swell. *In* "Mechanisms of Migration in Fishes" (J. D. McCleave, G. P. Arnold, J. J. Dodson and W. H. Neill, eds), pp. 79–101. Plenum Press, New York.
Cook, L. G. and Crisp, M. D. (2005). Directional asymmetry of long-distance dispersal and colonization could mislead reconstructions of biogeography. *Journal of Biogeography* **32**, 741–754.
Coombs, S. and Popper, A. N. (1979). Hearing differences among Hawaiian squirrel fishes (family Holocentridae) related to difference in peripheral auditory system. *Journal of Comparative Physiology* **132**, 203–207.
Corwin, J. T. (1981). Peripheral auditory physiology in the lemon shark negaprion-brevirostris evidence of parallel otolithic and nonotolithic sound detection. *Journal of Comparative Physiology* A **142**, 379–390.
Cowen, R. K., Lwiza, K. M., Su, M., Sponaugle, S., Paris, C. B. and Olson, D. B. (2000). Connectivity of marine populations: open or closed? *Science* **287**, 857–859.
Cury, P. (1994). Obstinate nature: an ecology of individuals. Thoughts on reproductive behaviour and biodiversity. *Canadian Journal of Fisheries and Aquatic Sciences* **51**, 1664–1673.
Danilowicz, B. S. (1996). Choice of coral species by naive and field-caught damselfish. *Copeia* **1996**, 735–739.
Denton, E. J. and Gray, J. A. B. (1985). Lateral-line–like antennae of certain of the Penaeidea (Crustacea, Decapoda, Natantia). *Proceedings of the Royal Society of London B* **226**, 249–261.
Dudley, B., Tolimieri, N. and Montgomery, J. C. (2000). Swimming ability of the larvae of some reef fishes from New Zealand waters. *Marine Freshwater Research* **51**, 783–787.
Dunkelberger, D. G., Dean, J. M. and Watabe, N. (1980). The ultrastructure of the otolithic membrane and otolith of juvenile mummichog. *Journal of Morphology* **163**, 367–377.

Elliott, J. K., Elliot, J. M. and Mariscal, R. N. (1995). Host selection, location, and association behaviors of anemonefishes in field settlement experiments. *Marine Biology* **122**, 377–389.

Fay, R. R. and Edds-Walton, P. L. (1997). Directional response properties of saccular afferents of the toadfish, *Opsanus tau*. *Hearing Research* **111**, 1–21.

Fay, R. R. and Megela Simmons, A. (1999). The sense of hearing in fishes and amphibians. *In* "Comparative Hearing: Fish and Amphibians" (R. R. Fay and A. N. Popper, eds), pp. 269–318. Springer Handbook of Auditory Research. Springer-Verlag, New York.

Fay, R. R. and Popper, A. N. (1980). Structure and function in teleost auditory systems. *In* "Comparative Studies of Hearing in Vertebrates" (A. N. Popper and R. R. Fay, eds), pp. 3–42. Springer-Verlag, New York.

Flamarique, I. N. and Hawryshyn, C. W. (1997). Is the use of underwater polarized light by fish restricted to crepuscular time periods. *Vision Research* **37**, 975–989.

Foote, A. D., Osborne, R. W. and Hoelzel, A. R. (2004). Whale-call response to masking boat noise. *Nature* **428**, 910.

Forward, R. B., Jr. and Tankersley, R. A. (2001). Selective-tidal stream transport of marine animals. *Oceanography and Marine Biology, an Annual Review* **39**, 305–353.

Forward, R. B. Jr., Tankersley, R. A., Smith, K. A. and Welch, J. M. (2003). Effects of chemical cues on orientation of blue crab, *Callinectes sapidus*, megalopae in flow: implications for location of nursery areas. *Marine Biology* **142**, 747–756.

Fraser, P. J. and Macdonald, A. G. (1994). Crab hydrostatic pressure sensors. *Nature* **371**, 383–384.

Fraser, P. J. and Shelmerdine, R. L. (2002). Dogfish hair cells sense hydrostatic pressure. *Nature* **415**, 495–496.

Fuiman, L. A., Smith, M. E. and Malley, V. N. (1999). Ontogeny of routine swimming speed and startle responses in red drum, with a comparison of responses to acoustic and visual stimuli. *Journal of Fish Biology* **55**, 215–226.

Gray, J. A. B. and Denton, E. J. (1979). The mechanics of the clupeid acoustico-lateralis system: low frequency measurements. *Journal of the Marine Biological Association of the United Kingdom* **59**, 11–26.

Hamilton, E. L. and Bachman, R. T. (1982). Sound velocity and related properties of marine sediments. *Journal of the Acoustical Society of America* **72**, 1891–1904.

Hawkins, A. D. and Myrberg, A. A. (1983). Hearing and sound communication under water. *In* "Bioacoustics: A Comparative Approach" (B. Lewis, ed.), pp. 347–405. Academic Press, London.

Higgs, D. M. (2002). Development of the fish auditory system: how do changes in auditory structure affect function? *Bioacoustics* **12**, 180–183.

Iwashita, A., Sakamoto, M., Kojima, T., Watanabe, Y. and Soeda, H. (1999). Growth effects on the auditory threshold of red sea bream. *Nippon Suisan Gakkaishi* **65**, 833–838.

James, M. K., Armsworth, P. R., Mason, L. B. and Bode, L. (2002). The structure of reef fish metapopulations: modelling larval dispersal and retention patterns. *Proceedings of the Royal Society of London* B: **269**, 2079–2086.

Jeffs, A. G., Chiswell, S. M. and Booth, J. D. (2001). Spiny lobster puerulus condition in the Wairarapa Eddy off New Zealand. *Marine and Freshwater Research* **52**, 1211–1216.

Jeffs, A. G., Diebel, C. E. and Hooker, S. H. (1997). Arrangement and significance of pinnate sensory setae on the antennae of the puerulus and post-puerulus of the spiny lobster *Jasus edwardsii* (Palinuridae). *Marine and Freshwater Research* **48**, 681–686.

Jeffs, A. G., Tolimieri, N., Haine, O. and Montgomery, J. C. (2003). Crabs on cue for the coast: the use of underwater sound for orientation by pelagic crab stages. *Marine and Freshwater Research* **54**, 841–845.

Johannes, R. E. (1978). Reproductive strategies of coastal marine fishes in the tropics. *Environmental Biology of Fishes* **3**, 65–84.

Jones, G. P., Caley, M. J. and Munday, P. L. (2002). Rarity in coral reef fish communities. *In* "Coral Reef Fishes" (P. F. Sale, ed.), pp. 81–101. Academic Press, San Diego.

Jones, G. P., Milicich, M. J., Emslie, M. J. and Lunow, C. (1999). Self-recruitment in a coral reef fish population. *Nature* **402**, 802–804.

Karlsen, H. E. (1992). Infrasound sensitivity in the plaice (*Pleuronectes platessa*). *Journal of Experimental Biology* **171**, 173–187.

Kennish, M. J. (2001). "Practical Handbook of Marine Science." CRC Press, Boca Raton, Florida.

Kenyon, T. N. (1996). Ontogenetic changes in the auditory sensitivity damselfishes (Pomacentridae). *Journal of Comparative Physiology* A **179**, 553–561.

Kenyon, T. N., Ladich, F. and Yan, H. Y. (1998). A comparative study of hearing ability in fishes: the auditory brainstem response approach. *Journal of Comparative Physiology A* **182**, 307–318.

Kingsford, M. (1997). Physical signals that may facilitate the detection of coral reefs by fish. *Journal of Fish Biology,* **51**(Suppl. A), 409 (Poster Abstract).

Kingsford, M. J. and Choat, J. H. (1986). Influence of surface slicks on the distribution and onshore movements of small fish. *Marine Biology* **91**, 161–171.

Kingsford, M. J. and Finn, M. (1997). The influence of phase of the moon and physical processes on the input of presettlement fishes to coral reefs. *Journal of Fish Biology* **51**, 176–205.

Kingsford, M. J., Leis, J. M., Shanks, A., Lindeman, K. C., Morgan, S. G. and Pineda, J. (2002). Sensory environments, larval abilities and local self recruitment. *Bulletin of Marine Science* **70**, 309–340.

Kingsland, S. and Macky, G. (1999). Canterbury directional wave buoy quarterly report—June to August 1999. National Institute of Water and Atmospheric Research Ltd., New Zealand.

Kinlan, B. P. and Gaines, S. D. (2003). Propagule dispersal in marine and terrestrial environments: A community perspective. *Ecology* **84**, 2007–2020.

Lecchini, D., Shima, J., Banaigs, B. and Galzin, R. (2005). Larval sensory abilities and mechanisms of habitat selection of a coral reef fish during settlement. *Oecologia* **143**, 326–334.

Leis, J. M. (2006). Are larvae of demersal fishes plankton or nekton? *Advances in Marine Biology* **51**, 57–141.

Leis, J. M., Sweatman, H. P. A. and Reader, S. E. (1996). What the pelagic stages of coral reef fishes are doing out in blue water: daytime field observations of larval behavioural capabilities. *Marine and Freshwater Research* **47**, 401–411.

Leis, J. M. and Carson-Ewart, B. M. (1998). Complex behaviour by coral-reef fish larvae in open-water and near-reef pelagic environments. *Environmental Biology of Fishes* **53**, 259–266.

Leis, J. M. and Carson-Ewart, B. M. (1999). *In situ* swimming and settlement behaviour of larvae in an Indo-Pacific coral-reef fish, the coral trout *Plectropomus leopardus* (Pisces: Serrandae). *Marine Biology* **134**, 51–64.

Leis, J. M. and Carson-Ewart, B. M. (2003). Orientation of pelagic larvae of coral-reef fishes in the ocean. *Marine Ecology Progress Series* **252**, 239–253.

Leis, J. M., Carson-Ewart, B. M. and Cato, D. H. (2002). Sound detection *in situ* by the larvae of a coral-reef damselfish (Pomacentridae). *Marine Ecology Progress Series* **232**, 259–268.

Leis, J. M., Carson-Ewart, B. M., Hay, A. C. and Cato, D. H. (2003). Coral-reef sounds enable nocturnal navigation by some reef-fish larvae in some places and at some times. *Journal of Fish Biology* **63**, 724–737.

Leis, J. M. and Lockett, M. I. (2005). Localization of reef sounds by settlement-stage larvae of coral-reef fishes (Pomacentridae). *Bulletin of Marine Science* **76**, 715–724.

Leis, J. M. and McCormick, M. I. (2002). The biology, behaviour and ecology of the pelagic larval stage of coral reef fishes. *In* "Coral Reef Fishes" (P. F. Sale, ed.), pp. 171–199. Academic Press, San Diego.

Levine, J. M. and Murrell, D. J. (2003). The community-level consequences of seed dispersal patterns. *Annual Reviews in Ecology Evolution and Systematics* **34**, 549–574.

Levin, S. A., Muller-Landau, H. C., Nathan, R. and Chave, J. (2003). The ecology and evolution of seed dispersal: a theoretical perspective. *Annual Reviews in Ecology Evolution and Systematics* **34**, 575–604.

Levin, L. E. and Gonzalez, O. (1994). Endogenous rectilinear guidance in fish: Is it adjusted by reference to the sun. *Behavioural Processes* **31**, 247–256.

Lewis, D. (1979). "We, the Navigators." University Press of Hawaii, Honolulu.

Lohmann, K. J. and Lohmann, C. M. F. (1992). Orientation to oceanic waves by green turtle hatchlings. *Journal of Experimental Biology* **171**, 1–13.

Lohmann, K. J., Swartz, A. W. and Lohmann, C. M. F. (1995). Perception of ocean wave direction by sea turtles. *Journal of Experimental Biology* **198**, 1079–1085.

Lohmann, K. J. and Lohmann, C. M. F. (1996). Orientation and open-sea navigation in sea turtles. *Journal of Experimental Biology* **199**, 73–81.

Lovella, J. M., Findlaya, M. M., Moateb, R. M. and Yanc, H. Y. (2005). The hearing abilities of the prawn *Palaemon serratus*. *Comparative Biochemistry and Physiology* **140A**, 89–100.

Lozano-Álvarez, E., Carrasco-Zanini, G. and Briones-Fourzán, P. (2002). Homing and orientation in the spotted spiny lobster, *Panulirus guttatus* (Decapoda, Palinuridae), towards a subtidal coral reef habitat. *Crustaceana* **75**, 859–873.

McCauley, R. D. and Cato, D. H. (2000). Patterns of fish calling in a nearshore environment in the Great Barrier Reef. *Philosophical Transactions of the Royal Society of London B* **35**, 1289–1293.

Macmillan, D. L., Phillips, B. F. and Coyne, J. A. (1992). Further observations on the antennal receptors of rock lobsters and their possible involvement in puerulus stage navigation. *Marine Behaviour and Physiology* **19**, 211–225.

Mann, D. A., Higgs, D. M., Tavolga, W. N., Souza, M. J. and Popper, A. N. (2001). Ultrasound detection by clupeiform fishes. *Journal of the Acoustical Society of America* **109**, 3048–3054.

Masters, W. M., Aicher, B., Tautz, J. and Markl, H. (1982). A new type of water vibration receptor on the crayfish antenna. II. Model of receptor function. *Journal of Comparative Physiology A* **149**, 409–422.

Medwin, H. and Clay, C. S. (1998). "Fundamentals of Acoustical Oceanography." Academic Press, Boston.

Metcalfe, J. D., Holford, B. H. and Arnold, G. P. (1993). Orientation of plaice (*Pleuronectes platessa*) in the open sea: Evidence for the use of external directional clues. *Marine Biology* **117**, 559–566.

Metz, J. A. J., de Jong, T. J. and Klinkhamer, P. G. L. (1983). What are the advantages of dispersing: A paper by Kuno explained and extended. *Oecologia* **57**, 166–169.

Miller, J. A. and Shanks, A. L. (2004). Evidence for limited larval dispersal in black rockfish (*Sebastes melanops*): implications for population structure and marine-reserve design. *Canadian Journal of Fisheries and Aquatic Sciences* **61**, 1723–1735.
Montgomery, J. C., Baker, C. F. and Carton, A. G. (1997). The lateral line can mediate rheotaxis in fish. *Nature* **389**, 960–963.
Montgomery, J. C., Carton, A. G., Voigt, R., Baker, C. F. and Diebel, C. (2000). Sensory processing of water currents by fish. *Philosophical Transactions of the Royal Society* **355**, 1–3.
Montgomery, J. C., Tolimieri, N. and Haine, O. (2001). Active habitat selection by pre-settlement reef fishes. *Fish and Fisheries* **2**, 261–277.
Mora, C. and Sale, P. F. (2002). Are populations of coral reef fish open or closed? *Trends in Ecology and Evolution* **17**, 422–428.
Myrberg, A. A. (1978). Ocean noise and the behaviour of marine animals: relationships and implications. *In* "Advanced Concepts in Ocean Measurements for Marine Biology" (F. P. Diemer, F. J. Vernberg and D. Z. Mirkes, eds), pp. 461–491. University of South Carolina Press, Columbia.
Myrberg, A. A. and Fuiman, L. A. (2002). The sensory world of coral reef fishes. *In* "Coral Reef Fishes" (P. F. Sale, ed.), pp. 81–101. Academic Press, San Diego.
Myrberg, A. A. J. and Spires, J. Y. (1980). Hearing in damselfishes: An analysis of signal detection among closely related species. *Journal of Comparative Physiology* **140**, 135–144.
Myrberg, A. A. J., Mohler, M. and Catala, J. D. (1986). Sound production by males of a coral reef fish (*Pomacentrus partitus*): Its significance to females. *Animal Behaviour* **34**, 913–923.
Nishida, S. and Kittaka, J. (1992). Integumental organs of the phyllosoma larva of the rock lobster *Jasus edwardsii* (Hutton). *Journal of Plankton Research* **14**, 563–573.
Palumbi, S. R. (2003). Population genetics, demographic connectivity, and the design of marine reserves. *Ecological Applications* **13**, S146–S158.
Paris, C. B. and Cowen, R. K. (2004). Direct evidence of a biophysical retention mechanism for coral reef fish larvae. *Limnology and Oceanography* **49**, 1964–1979.
Phillips, B. F. and Sastry, A. N. (1980). Larval ecology. *In* "The Biology and Management of Lobsters" (J. S. Cobb and B. F. Phillips, eds). Vol. II, pp. 11–57. Academic Press, New York.
Phillips, B. F. and Macmillan, D. L. (1987). Antennal receptors in puerulus and postpuerulus stages of the rock lobster *Panulirus cygnus* (Decapoda: Palinuridae) and their potential role in puerulus navigation. *Journal of Crustacean Biology* **7**, 122–135.
Phillips, B. F. and Penrose, J. D. (1985). The puerulus stage of the spiny (rock) lobster and its ability to locate the coast. *School of Physics and Geosciences, Western Australian Institute of Technology, Report SPG 374/1985/AP92.* p. 48.
Pittman, S. J. and McAlpine, C. A. (2003). Movements of marine fish and decapod crustaceans: process, theory and application. *Advances in Marine Biology* **44**, 205–294.
Platt, C. and Popper, A. N. (1984). Variation in lengths of ciliary bundle on hair cells along the macula of the sacculus in two species of teleost fishes. *Scanning Electron Microscopy* **1984**, 1915–1924.
Popper, A. N. (1971). The effects of fish size on auditory capacities of the goldfish. *Journal of Auditory Research* **11**, 239–247.
Popper, A. N. (2003). Effects of anthropogenic sounds on fishes. *Fisheries* **28**, 24–31.

Popper, A. N. and Fay, R. R. (1999). The auditory periphery in fishes. *In* "Comparative Hearing: Fish and Amphibians" (R. R. Fay and A. N. Popper, eds), pp. 43–100. Springer-Verlag, New York.

Popper, A. N. and Platt, C. (1983). Sensory surface of the saccule and lagena in the ears of ostariophysan fishes. *Journal of Morphology* **176**, 121–129.

Popper, A. N., Salmon, M. and Horch, K. W. (2001). Acoustic detection and communication by decapod crustaceans. *Journal of Comparative Physiology* **187**, 83–89.

Queiroga, H. and Blanton, J. (2004). Interactions between behaviour and physical forcing in the control of horizontal transport of decapod crustacean larvae. *Advances in Marine Biology* **47**, 107–214.

Roberts, W. M., Howard, J. and Hudspeth, A. J. (1988). Hair cells: transduction, tuning and transmission in the inner ear. *Annual Reviews of Cell Biology* **4**, 63–92.

Rodgers, P. H. and Cox, M. (1988). Underwater sound as a biological stimulus. *In* "Sensory Biology of Aquatic Animals" (J. Atema, R. R. Fay, A. N. Popper and W. N. Tavolga, eds), pp. 131–149. Springer-Verlag, New York.

Ross, D. (1993). On ocean underwater ambient noise. *Institute of Acoustics Bulletin* **18**.

Saidel, W. M. and Popper, A. N. (1987). Sound reception in two anabatid fishes. *Comparative Biochemistry and Physiology* **88A**, 37–44.

Sale, P. F. and Kritzer, J. P. (2003). Determining the extent and spatial scale of population connectivity: Decapods and coral reef fishes compared. *Fisheries Research* **65**, 153–172.

Sale, P. F. (2004). Connectivity, recruitment variation, and the structure of reef fish communities. *Integrative and Comparative Biology* **44**, 390–399.

Sand, O. and Karlsen, H. E. (2000). Detection of infrasound and linear acceleration in fishes. *Philosophical Transactions of the Royal Society of London B* **355**, 1295–1298.

Sekiguchi, H. and Terazawa, T. (1997). Statocyst of *Jasus edwardsii* pueruli (Crustacea, Palinuridae), with a review of crustacean statocysts. *Marine and Freshwater Research* **48**, 715–719.

Simpson, S. D., Meekan, M. G., McCauley, R. D. and Jeffs, A. G. (2004). Attraction of settlement-stage coral reef fishes to reef noise. *Marine Ecology Progress Series* **276**, 263–268.

Simpson, S. D., Yan, H. Y., Wittenrich, M. L. and Meekan, M. G. (2005). Response of embryonic coral reef fishes (Pomacentridae: *Amphiprion* spp.) to noise. *Marine Ecology Progress Series* **287**, 201–208.

Smith, R. J. F. and Smith, M. J. (1998). Rapid acquisition of directional preferences by migratory juveniles of two amphidromous Hawaiian gobies (*Awaous guamensis* and *Sicyopterus stimpsoni*). *Environmental Biology of Fishes* **53**, 275–282.

Smith, W. L., Webb, J. F. and Blum, S. D. (2003). The evolution of the laterophysic connection with a revised phylogeny and taxonomy of butterflyfishes (Teleostei: Chaetodontidae). *Cladistics* **19**, 287–306.

Sponaugle, S., Cowen, R. K., Shanks, A., Morgan, S. G., Leis, J. M., Pineda, J., Boehlert, G. W., Kingsford, M. J., Lindeman, K. C., Grimes, C. and Munro, J. L. (2002). Predicting self-recruitment in marine populations: biophysical correlates and mechanisms. *Bulletin of Marine Science* **70** (Suppl.), 431–375.

Steele, C. W. (1989). Evidence for use of continuous angle compensation as a mechanism for inertial guidance by sea catfish *Arius felis*. *American Midland Naturalist* **122**, 183–192.

Stobutzki, I. C. and Bellwood, D. R. (1994). An analysis of the sustained swimming abilities of pre-settlement and postsettlement reef fishes. *Journal of Experimental Marine Biology and Ecology* **175**, 275–286.
Stobutzki, I. C. and Bellwood, D. R. (1998). Nocturnal orientation to reefs by late pelagic stage coral reef fishes. *Coral Reefs* **17**, 103–110.
Strathmann, R. R., Hughes, T. P., Kuris, A. M., Lindeman, K. C., Morgan, S. G., Pandolfi, J. M. and Warner, R. R. (2002). Evolution of local-recruitment and its consequences for marine populations. *Bulletin of Marine Science* **70** (Suppl.), 377–396.
Sugihara, I. and Furukawa, T. (1989). Morphological and functional aspects of two different types of hair cells in the goldfish sacculus. *Journal of Neurophysiology* **62**, 1330–1343.
Swearer, S. E., Caselle, J. E., Lea, D. W. and Warner, R. R. (1999). Larval retention and recruitment in an island population of a coral-reef fish. *Nature* **402**, 799–802.
Sweatman, H. (1988). Field evidence that settling coral reef fish larvae detect resident fishes using dissolved chemical cues. *Journal of Experimental Marine Biology and Ecology* **124**, 163–174.
Tait, R. I. (1962). "The Evening Chorus: A Biological Noise Investigation." Naval Research Laboratory, MHNZ Dockyard, Auckland, New Zealand.
Tautz, J., Masters, W. M., Aicher, B. and Markl, H. (1981). A new type of water vibration receptor on the crayfish antenna. I. Sensory physiology. *Journal of Comparative Physiology* A **144**, 533–541.
Tazaki, K. and Ohnishi, M. (1974). Responses from tactile receptors in the antenna of the spiny lobster *Panulirus japonicus*. *Comparative Biochemistry and Physiology* A **47**, 1323–1327.
Thorrold, S. R., Latkoczy, C., Swart, P. K. and Jones, C. M. (2001). Natal homing in a marine fish metapopulation. *Science (Washington D C)* **291**(5502), 297–299.
Tolimieri, N., Haine, O., Jeffs, A. G., McCauley, R. and Montgomery, J. C. (2004). Directional orientation of pomacentrid larvae to ambient reef sound. *Coral Reefs* **24**, 184–191.
Tolimieri, N., Haine, O., Montgomery, J. C. and Jeffs, A. G. (2002). Ambient sound as a navigational cue for larval reef fish. *Bioacoustics* **12**, 214–217.
Tolimieri, N., Jeffs, A. G. and Montgomery, J. C. (2000). Ambient sound as a cue for navigation by the pelagic larvae of reef fishes. *Marine Ecological Progress Series* **207**, 219–224.
Urick, R. J. (1983). "Principles of Underwater Sound." McGraw-Hill, New York.
van den Berg, A. V. and Schuijf, A. (1983). Discrimination of sounds based on the phase difference between particle motion and acoustic pressure in the shark *Chiloscyllium griseum*. *Proceedings of the Royal Society of London* B **218**, 127–134.
Vedel, J. P. (1985). Cuticular mechanoreception in the antennal flagellum of the rock lobster *Palinurus vulgaris*. *Comparative Biochemistry and Physiology* A **80**, 151–158.
Villanueva, R., Nozais, C. and Boletzky, S. V. (1997). Swimming behavior and food searching in planktonic *Octopus vulgaris* Cuvier from hatching to settlement. *Journal of Experimental Marine Biology and Ecology* **208**, 169–184.
Walton, A. S. and Herrnkind, W. F. (1977). Hydrodynamic orientation of spiny lobster, *Panulirus argus* (Crustacea: Palinuridae): wave surge and unidirectional currents, pp. 184–211. Memorial University of Newfoundland Marine Sciences Research Laboratory. Technical Report 20.

Webb, J. F. and Smith, W. L. (2000). The laterophysic connection in chaetodontid Butterfly fish: morphological variation and speculations on sensory function. *Philosophical Transactions of the Royal Society of London B* **355**, 1125–1129.
Wilkens, L. A., Schmitz, B. and Herrnkind, W. F. (1996). Antennal responses to hydrodynamic and tactile stimuli in the spiny lobster *Panulirus argus*. *Biological Bulletin* **191**, 187–198.
Wright, K. J., Higgs, D. M., Belanger, A. J. and Leis, J. M. (2005). Auditory and olfactory abilities of pre- and post-settlement juveniles of a coral damselfish (Pisces: Pomacentridae). *Marine Biology* **147**, 1425–1434.
Wubbels, R. J. and Schellart, N. A. M. (1997). Neuronal encoding of sound direction in the auditory midbrain of the rainbow trout. *Journal of Neurophysiology* **77**, 3060–3074.
Wysocki, L. E. and Ladich, F. (2001). The ontogenetic development of auditory sensitivity, vocalization and acoustic communication in the labyrinth fish *Trichopsis vittata*. *Journal of Comparative Physiology A* **187**, 177–187.

Crustacea in Arctic and Antarctic Sea Ice: Distribution, Diet and Life History Strategies

Carolin E. Arndt*,†,§ and Kerrie M. Swadling*,‡

**School of Zoology, University of Tasmania, Hobart, Tasmania, Australia*
†University Center on Svalbard, Longyearbyen, Norway
‡Tasmanian Aquaculture and Fisheries Institute, University of Tasmania, Hobart, Tasmania, Australia
§World Climate Research Programme, Geneva, Switzerland

ADVANCES IN MARINE BIOLOGY VOL 51

0065-2881/06 $35.00
DOI: 10.1016/S0065-2881(06)51004-1

This review concerns crustaceans that associate with sea ice. Particular emphasis is placed on comparing and contrasting the Arctic and Antarctic sea ice habitats, and the subsequent influence of these environments on the life history strategies of the crustacean fauna. Sea ice is the dominant feature of both polar marine ecosystems, playing a central role in physical processes and providing an essential habitat for organisms ranging in size from viruses to whales. Similarities between the Arctic and Antarctic marine ecosystems include variable cover of sea ice over an annual cycle, a light regimen that can extend from months of total darkness to months of continuous light and a pronounced seasonality in primary production. Although there are many similarities, there are also major differences between the two regions: The Antarctic experiences greater seasonal change in its sea ice extent, much of the ice is over very deep water and more than 80% breaks out each year. In contrast, Arctic sea ice often covers comparatively shallow water, doubles in its extent on an annual cycle and the ice may persist for several decades. Crustaceans, particularly copepods and amphipods, are abundant in the sea ice zone at both poles, either living within the brine channel system of the ice-crystal matrix or inhabiting the ice–water interface. Many species associate with ice for only a part of their life cycle, while others appear entirely dependent upon it for reproduction and development. Although similarities exist between the two faunas, many differences are emerging. Most notable are the much higher abundance and biomass of Antarctic copepods, the dominance of the Antarctic sea ice copepod fauna by calanoids, the high euphausiid biomass in Southern Ocean waters and the lack of any species that appear fully dependent on the ice. In the Arctic, the ice-associated fauna is dominated by amphipods. Calanoid copepods are not tightly associated with the ice, while harpacticoids and cyclopoids are abundant. Euphausiids are nearly absent from the high Arctic. Life history strategies are variable, although reproductive cycles and life spans are generally longer than those for temperate congeners. Species at both poles tend to be opportunistic feeders and periods of diapause or other reductions in metabolic expenditure are not uncommon.

1. INTRODUCTION

Variable coverage by ice characterises marine environments at high latitudes. At its winter maximum, sea ice covers up to 6% of the earth's surface

(Dieckmann and Hellmer, 2003), and the high albedo of sea ice gives it a pivotal role in the global energy balance as it influences important interactions between the ocean and the atmosphere. The seasonality in ice coverage and freeze-melt episodes, coupled with the annual cycle in irradiance changing from months of almost total darkness to months of continuous daylight, regulates biological cycles in ice-covered waters. If, as predicted, sea ice at both poles diminishes in extent over the next few decades (e.g., Budd and Wu, 1998; Comiso, 2003), there will be profound changes to the marine environment. The vulnerability of the polar ecosystems has become apparent via the discoveries of an apparent mid-twentieth century decline in Antarctic sea ice extent (de la Mare, 1997), the decrease in the proportion of old ice in the Arctic (Parkinson, 2000; Comiso, 2002), the thinning of sea ice in the Arctic (Rothrock *et al.*, 1999, but see Holloway and Sou, 2001), and the collapse of ice shelves at both poles (Vaughan and Doake, 1996; Mueller *et al.*, 2003).

Sea ice harbours high standing stocks of bacteria, algae, protists and invertebrates. In ice-covered seas, ice biota form a major energetic link to higher trophic levels, including sea birds and mammals. Crustaceans are an integral component of the sea ice biota in both the Arctic and the Antarctic and often dominate the metazoan assemblages in abundance and biomass. Their roles in sea ice food webs are gradually being elucidated, although many questions remain. In particular, responses of individual species to the extreme seasonality of the polar habitat are still poorly understood.

The idea that sea ice harbours a flourishing assemblage of small organisms is certainly not recent. Joseph Hooker was the first to collect particulate material from Antarctic sea ice, which was subsequently examined by C. G. Ehrenberg and found to contain diatoms (Ross, 1847). During early exploration in the Arctic, Nansen (1897) discovered that diatoms present in ice from the Bering Sea were similar to those found in Greenland sea ice.

Once it was established that substantial primary production was occurring within sea ice (Bunt, 1963; Bunt and Wood, 1963), attention turned to verifying the presence of an ice-associated fauna. In the Antarctic, Ray (1966) photographed seals under the ice at McMurdo Sound and recorded the presence of "little shrimplike amphipods" living amongst ice crystals at the under-ice surface. Gruzov *et al.* (1967) undertook several diving trips under the ice near Mirny Station (66°33′S, 93°01′E) in East Antarctica in the early 1960s. As part of their observations, they recorded the presence of calanoid copepods (no further information was given), three harpacticoid copepod genera (*Tisbe, Harpacticus* and *Dactylopodia*), the cyclopoid copepod *Oithona* sp. and two species of amphipods, *Bovallia walkeri* (*Paramoera walkeri* Stebbing) and *Orchomenopsis* sp. They also noted the lack of *Euphausia crystallorophias* Holt and Tattersall, a common inshore euphausiid that was assumed to live in association with the ice. In a related publication, Gruzov *et al.* (1968) recorded that a "rather rich fauna" was growing under the ice, and that this fauna consisted of copepods, amphipods and fish. From the same sampling efforts, 12 taxa, including polychaetes, copepods, amphipods and fish, were described as having strong association with the bottom layers of Antarctic fast ice near Mirny Station; in particular *Oithona* sp. was found in ice samples in numbers up to several hundreds per square metre, whereas the three harpacticoid genera listed above occurred in abundances of tens to hundreds per square metre (Andriashev, 1968). This pioneering diving work under sea ice was carried out by scientists from the USSR almost 40 years ago and included the use of a quantitative sampler designed to scrape the bottom of the ice. Yet, apart from some detailed studies of amphipods in the early 1970s (see below) and the work by the Japanese in the early 1980s (see below), few other studies of small Antarctic ice-associated fauna were undertaken during the next 2 decades. Much greater attention was given to ice-associated primary producers and larger grazers (i.e., euphausiids) during that time.

In the Arctic, Golikov and Scarlato (1973) were pioneers in describing trophic relationships of ice biota, while Carey (1985) was the first to develop a classification system for ice fauna. He catalogued >10 copepod and 13 amphipod species found in Arctic sea ice and defined different fast- and drift-ice communities with respect to underlying water depth and age of ice. Comprehensive quantitative studies of Arctic and Antarctic fauna added turbellarians, nematodes, rotifers and euphausiids to the species list (Carey and Montagna, 1982; Cross, 1982; Kern and Carey, 1983; Grainger and Hsiao, 1990; Garrison, 1991). The complex physical structure of sea ice provides these animals with a refuge from predation. Furthermore, the extended growing season of ice algae compared to phytoplankton at high latitudes (e.g., Garrison and Buck, 1989) means that the foraging potential provided by the ice habitat is prolonged. Some crustacea live and feed within the interstitial brine channels (Kern and Carey, 1983; Hoshiai *et al.*, 1987; Kurbjeweit *et al.*, 1993), whereas others congregate to feed at the ice–water interface (Gulliksen, 1984; Conover *et al.*, 1986; Runge and Ingram, 1988; Daly, 1990).

One difficulty in studying polar crustaceans is that their life cycles are long, usually taking ≥1 yr to complete. Furthermore, polar science often involves a trade-off between the good spatial but poor temporal coverage provided by ship-based studies, and the good temporal but less rigorous spatial coverage of station-based research. In spite of some logistical difficulties, our knowledge of ice-associated crustaceans has progressed significantly in the last 15 years, so it is timely to review current knowledge of the two polar systems, emphasising their similarities and differences. Such analysis will facilitate understanding how changes to the physical environment might affect key species.

While there have been some important reviews of the ecology of sea ice in the last few years (e.g., Brierley and Thomas, 2002; chapters in Thomas and

Dieckmann, 2003), they have by design been very broad ranging in their subject matter. As crustaceans play a significant role in the sea ice habitat, a comprehensive review of their biology and ecology is justified. Defining the relationships of the different species associated with sea ice will help us infer possible future impacts of climate change on the sea ice habitat. Here, we present a comparative analysis of crustaceans exploiting Arctic and Antarctic sea ice environments, focusing on those species that associate with sea ice at some point, however briefly, in their life cycle.

1.1. Definitions

1.1.1. Sea ice

For the purposes of this review, we have distinguished the following ice types and features (Figure 1): (1) the sea ice proper (i.e., interstitial ice), consisting of the ice crystal matrix interspersed with the brine channel system, (2) the ice–water interface, where seawater flows across the bottom of the ice and

Figure 1 Ice in the Southern Ocean (A, B) and the Arctic Ocean (C, D) showing (A) pancakes in the pack ice, with newly forming ice between, (B) heavily rafted ice with thick snow covering, (C) consolidation of ice crystals to a translucent surface layer, (D) surface melt ponds on old rafted ice in summer. (Photos courtesy K. Swadling, L. Gurney, with permission, C. Arndt.)

there is usually free exchange of nutrients and gases between the top of the water column and the bottom few centimetres of ice, (3) platelet ice, common in the Weddell Sea, which consists of loose layers of ice forming deep aggregations under the solid ice cover, and (4) slush ice, those regions, usually found in summer, where seawater has infiltrated middle layers of ice, creating more liquid environments. (5) Pack ice is drift ice that is either annual and reaches a maximum of ~2 m thick, more or less level with recently broken edges, or perennial with a thickness of 3–4 m and rough deformed surfaces. In contrast to (6) land-fast ice, pack ice moves according to the whims of wind and currents. Quiescent conditions favour the formation of (7) congelation ice that forms by the growth of large elongated (columnar) crystals perpendicular to the underside of the ice sheet (Horner *et al.*, 1992). Alternatively, turbulent conditions promote consolidation of agglomerations of (8) frazil ice crystals into sheets of granular ice (Eicken and Lange, 1989). Where necessary, we will use these terms to clarify a taxon's association with sea ice.

1.1.2. Crustaceans

Sympagic (i.e., ice-associated) macrofauna have been classified as those species that complete their entire life cycle within the sea ice (autochthonous fauna) or those that spend only part of their life cycle associated with ice (allochthonous fauna) (Melnikov and Kulikov, 1980). Autochthonous species have rarely, if ever, been recorded from the Antarctic, as >80% of Antarctic sea ice is seasonal and must be colonised anew each year. In contrast, Arctic sea ice is mainly perennial (Spindler, 1990), so there is a higher probability that autochthonous species have evolved. These species rely on year-round ice cover and a drift pattern of ice that maintains the floe within the perennial ice zone. However, as ice is eventually advected out of this zone, autochthonous species must be able to spread laterally from floe to floe (Gulliksen and Lønne, 1991). Arctic ice can support species with life spans of >1 yr. For example, the amphipod *Gammarus wilkitzkii* Berula can live for at least 6 yr (Poltermann, 2000) and sustains a population in the ice all year round. In contrast, in the Antarctic, the ice habitat has a much more ephemeral character, so most ice-dwelling copepod species have life spans of ≤1 yr, equal to the life spans of their habitat (Dahms *et al.*, 1990; Schnack-Schiel *et al.*, 1995; Tanimura *et al.*, 1996).

Some species that have been classified as autochthonous (Gulliksen and Lønne, 1991) seem to survive well in the absence of sea ice and have been recorded from open water down to the abyss; these include the amphipods *Apherusa glacialis* Hansen (Barnard, 1959) and *G. wilkitzkii* (Steele and Steele, 1974; Arndt and Pavlova, 2005; Arndt *et al.*, 2005a). For other amphipod species, no reproductive stage has ever been sampled in sea ice (reviewed in Arndt and Beuchel, 2006), which further raises questions about the concept of autochthonous ice species. The ice environment has been described as "upside-down benthic habitat" (Mohr and Tibbs, 1963), suggesting that all organisms living in association with ice are by nature benthic organisms with a strong affiliation to surfaces, irrespective of the type of substrate. Because many aspects of the biology of sympagic species are not yet known, it is not feasible to call them "permanent residents" of sea ice, so we have set aside the dichotomy of autochthonous versus allochthonous species and broadened our focus to include all crustaceans that are regularly or occasionally found associated with sea ice.

2. THE SEA ICE ENVIRONMENT

2.1. Movement and coverage

Although superficially similar, there are substantial differences between the Arctic and Southern Oceans (Figures 2 and 3). The Arctic Ocean, a basin surrounded by continents, has limited water exchange with other oceans via several narrow passages. A high proportion of the Arctic Ocean covers shallow shelf areas. Conversely, the Southern Ocean, about half of which is influenced by ice cover, consists of deeper water circulating around a large ice-covered continent. The shelf regions are generally deep and narrow and the Southern Ocean has the capacity to exchange with all other global oceans except the Arctic (Johnson, 1990). Sea ice coverage at both poles is seasonal in its extent. In Antarctica, it undergoes a fivefold increase in area from a summer minimum of 4×10^6 km^2 to a winter maximum of 19×10^6 km^2. In contrast, sea ice coverage of the Arctic Ocean undergoes less than a doubling in area from a summer minimum of 9×10^6 km^2 to a winter maximum of 16×10^6 km^2 (Comiso, 2003).

The summer ice extent generally corresponds to the amount of multiyear pack ice (Gloersen *et al.*, 1993). In the Arctic, >50% of the sea ice is 5- to 7-yr-old perennial ice (Dieckmann and Hellmer, 2003) of 3- to 5-m thickness (Wadhams, 2000). Sea ice circulating in the cyclonic Beaufort Gyre in the Canada Basin may be more than a decade old (Rigor *et al.*, 2002). In contrast, the Transpolar Drift Stream is a rapid conveyer belt for young sea ice that has formed along the Siberian Coast and is transported within 3–5 yr towards the Greenland Sea, where it eventually melts (Rigor *et al.*, 2002). The mean ice thickness in Antarctica is 0.5–0.6 m and >80% of the Southern Ocean pack ice melts each summer (Thomas and Dieckmann, 2003). Residence times of the remaining 20% in the Southern Ocean are rarely >2 yr (Comiso, 2003), with multiyear ice persisting only in restricted

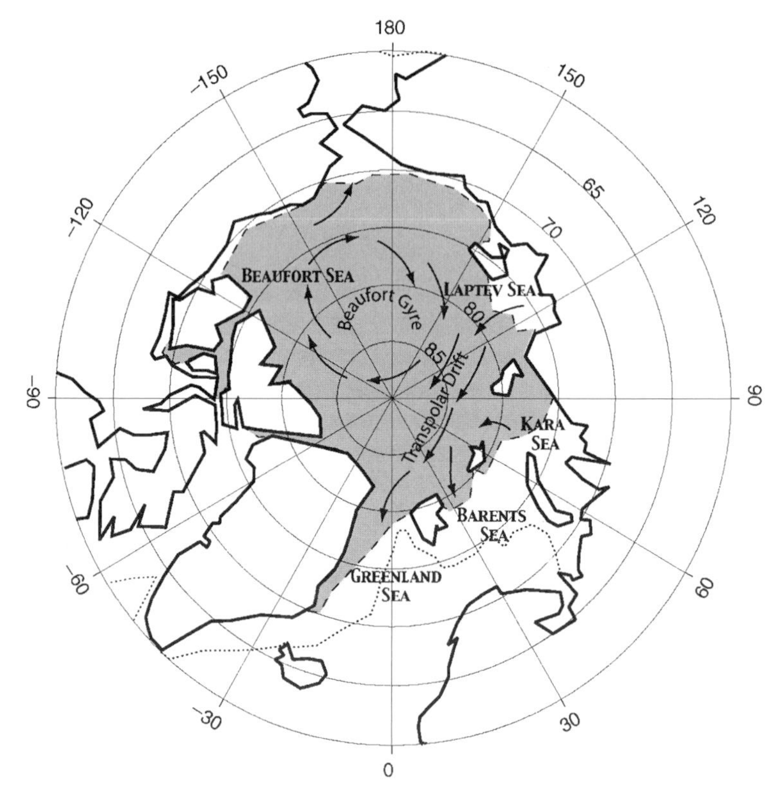

Figure 2 Oceanographic map of the Arctic indicating the main surface currents and sea ice zones; shaded area: perennial sea ice zone; dotted line: southern-most extension of the seasonal sea ice zone. (After Maykut, 1985, and Wadhams, 2000.) Note that parts of the southern limit of winter ice (seasonal ice) are outside the area of the map.

regions such as the northwestern Weddell Sea and some localised regions of coastal fast ice. For example, near Hut Point, McMurdo Sound (77°51′S, 166°02′E), 2-yr-old ice is not uncommon and can reach 5 m in thickness (El-Sayed, 1971).

The wind system that drives ice flow in the Arctic arises because of the presence of the two polar highs over the centre of the Beaufort Sea and over Siberia. The high pressure cells are balanced by the Icelandic Low and the Aleutian Low (Wadhams, 2000). The wind-driven motion in the central Arctic alternates between anticyclonic and cyclonic circulation, with each regimen persisting for 5–7 yr. Shifts from one regimen to another are forced by changes in the location and intensity of the Icelandic Low and the Siberian High (Proshutinsky and Johnson, 1997). The resulting changes in

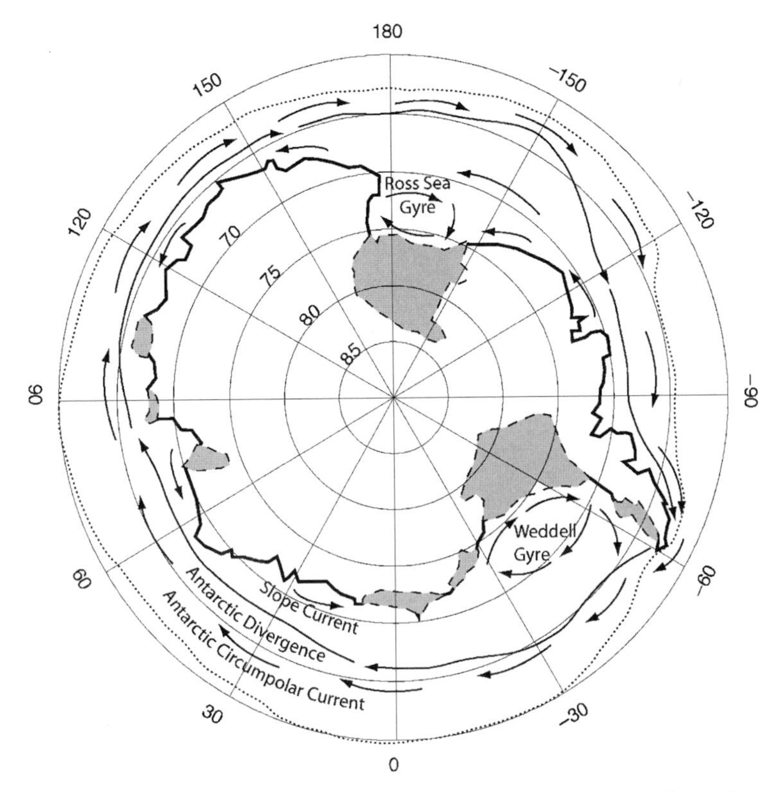

Figure 3 Oceanographic map of the Antarctic indicating the main surface currents and sea ice zones; shaded areas: perennial sea ice zones; dotted line: northernmost extension of the seasonal sea ice zone. (After Maykut, 1985, and Wadhams, 2000.)

ice drift pattern affect the origin and distribution of different ice types, the ice concentration and extent, the local ice thickness and flux rates.

Major features of the mean pattern (climatic mean) of sea ice drift in the Southern Ocean are the dominance of westerly (eastward) flow within the Antarctic Circumpolar Current (ACC) (Figure 3) and a more easterly (westward) flow closer to the coast (i.e., within the near-coastal East Wind Drift). The Antarctic Divergence approximately separates these two regimens. These patterns reflect the large-scale/long-term ocean circulation, which is itself a response to the surface wind field (i.e., prevailing easterly winds close to the coast and westerlies to the north). Atmospheric pressure patterns are dominated by high-pressure systems over the continent surrounded by a ring of low-pressure systems situated between 60° S and 70° S. Semi-permanent and intense low-pressure systems are encountered over the Ross and Weddell Seas and off East Antarctica in winter. The tendency of the ice to drift slightly to the left of the westerlies (due to the Coriolis force), that is, to the north, results in a general northward expansion of the sea ice. This means that the Antarctic sea ice is on average a divergent ice cover. In the Ross and Weddell Seas, the combination of the easterly drift along the coast and the ACC causes two incomplete gyres, and so ice cannot continue to recirculate for years. Note that there are seasonal (and indeed interannual) oscillations in these pressure patterns, and that the distribution of sea ice is also closely tied to the seasonal temperature regimen (Lubin and Massom, 2005).

2.2. Production in sea ice

Although rates of primary production by sea ice algae are generally low compared to their phytoplankton counterparts, they are often virtually the sole source of fixed carbon for higher trophic levels in ice-covered waters. Together the biota of the polar oceans may produce ~70 Tg of biogenic carbon each year, with both poles contributing similar amounts (Arrigo, 2003). This is approximately equivalent to 10% of total phytoplankton production in the same regions, suggesting that primary production in sea ice is of much less importance than production in the water column. However, the value of ice-algal production for crustacean grazers lies primarily in its extended growing season. In spring fast ice near Davis Station (68°35′S, 77°58′E), east Antarctica, algae were thriving in the ice at least 1 mo before measurable quantities of phytoplankton appeared in the water column, providing a potential source of nutrition for grazers that had experienced winter deprivation (Swadling *et al.*, 2004). Similarly, in autumn the growing season in the ice extended well beyond the time when phytoplankton cell abundance decreased in the water column. This tendency of ice algae to prolong their growing season relative to that of phytoplankton is evident from chlorophyll *a* measurements taken from pack and fast ice regions at both poles (Horner, 1985, and references therein; Garrison and Buck, 1989; Swadling *et al.*, 1997a).

Annual sea ice primary production is similar in magnitude at both poles, but the variability is greater in Antarctic ice. Production values range from 5 to 15 g C m^{-2} yr^{-1} in the Arctic and from 0.3 to 38 g C m^{-2} yr^{-1} in the Antarctic (Arrigo, 2003). In the Antarctic, production associated with platelet ice alone ranges from 0.26 to 14 g C m^{-2} yr^{-1}. In comparison to these figures, only a relatively small percentage of the carbon produced by autotrophs is transferred to secondary producers. For example, in the ice-covered domains of the Barents and Bering Seas in the Arctic, it is <10% (Sakshaug and Walsh, 2000). In the Greenland Sea, total "interstitial biomass" was 196 mg C m^{-2} in pack ice, with 43% being autotrophs, 51% bacteria and heterotrophic flagellates and 4% meiofauna (~8 mg C m^{-2}) (Gradinger *et al.*, 1999). In the annual ice in Northern Baffin Bay, meiofaunal carbon biomass ranged between 0 and 19.4 mg C m^{-2} (Nozais *et al.*, 2001). In the Weddell Sea, Antarctica, integrated biomass ranged between 0 and 118 mg C m^{-2}, with turbellarians contributing 25–41%, crustaceans 35–62% and foraminiferans 6–28% (Gradinger, 1999). In perennial ice, nematodes and turbellarians often dominate the total biomass (Friedrich, 1997; Gradinger *et al.*, 1999), while in first-year ice, it is the copepods (Nozais *et al.*, 2001).

2.3. Sea ice communities

Several distinct habitats, with their associated biota, have been described for sea ice. At the surface, "snow ice" results when snow loading reaches a point where it depresses the ice below sea level, thereby becoming infiltrated with seawater. Snow ice, which covers much of the Weddell Sea, is characterised by a mixed diatom-flagellate community and supports high concentrations of chlorophyll *a* (100–400 mg m^{-3}) (Ackley and Sullivan, 1994). Deformation communities, also described from Antarctica, arise via seawater infiltration of pressure ridges or when ice deflected below the surface is flooded by seawater that then collects into ponds. A diverse range of autotrophic and heterotrophic microorganisms have been described from these habitats (Garrison, 1991). Finally, melt pool communities form via flooding and/or thawing of the surface. These communities, often consisting of diatoms, flagellates and ciliates, are particularly common in the Arctic where they may cover up to 50–60 % of the sea ice (Maykut, 1985).

Freeboard communities develop in the ice, just below sea level, when surface temperatures warm the ice and cause partial brine drainage from the upper layers. At the same time, algal growth increases, heat is trapped and the ice melts (Horner *et al.*, 1992). Solid layers of ice occur above and below the freeboard layer. The salinity of this layer is often much higher than observed in other sections of the ice, and chlorophyll *a* concentrations up to 425 mg m^{-3} have been reported (Ackley and Sullivan, 1994). The brine channel systems that branch throughout the ice constitute the most common interior habitat. Organisms are either concentrated in narrow bands or spread diffusely throughout the ice (Horner *et al.*, 1992).

Bottom ice communities (Figure 4) have been the most studied. The lower layers of congelation ice have high stability, a high probability of colonisation from pelagic and benthic habitats and other ice communities and free exchange of nutrients with underlying seawater. Approximately 99% of the primary productivity of congelation ice in McMurdo Sound,

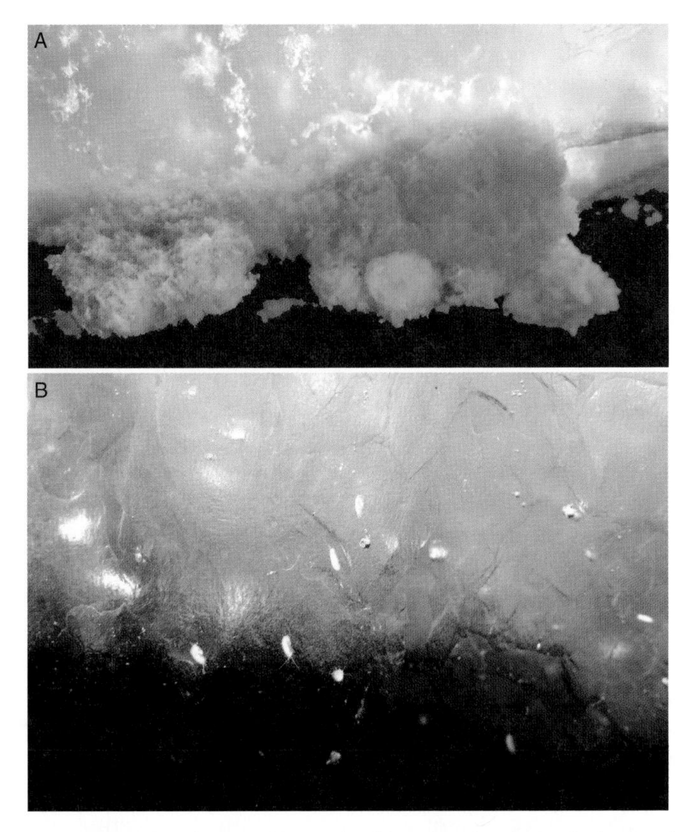

Figure 4 Rich life flourishes at the underside of sea ice, in both the Arctic and the Antarctic. (A) Sea ice diatoms grow in mats at the lower surface of Arctic multiyear ice. (B) Ice amphipods, probably *Apherusa glacialis* (maximal body length 14 mm), gather in large numbers along the edges of Arctic sea ice. (Photos courtesy C. Arndt.)

Antarctica, was measured in the bottom 20 cm, where chlorophyll *a* concentrations up to 310 mg m^{-2} were recorded (Palmisano and Sullivan, 1983). In the Arctic, maximal chlorophyll *a* concentrations were 10 mg m^{-2} in pack ice and 40 mg m^{-2} in fast ice (Gradinger *et al.*, 1991, and references cited therein). Furthermore, the highest densities of ice-associated metazoans are usually found in these bottom layers of sea ice (Friedrich, 1997; Swadling *et al.*, 1997a, 2000a; Gradinger *et al.*, 1999; Schnack-Schiel *et al.*, 2001a). Larger taxa such as amphipods and euphausiids are generally excluded from the narrow-channelled ice matrix and tend to concentrate at the ice–water interface (Carey, 1985). The sub-ice platelet layer typically occurs in the Antarctic and is also a region of high algal growth. High concentrations of chlorophyll *a* (2120 mg m^{-2}) were measured in the platelet layer in McMurdo

Sound (Ackley and Sullivan, 1994). An extension of these under-ice communities is the strand communities (McConville *et al.*, 1985; Watanabe, 1988), largely composed of chains of diatoms, which form in spring and are possibly derived from melting of the platelet layer (McGrath-Grossi *et al.*, 1987).

3. SAMPLING CRUSTACEANS FROM SEA ICE

In the Arctic, most of the research was conducted on land-fast ice where access and logistics are easy due to land-based research stations (Figure 5). A few drift stations on ice successfully operated for several years in the perennial ice pack of the Arctic. The long-term records and observations of these drift stations are still of unique value to ice researchers. In the Southern Ocean, the bulk of research was conducted in the Weddell Sea pack ice using ships as operating platforms (Figure 6). Research vessels generally operate along the marginal ice zone where disturbances are greatest to the ice ecosystem. In the inner ice pack, research is considered cost intensive and logistically difficult.

Various methods have been used to sample sea ice and its associated organisms (Figure 7). Those fauna that are usually found in association with the ice matrix proper have generally been sampled with a simple barrel corer, either of the SIPRE (Snow, Ice, and Permafrost Research Establishment) or CRREL (Cold Regions Research and Engineering Laboratory) types. These corers take a discrete unit of ice that is then melted and the organisms extracted. Knowing the diameter of the corer and the thickness of the sea ice facilitates the conversion from numbers per core to numbers per square metre or cubic metre. Although Horner *et al.* (1992) recommended that ice fauna be reported as "number m^{-2}," this is not always the case in the literature. For easy translation between numbers m^{-2} and numbers m^{-3}, we recommend that the thickness of each ice core be reported.

The distribution of sea ice organisms can be extremely patchy, both horizontally and vertically (Swadling *et al.*, 1997a; Brierley and Thomas, 2002), and it is likely that the ecologically meaningful scales for many sea ice organisms are of the order of millimetres. Therefore, while ice corers are a relatively cost-effective method for sampling sea ice, they are probably failing to sample the fauna at appropriate biological scales. The sea ice mesocosms developed by Krembs *et al.* (2001, 2002) show much potential for experimental studies of ice-associated grazers, especially smaller organisms such as copepods. In particular, it should be possible to examine their small-scale interactions with ice algae. However, this approach is not suitable for routine examination of abundance and distribution of organisms in sea ice, so a combination of methods is usually required.

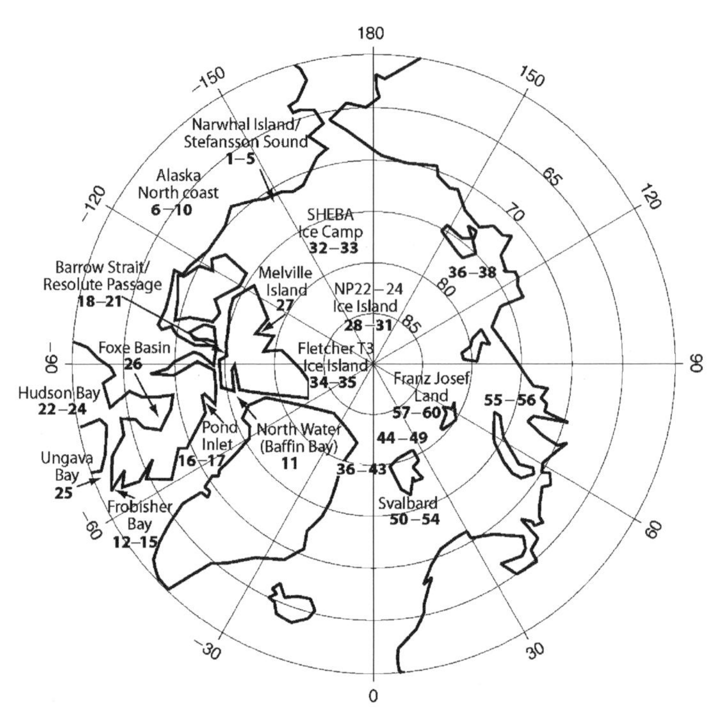

[Alaska] 1: Montagna and Carey, 1978; **2:** Carey and Montagna, 1982; **3:** Kern and Carey, 1983; **4:** Carey, 1992; **5:** Dunton *et al.*, 1982; **6:** Gradinger *et al.*, 2005; **7:** MacGinitie, 1955; **8:** Mohr and Tibbs, 1963; **9:** Horner, 1972; **10:** Griffiths and Dillinger,1981; **[Canada] 11:** Nozais *et al.*, 2001; **12:** Grainger *et al.*, 1985; **13:** Grainger and Mohammed, 1986; **14:** Grainger and Hsiao, 1990; **15:** Grainger and Mohammed, 1991; **16:** Cross, 1982; **17:** Bradstreet and Cross, 1982; **18:** Fortier *et al.*, 1995; **19:** Fortier *et al.*, 2001; **20:** Green and Steele, 1977; **21:** Pike and Welch, 1990; **22:** Runge and Ingram, 1988; **23:** Runge and Ingram, 1991; **24:** Siferd *et al.*, 1997; **25:** Dunbar, 1954; **26:** Grainger, 1962; **27:** Buchanan *et al.*, 1977; (cited in Grainger *et al.*, 1985) **[Arctic Ocean] 28:** Melnikov and Kulikov, 1980; **29:** Pavshtiks, 1980; **30:** Melnikov, 1989; **31:** Melnikov, 1997; **32:** Melnikov *et al.*, 2001; **33:** Melnikov *et al.*, 2002; **34:** Barnard, 1959; **35:** George and Paul, 1970; [Greenland/Svalbard] **36:** Friedrich, 1997; **37:** Werner, 1997; **38:** Werner and Arbizu, 1999; **39:** Werner, 2000; **40:** Werner *et al.*, 2002b; **41:** Gradinger*et al.*, 1991; **42:** Gradinger *et al.*, 1992; **43:** Gradinger *et al.*, 1999; **44:** Gulliksen, 1984; **45:** Lønne and Gulliksen, 1991a; **46:** Lønne and Gulliksen, 1991b; **47:** Beuchel *et al.*, 1998; **48:** Hop *et al.*, 2000; **49:** Arndt and Pavlova, 2005; **50:** Klekowski and Weslawski, 1991; **51:** Weslawski *et al.*, 1993; **52:** Hop *et al.*, 2002; **53:** Werner *et al.*, 2004; **54:** Arndt *et al.*, 2005; **[Russia] 55:** Gurjanova, 1936; **56:** Birula, 1937; **57:** Golikov and Scarlato, 1973; **58:** Golikov and Averintzev, 1977; **59:** Averintzev, 1993; **60:** Poltermann, 1998.

Figure 5 Sampling locations and distributional references in the Arctic; the numbering in Table 1 refers to this figure.

Sampling the ice–water interface, including platelet ice and brackish ice, is more problematical. Commonly these environments have been sampled by divers scraping nets and other equipment along the underice surface (e.g., Menshenina and Melnikov, 1995). In many cases, the samples are qualitative only, which, though adding useful information about species' distributions, cannot be used to determine their abundances accurately. Under-ice pumping

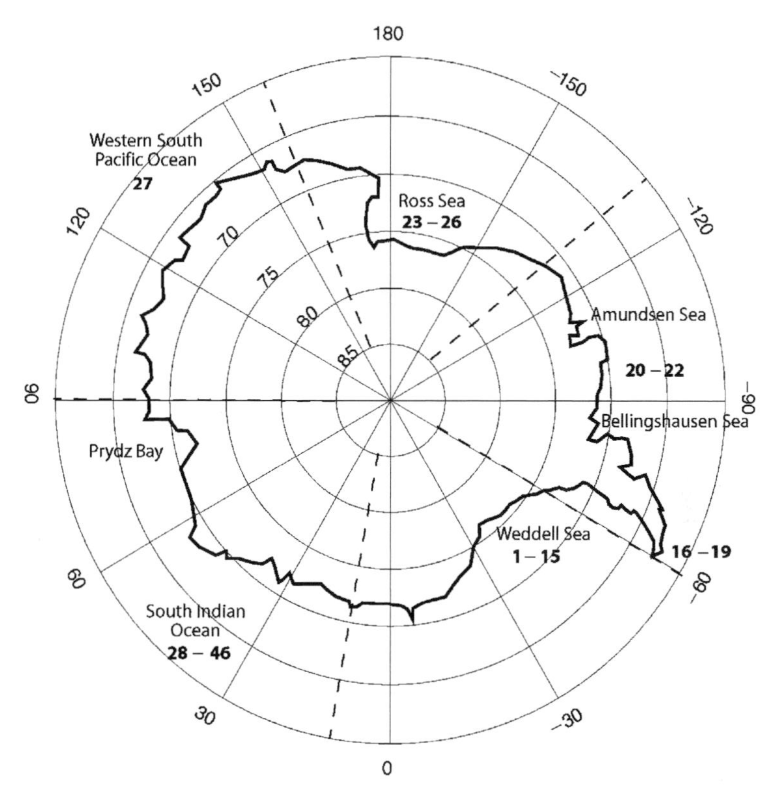

[Weddell Sea] 1: Schnack-Schiel *et al.*, 2001a; **2:** Dahms *et al.*, 1990; **3:** Schnack-Schiel *et al.*, 1995; **4:** Menshinina and Melnikov, 1995; **5:** Dahms and Schminke, 1992; **6:** Kurbjeweit *et al.*, 993; **7:** Schnack-Schiel *et al.*, 2001b; **8:** Günther *et al.*, 1999; **9:** Bergmans *et al.*, 1991; **10:** Brierley *et al.*, 2002; **11:** Aarset, 1987; **12:** Marschall, 1988; **13:** Melnikov and Spiridonov, 1996; **14:** Daly, 1990; **15:** Daly and McCauley, 1991; **[Peninsula] 16:** Hamner *et al.*, 1989; **17:** Richardson and Whitaker, 1979; **18:** Whitaker, 1977; **19:** Ito and Fukuchi, 1978; **[Bellingshausen and Amundsen Seas] 20:** Schnack-Schiel *et al.*, 1998; **21:** Thomas *et al.*, 1998; **22:** Bradford, 1978; **[Ross Sea] 23:** Waghorn, 1979; **24:** Waghorn and Knox, 1988; **25:** Costanzo *et al.*, 2002; **26:** Sagar, 1980; **[Western South Pacific Ocean] 27:** Gruzov *et al.*, 1967; **[South Indian Ocean] 28:** Taninura *et al.*, 2002; **29:** Tanimura *et al.*,1996; **30:** Tanimura *et al.*,1984a; **31:** Tanimura *et al.*,1984b; **32:** Kirkwood and Burton, 1987; **33:** Hoshiai and Tanimura, 1986; **34:** Tucker and Burton, 1988; **35:** Swadling *et al.*, 1997a; **36:** Swadling *et al.*, 2000a; **37:** Swadling *et al.*, 2000b; **38:** Swadling, 2001; **39:** Swadling *et al.*, 2004; **40:** Rakusa-Suszczewski, 1972; **41:** Rakusa-Suszczewski and Klekowski, 1973; **42:** Klekowski *et al.*, 1973; **43:** Rakusa-Suszczewski and Dominas, 1974; **44:** Opalinski, 1974; **45:** Ikeda and Kirkwood, 1989; **46:** Harrington and Thomas, 1987.

Figure 6 Sampling locations and distributional references in the Antarctic; the numbering in Table 2 refers to this figure.

systems (e.g. the ADONIS sampler [Dieckmann *et al.*, 1992]) can be used quantitatively, although there are limitations to these devices. Alternatively, diver-operated suction samplers (Lønne, 1988) can be used for quantitative samples. For larger-scale studies involving bigger animals (e.g., Antarctic

Figure 7 Sampling the underside of sea ice and the underlying water column using (A) fine-meshed zooplankton nets or (B) divers-held rectangular nets and battery-driven suction pumps. (C) Sampling the pack ice from the Australian ice-breaker, *RV Aurora Australis*. (Photos Courtesy K. Swadling, B. Gulliksen, with permission, S. Forbes, with permission.)

krill), remotely operated vehicles are showing much promise. Early results from Autosub 2 (Brierley *et al.*, 2002) indicate the usefulness of this approach. The slush ice habitat common in the interior of sea ice during the summer months (Schnack-Schiel *et al.*, 1998; Thomas *et al.*, 1998) is also difficult to sample quantitatively. Nevertheless, qualitative samples are preferable to no samples at all, so we encourage the continued exploration of these habitats.

4. CHARACTERISTIC SPECIES OF SEA ICE: DISTRIBUTION AND BIOMASS

The crustacean groups that associate most strongly with sea ice are copepods, amphipods and euphausiids (Tables 1 and 2). These taxa possess characteristics, such as generally small size, trophic flexibility and motility, that predispose them to take advantage of the sea ice habitat. However, as will be shown, there are important differences between the main groups inhabiting the two polar regions.

4.1. Arctic species

Copepods and amphipods are the most prominent crustaceans found in Arctic sea ice, although occasionally decapods, mysids and cirriped larvae occur (Table 1, Figure 5). Amphipods are by far the best studied organisms in Arctic sea ice; the routine use of SCUBA diving in the Arctic has certainly facilitated this work. Copepods have gained much less attention. In particular, biomass values are sparse and there are few studies in which copepods were assigned to species level.

4.1.1. Copepods

Harpacticoid and cyclopoid copepods occur in numbers of up to tens of thousands of individuals m^{-2} in the interstitial spaces of Arctic sea ice (Cross, 1982; Kern and Carey, 1983; Friedrich, 1997) and are several orders of magnitude more abundant in sea ice compared to the ice–water interface (Grainger *et al.*, 1985; Grainger and Hsiao, 1990; Werner *et al.*, 2002a). Maximum carbon biomass values of $\sim$20 mg C m^{-2} have been recorded from the Laptev Sea (Friedrich, 1997) and the Northwater Polynya (Nozais *et al.*, 2001). Harpacticoids and cyclopoids are often among the most abundant and widespread meiofaunal elements in the benthos (Hicks and Coull, 1983; Gulliksen *et al.*, 1999), so it is not surprising to find higher abundances in (and directly beneath) land-fast and annual coastal drift ice and few individuals in drift ice above deep water (Tables 3 and 4). Highest copepod

Table 1 List of Arctic crustacea sampled within or beneath the ice; common species are referred to as "character species." Note that references listed in Figure 5 but omitted in this table do not refer to species but to taxa

Species	Affiliation[a]	Habitat in ice[b]	Abundance[c]	Location/reference[d]
Arctic				
Copepods				
Harpacticoids				
Harpacticus sp.	b	i,w	c	1,3,4,11
Harpacticus superflexus	b	i,w	c	1,2,3,12,13,14,30,31,36,37,38,40
Halectinosoma sp.	b	i	c	3,12,13,14,16,36,37
Halectinosoma neglectum	b	i,w	c	2,4,14,30,36,37,40
Hal. finmarchicum	b	i,w	c	4,14,36,37,40
Tisbe sp.	b	w	c	14,37,40
Tisbe furcata	b	i,w	c	1,12,15,16,31,36,37,40
Dactylopodia signata	b	w	r-o	3,14
Pseudobradya sp.	b	i,w	o	2,4,36,37
Cyclopoids				
Cyclopina sp.	b	i	c	12,31,36,40
Cyclopina gracilis	b	i	c	2,3,4
C. schneideri	b	i,w	c	3,13,14,16,31,36,37,40
Oithona similis	p	i,w	o	12,16,19,36,37,38,40
Arctocylopina pagonasta	b	i	o	14
Oncaea borealis	p	i,w	o	3,12,19,36,37,38,40
Calanoids				
Calanus glacialis	p	w	r-o	12,19,23,40
C. hyperboreus	p	w	r-o	12,19,40
Microcalanus pygmaeus	p	w	r-o	19,37
Pseudocalanus sp.	p	w	o	12,18,19,22,37,40,51
Metridia longa	p	w	r-o	12,19,37
Ectinosoma sp.	p	w	r	12,16,31
Acartia longiremis	p	w	r	12
Jaschnovia brevis	b-p	w	r-o	31,36,38,40

Amphipods				
Gammaroids				
Gammarus wilkitzkii	b	w	c	12,16,25,27,28,33,34,37,45,46,47,48,49,54,58,59,60
G. setosus	b	w	r-o	4,5,8,20,21,24,25,27,31,50,52,56,57,58,59
Weyprechtia pinguis	b-p	w	r-o	4,5,10,21,24,25,28,46,50,55
Apherusa glacialis	b	w/i?	c	4,7,12,16,25,26,27,28,33,34,37,44,45,46,47,48, 49,57,58,59,60
A. megalops (=A. sarsii)	b	w	r	10,24,46,57
Onisimus nanseni	b	w	c	16,25,26,27,28,33,34,44,48,49,60
O. glacialis	b	w/i?	c	16,25,26,27,28,48,49,60
O. litoralis	b	w	o	4,5,9,20,21,25,27,50,55,59
O. edwardsi	b	w	r-o	10,17,20
Anonyx nugax	b	w	r	20,25,28,45,50,53
A. sarsi	b	w	r	53
Gammaracanthus loricatu	b	w	r-o	5,9,10,20,21,24,25,27,31,34,45,46,50,55,59
Gammarellus homari	b	w	r-o	24
Ischyrocerus anguipes	b	w	o	10,16,21,24,25,45,50,55
Pontogenia inermis	b	w		24
Eusirus holmi	b	w		10,16,29,31,34
E. cuspidatus	b	w		17,20,55
Melita formosa	b	w	r	5
Metopa wiesei	b	w	r	31
Neopleustes sp.	b	w	r	31
Hyperiids				
Parathemisto libellula	p	w	o	10,12,17,19,20,21,24,25,31,34,35,45,46,50,55
Mysids				
Mysis polaris	p	w	r	16,31
Decapods				
Eualus gaimardii	b	w	r	28,31

[a]p: pelagic, b: benthic.
[b]i: interstitial, w: ice-water interface.
[c]c: common/frequently recorded, o: (occasionally) abundant, r: rare/single ind. recorded.
[d]See Figure 5.

Table 2 List of Antarctic crustacea sampled within or beneath the ice; common species are referred to as "character species"

Species	Affiliation[a]	Habitat in ice[b]	Abundance[c]	Location/reference[d]
Antarctic				
Copepods				
Calanoids				
Stephos longipes	p	i,w	c	1,3,4,6,7,8,20,25,34,35,36
Paralabidocera antarctica	p	i,w	c	4,8,28,29,30,31,32,33,34,35,36,37,38,39
Calanus propinquus	p	w	o/c?	34
Ctenocalanus citer	p	w	o/c?	4,33,34,35
Paralabidocera grandispina	p	i,w	o	23,24
Metridia gerlachei	p	w	o	34
Microcalanus pygmaeus	p	w	o	4
Cyclopoids				
Oithona similis	p	w/i?	c	4,8,32,33,34,35
Oncaea curvata	p	w	c	8,31,33,34,35
Pseudocyclopina belgicae	p	w/i?	o	4,8,24
Harpacticoids				
Drescheriella glacialis	p/b?	i	c	1,2,7,8,9,20,36,38
Harpacticus furcifer	b	w/i?	o	2,5,8

Harpacticus furcatus	b	w/i?	o	22,32
Microsetella norvegica	p	w/i?	r-o	2
Hastigerella antarctica	b	w/i?	r	5,8
Drescheriella racovitzai	b	w/i?	r	5,8
Idomene antarctica	b	w/i?	r	5
Drescheriella sp.	b	w/i?	r	2
Idomene sp.	b	w/i?	r	8
Hastigerella sp.	b	w/i?	r	2
Tisbe prolata	b	w/i?	r-o	8,23,24
Tisbe sp.	b	w/i?	r	2,8
Dactylopodella sp.	b	w/i?	r	2
Nitrocra gracilimana	b	w/i?	r	8
F. Ameriridae (3 spp)	b	w/i?	r	8
F. Ectinosomatidae (5 spp)	b	w/i?	r	8
Amphipods				
Gammarids				
Paramoera walkeri	b	w	c	26,27,32,34,40,41,42,43,44
Orchomene cf. *plebs*	b	w	c	34
Pontogenia antarctica	b	w	o	17,18
Eusirus antarcticus	b	w	o	11
Oradarea sp.	b	w	r	8
Probolisca sp.	b	w	r	8

(Continued)

Table 2 (Continued)

Species	Affiliation[a]	Habitat in ice[b]	Abundance[c]	Location/reference[d]
F. Lysianassidae	b	w	r	8
Orchomene sp.	b	w	r	34
Eusirus cf. fragilis	b	w	r	34
Cheirimedon femoratus	b	w	o	19
Orchomene cavimanus	b	w	r	27
Hyperiids				
Hyperia macrocephalus	p	w	r	34
Mysids				
Mesaiokeras sp.	p	w	r	8
Euphausids				
Euphausia superba	p	w	c	4,10,11,12,13,14,15,16,43
Euphausis crystallorophias	p	w?	c	43,44

[a]p: pelagic, b: benthic.
[b]i: interstitial, w: ice-water interface.
[c]c: common/frequently recorded, o: (occasionally) abundant, r: rare/single ind. recorded.
[d]See Figure 6.

Table 3 Abundance (ind. m^{-2}) of interstitial copepods and nauplii and carbon biomass (mg Cm^{-2}) in Arctic sea ice; mean ± SD (min-max) conversion factors after Friedrich (1997), mean carbon content for copepods: 0.6 μg C $specimen^{-1}$, and nauplii: 0.02 μg C $specimen^{-1}$

Location (ice type)/date	Copepodites		Nauplii		Reference
	ind m^{-2}	mg Cm^{-2}	ind m^{-2}	mg Cm^{-2}	
Greenland Sea (MYI)					
Jul-Aug 1994	1643 (145–7694)	0.99 (0.08–4.62)[b]	81 (0–544)	0.002 (0–0.01)[b]	Friedrich, 1997 (Table 4.3)
May 1988	(150–300)	(0.09–0.15)[b]	(300–3900)	(0.006–0.08)[b]	Gradinger *et al.*, 1991 (Table 2)
Jul-Aug 1994/Sep-Oct 1995	2400	1.22[b]			Gradinger *et al.*, 1999 (p. 1464)
Barents Sea (MYI)					
Aug 1993	1916 (0–8639)	1.15 (0–5.18)[b]	2109 (0–9125)	0.04 (0–0.18)[b]	Friedrich, 1997 (Table 4.3)
Laptev Sea/Russia (FYI)					
Aug-Sep 1993	4493 (0–33,438)	2.70 (0–20.06)[b]	11,670 (0–127,954)	0.033 (0–2.56)[b]	Friedrich, 1997 (Table 4.3)
Near Resolute/Alaska (MY)					
Aug 2002/Aug-Sep 2003	(0–7300)	(0–4.2)[b]			Gradinger *et al.*, 2004 (Figure 5)

(Continued)

Table 3 (Continued)

Location (ice type)/date	Copepodites		Nauplii		Reference
	ind m^{-2}	mg Cm^{-2}	ind m^{-2}	mg Cm^{-2}	
Narwhal island/ Alaska (FY-F)					
Mar 1979	120 ± 3.4	0.07 ± 0.002[b]	1770 ± 1940	0.035 ± 0.039[b]	Carey and Montagna, 1982 (Table 2)
May 1979	810	0.49[b]	230 ± 460	0.005 ± 0.009[b]	Carey and Montagna, 1982 (Table 2)
Apr 1980	5094 ± 2813	3.06 ± 1.69[b]			Kern and Carey, 1983 (Table 3)
May 1980	10,971 ± 5663	6.58 ± 3.40[b]			Kern and Carey, 1983 (Table 3)
Jun 1980	6987 ± 3651	4.19 ± 2.19[b]			Kern and Carey, 1983 (Table 3)
Frobisher Bay/Canada (FY-F)					
Feb 1981	2450	1.47[b]	90,920	1.82[b]	Grainger *et al.*, 1985 (Table 1-2)
Mar 1981			26,080	0.52[b]	Grainger *et al.*, 1985 (Table 1-2)

May 1981[a]	252	0.15[b]	4200	0.08[b]	Grainger *et al.*, 1985 (Table 1-2)
Jun 1981			800	0.02[b]	Grainger *et al.*, 1985 (Table 1-2)
Feb 1982	3716	2.23[b]	48,156	0.96[b]	Grainger *et al.*, 1985 (Table 1-2)
Mar 1982			7886	0.16[b]	Grainger *et al.*, 1985 (Table 1-2)
May 1982	109	0.07[b]	8891	0.18[b]	Grainger *et al.*, 1985 (Table 1-2)
Pond Inlet/Canada (FY-F)					
May 1979	(50–20,470)	(0.03–12.28)[b]			Cross, 1982 (p. 23–24)
Jun-Jul 1979	(0–1410)	(0–0.85)[b]			Cross, 1982 (p. 23–24)
Northwater/Canada (FYI)					
Apr-May 1998	(0–11,080)	(0–19.15) (0–6.65)[b]	(0–4090)	(0–0.79) (0–0.08)[b]	Nozais *et al.*, 2001 (Table 2)

[a]Data from 1st May used.
[b]Based on conversion factors.
Note: FYI: first year (pack) ice, FY-F: first year fast ice, MYI: multiyear ice.

Table 4 Abundances (ind m^{-3}) of under-ice copepods and nauplii and carbon content (mg Cm^{-3}) in Arctic sea ice; mean (min-max); for conversion factors see Table 3

Location (Ice type)/date	Copepodites		Nauplii		Reference
	ind m^{-3}	mg Cm^{-3}	ind m^{-3}	mg Cm^{-3}	
Greenland Sea (MYI)					
Jul-Aug 1994	(117–754)	(0.07–0.45)[b]	635 (151–2533)	0.01 (0–0.05)[b]	Werner, 1997a (Table 13)
Sep-Oct 1995	(282–1757)	(0.17–1.05)[b]	2869 (1509–2865)	0.06 (0.03–0.06)[b]	Werner, 1997a (Table 13)
Laptev Sea/Russia (FYI)					
Jul-Sep 1995	(104–6307)	(0.06–3.78)[b]	2011 (130–23,911)	0.04 (0–0.48)[b]	Werner, 1997a (Table 13)
Frobisher Bay/ Canada (FY-F)					
Feb 1981	2450	1.47[b]	90,920	1.82[b]	Grainger *et al.*, 1985 (Table 1-2)[c]
Mar 1981			26,080	0.52[b]	Grainger *et al.*, 1985 (Table 1-2)[c]
May 1981[a]	252	0.15[b]	4200	0.08[b]	Grainger *et al.*, 1985 (Table 1-2)[c]
Jun 1981			800	0.02[b]	Grainger *et al.*, 1985 (Table 1-2)[c]

[a]Data from 1st May used.
[b]Based on conversion factors.
[c]Values recalculated from a 50-m water layer.
Note: FYI: first year (pack) ice, FY-F: first year fast ice, MYI: multiyear ice.

densities generally coincide with chlorophyll *a* maxima at the underside of sea ice (e.g., Friedrich, 1997; Gradinger *et al.*, 1999).

The young stages of copepods (eggs and nauplii) may accumulate in sea ice in numbers of more than 100,000 m^{-2} (Friedrich, 1997). The eggs and nauplii do not necessarily stem only from those harpacticoid and cyclopoid copepods that inhabit the ice (see below), but also from pelagic organisms such as calanoid copepods. For example, the eggs of *Calanus hyperboreus* Krøyer and *Calanus glacialis* Jaschnov are positively buoyant (Conover *et al.*, 1991), and while developing, they ascend slowly to the surface to time their arrival with the onset of the phytoplankton bloom. It is believed that eggs are the main dispersal mechanism for benthic copepods to colonise the ice environment. Further means for entering the ice ecosystem are advection by currents (and, of course, drifting ice floes), anchor ice lift (Carey and Montagna, 1982), suspension freezing (Reimnitz *et al.*, 1993) and swimming (Carey and Montagna, 1982; Ólafsson *et al.*, 2001). Although primarily benthic species, the harpacticoids *Tisbe furcata* (Baird) and *Harpacticus superflexus* Willey are also common constituents of the pelagos and have been sampled in the water column of the polar basins down to 2000 m (Montagna and Carey, 1978; Mumm, 1991).

The copepod genera *Harpacticus, Halectinosoma, Tisbe* (all Harpacticoida) and *Cyclopina* (Cyclopoida) have circum-Arctic distributions and are regularly found in association with the ice interior or the ice–water interface (Tables 5 and 6). *H. superflexus* and other *Harpacticus* species appear to be virtually absent from the interstices of perennial ice in the Arctic Ocean and the northern Barents and Greenland Seas (Melnikov, 1989; Friedrich, 1997) and are also scarce underneath old ice (Werner, 1997a). In contrast, in the fast ice of Frobisher Bay in Canada, *H. superflexus* reached abundances between 3 and >380 individuals m^{-2} (Grainger *et al.*, 1985). In the shear zone between the fast ice belt off the north coast of Alaska and the Beaufort Sea ice pack, near Narwhal Island, several thousands of individuals of *Harpacticus* sp. were found in the bottom of the ice (Kern and Carey, 1983). *Halectinosoma* spp. were also numerically important constituents of the interstitial copepod fauna of that region. This genus is found more regularly in the multiyear ice north of Svalbard and in the Greenland Sea than are *Harpacticus* spp. (Friedrich, 1997) and is more abundant in the ice–water interface than any other ice-associated copepod (Werner, 1997a). Two species, *Halectinosoma neglectum* (Sars) and *Halectinosoma finmarchicum* (Scott), often co-occur in the interstitial and the ice–water habitats. Only single individuals of the harpacticoid *T. furcata* were found in the multiyear ice north of Svalbard and the Greenland Sea (Friedrich, 1997). Werner (1997a) observed several species of *Tisbe* but did not assign them to species level. However, in fast ice, *T. furcata* seems to be a numerically important component of the copepod fauna, as observed at

Table 5 Abundances (ind m^{-2}) of interstitial copepods in Arctic sea ice, sorted by genera; mean ±SD

Location (ice type)/date	*Harpacticus* sp. ind m^{-2}	*Halectinosoma* sp. ind m^{-2}	*Tisbe* sp. ind m^{-2}	*Cyclopina* sp. ind m^{-2}	Reference
Narwhal island/ Alaska (FY-F)					
Apr 1980	1039 ± 796	1552 ± 1141		2297 ± 1326	Kern and Carey, 1983 (Table 3)
May 1980	3156 ± 1516	1405 ± 1238		6078 ± 3424	Kern and Carey, 1983 (Table 3)
Jun 1980	2059 ± 1345	808 ± 501		3465 ± 2727	Kern and Carey, 1983 (Table 3)
Frobisher Bay/ Canada (FY-F)					
Feb 1981	20		255	2175	Grainger *et al.*, 1985 (Table 1-2)

Note: FY-F: first year fast ice.

Table 6 Abundances (ind m^{-3}) of under-ice copepods in Arctic sea ice, sorted by genera; mean (min-max)

Location (ice type)/date	*Harpacticus superflexus* ind m^{-3}	*Halectinosoma* sp. ind m^{-3}	*Tisbe* sp. ind m^{-3}	*Cyclopina* sp. ind m^{-3}	Reference
Greenland Sea (MYI)					
Jul-Aug 1994	2	12 (1–41)	5 (2–448)	1 (0–45)	Werner, 1997 (Table 13)
Sep-Oct 1995	<1	4 (4–7)	4 (0–6)	2 (<1–10)	Werner, 1997 (Table 13)
Laptev Sea/ Russia (FYI)					
Jul-Sep 1995		51 (2–481)	11 (1–98)	6 (<1–61)	Werner, 1997 (Table 13)
Frobisher Bay/ Canada (FY-F)					
Feb 1981	0.24		42.16	0	Grainger *et al.*, 1985 (Table 1-2)[b]
Mar 1981	0		34.92	2.28	Grainger *et al.*, 1985 (Table 1-2)[b]
May 1981[a]	0		10.74	2.28	Grainger *et al.*, 1985 (Table 1-2)[b]
Jun 1981	1.52		5.2	6.64	Grainger *et al.*, 1985 (Table 1-2)[b]

[a]Data from 1st May used.
[b]Values recalculated from a 50-m water layer.
Note : FYI: first year (pack) ice, FY-F: first year fast ice, MYI: multiyear ice.

Pond Inlet (Cross, 1982) and Frobisher Bay (Canada), where up to 1180 individuals m^{-2} have been found in the bottom ice (Grainger and Hsiao, 1990). Beneath the old ice in the Greenland Sea, abundances of *Tisbe* sp. ranged between zero and 448 individuals m^{-2} (Werner, 1997a). Only single individuals were sampled of *Dactylopodia signata* (Willey) and *Dactylopodia vulgaris* (Sars) under fast ice north of Alaska (Kern and Carey, 1983) and in Frobisher Bay (Grainger and Hsiao, 1990). These species have not been observed near Svalbard. *D. signata* is not part of the benthic meiofauna of the Svalbard archipelago, while *D. vulgaris* is present (Kotwicki, 2002).

The most conspicuous copepods in land-fast ice, however, appear to be two species of *Cyclopina: Cyclopina gracilis* Claus and *Cyclopina schneideri* Scott. Near Narwhal Island, their numbers varied between 2300 and >6000 individuals m^{-2} (Kern and Carey, 1983), and during continuous winter sampling in Frobisher Bay, the abundance of *Cyclopina* spp. peaked in March 1982 when >9300 individuals m^{-2} were found (Grainger and Hsiao, 1990). At the ice–water interface, *C. schneideri* generally occurs as single individuals (Werner, 1997a).

It is interesting to note that none of the aforementioned typical "ice copepods" has been recorded from the SHEBA Ice Camp, a multiyear ice floe in the Canadian Basin that was followed on its drift from October 1997 until October 1998. The ice interstices only contained skeletal parts and exuviae from benthic copepods, and *T. furcata* was the only species sampled from directly beneath the ice (Melnikov *et al.*, 2002).

Calanoid copepods do not appear to colonise the sea ice itself but may be advected to the ice sheet during the egg and naupliar stages (Cross, 1982). Calanoids have been observed utilising the rich food supply in the bottom layers of sea ice. Grazers of ice algae, including *Calanus glacialis, Calanus hyperboreus, Pseudocalanus acuspes* (Giesbrecht), *Metridia longa* (Lubbock) (Conover *et al.*, 1988; Runge and Ingram, 1991) and, to a minor extent, *Oithona similis* Claus, *Oncaea borealis* Sars and *Microcalanus pygmaeus* (Sars), perform diel vertical migrations to the ice surface at dusk (Fortier *et al.*, 2001). Grainger *et al.* (1985) pointed out that several copepod species found beneath sea ice are small enough to inhabit the ice "but seem to do so either only rarely or not at all." Numbers confirm this observation; in the ice–water boundary layer of drift ice in the Laptev Sea, *O. similis* dominated the copepod assemblage, with up to 4,660 individuals m^{-3} (Werner and Arbizu, 1999), but in the lowermost centimetre of Svalbard fast ice, a maximum of 80 individuals m^{-3} was observed (Weslawski *et al.*, 1993). Given the relative abundances of planktonic *Oithona* versus ice core observations, this species seems more likely to be an occasional visitor to the ice ecosystem in search of food but is not associated strongly with ice during its life cycle. Similarly, *Acartia longiremis* (Lilljeborg) and *O. borealis* appear to be independent from sea ice for reproduction but utilise it as a food source (Grainger *et al.*, 1985). According to Grainger *et al.* (1985), these species show "accidental presence in the ice... at times when great numbers of the same species are present in the water below."

Biomass estimates for ice-associated copepods are also highly variable, ranging between zero and 20 mg Cm^{-2} (Friedrich, 1997; Nozais *et al.*, 2001) (Tables 3 and 4). Note that only the study of Nozais *et al.* (2001) published *in situ* values on copepod biomass in sea ice. Several studies (e.g., Gradinger *et al.*, 1991, 1999; Gradinger and Bluhm, 2004) have estimated ice-associated biomass on the basis of individual conversion factors for different copepod species and nauplii determined by Friedrich (1997). In Table 3, we used the same approach. However, in the study by Nozais *et al.* (2001), the comparison of observed versus calculated biomass shows great discrepancies; the carbon values for copepods are nearly 3 times higher and those for nauplii 10 times higher from *in situ* measurements compared to the calculated estimates. Therefore, many of the biomass values listed in Tables 3 and 4 are probably gross underestimations.

4.1.2. Amphipods

Amphipods tend to be excluded from the ice interior because of their large size but inhabit the subsurface of the ice, where they penetrate into larger holes, crevices and channels that have widened during the summer melt. Amphipods (Figure 8) have been sampled from the ice all over the Arctic and most of the species seem to have a circum-Arctic distribution. However, there are large differences in their abundances when comparing land-fast ice and drift ice above shallow water with the ice pack that has been drifting above deep water for many years (Carey, 1992). The species *Gammarus wilkitzkii* (Gammaridea), *Apherusa glacialis* (Calliopiidae), *Onisimus nanseni* Sars and *Onisimus glacialis* Sars (Lysianassidae) occur in any ice type throughout those Arctic seas that are regularly impacted by perennial ice from the Arctic Ocean. Except for *Onisimus* spp., these amphipods are scarce in the narrow bays and straits of the Canadian archipelago, where ice is locally produced anew every winter (e.g., Grainger and Hsiao, 1990; Siferd *et al.*, 1997). In Barents Sea drift ice, these four species generally occur in numbers <100 individuals m^{-2} (Table 7). Maximum abundances for the area north of Svalbard are 280 individuals m^{-2}, which corresponds to nearly 8 g WM (met mass) m^{-2} (Lønne and Gulliksen, 1991a) and is equivalent to 1.3 g C m^{-2}. *G. wilkitzkii* and *A. glacialis* are equally abundant, but because of the high individual body weight of *G. wilkitzkii*, total biomass values depend heavily on the relative contribution of this species. It is interesting to note that its congener *Gammarus setosus* Dementiera has been observed

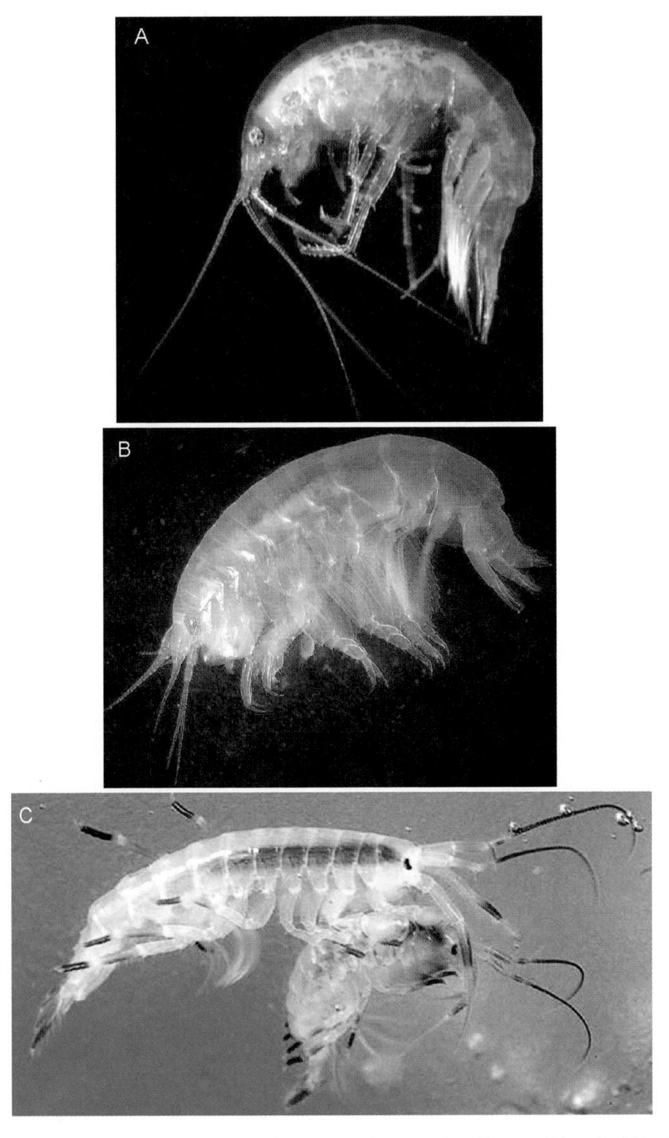

Figure 8 Examples of Arctic sea ice macrofauna. Arctic amphipods (A) *Apherusa glacialis* (maximal body length BL = 14 mm), (B) *Onisimus nanseni* (BL = 25 mm) and (C) *Gammarus wilkitzkii* (BL = 55 mm). (Photos courtesy C. Arndt, E. Svensen, with permission.)

Table 7 Abundances (ind m^{-2}) of Arctic sympagic amphipods, wet mass (g WM m^{-2}) and estimated carbon content (mg Cm^{-2}) based on the conversion factor for amphipods given by Ricciardi and Bourget (1998): AFDM/WM = 16%; *p*: single individuals present; mean values

Location (ice type)/date (replicate)	Amphipods			*Gammarus wilkitzkii*		*Apherusa glacialis*		*Onisimus* spp.[a]		Reference
	ind m^{-2}	g WM m^{-2}	mg C m^{-2}	ind m^{-2}	g WM m^{-2}	ind m^{-2}	g WM m^{-2}	ind m^{-2}	g WM m^{-2}	
Barents Sea (MYI/FYI)										
1996 (1)	100.9	3.001	495.1	88.1	2.900	12.4	0.081	0.4	0.02	Hop *et al.*, 2000[b]
1996 (2)	89.8	3.216	530.6	75.2	3.089	13.9	0.083	0.7	0.044	Hop *et al.*, 2000[b]
1996 (3)	24.8	1.379	227.5	18.9	1.249	5.1	0.034	0.8	0.096	Hop *et al.*, 2000[b]
1996 (4)	59.1	2.086	344.2	50.1	1.993	8.5	0.058	0.5	0.035	Hop *et al.*, 2000[b]
May-Jun 1983	10.4	0.035	5.8	0.1	0.007	10.0	0.019	0.3	0.009	Lønne and Gulliksen, 1991a
Jul 1986	103.4	2.048	337.9	33.4	0.848	46.3	0.366	23.6	0.834	Lønne and Gulliksen, 1991a
May 1988	6.5	0.067	11.1	0.5	0.02	4.4	0.026	1.6	0.021	Lønne and Gulliksen, 1991a
Jul-Aug 1986	138.5	7.207	1189.1	32.6	4.374	94.8	0.478	11.1	0.388	Lønne and Gulliksen, 1991b
Sep 1988	278.2	7.85	1295.3	69.7	6.34	186.6	0.86	21.9	0.65	Lønne and Gulliksen, 1991b
Aug 1982	19.7	0.308	50.8	2.9	0.152	15.7	0.097	1.1	0.059	Gulliksen, 1984
Aug 1982	0.4	0.075	12.4	0.3	0.07	0.0	0	0.1	0.005	Gulliksen, 1984
Sep 2000 (1)	12.8	0.5	82.5	2.4	0.1	10.1	0	0.2	0	Arndt *et al.*, 2000[c]
Sep 2000 (2)	14.8	0.6	99.0	4.0	0.5	9.6	0	1.2	0	Arndt *et al.*, 2000[c]
Sep 2000 (3)	32.2	0.2	33.0	3.0	0.1	28.8	0.1	0.3	0	Arndt *et al.*, 2000[c]
Sep 1998 (1)	29.0	0.668	110.3	16.0	0.587	12.6	0.066	0.4	0.016	Beuchel *et al.*, 1998[c]
Sep 1998 (2)	19.1	0.643	106.1	11.5	0.607	7.6	0.033	0.0	0.003	Beuchel *et al.*, 1998[c]
Sep 1998 (3)	16.4	0.557	91.9	6.9	0.452	8.1	0.059	1.4	0.046	Beuchel *et al.*, 1998[c]
Sep 2002 (1)	9.5	0.2	33.0	4.5	0.1	3.8	0	1.3	0	Arndt and Pavlova, 2005

(Continued)

Table 7 (Continued)

Location (ice type)/date (replicate)	Amphipods			*Gammarus wilkitzkii*		*Apherusa glacialis*		*Onisimus* spp.[a]		Reference
	ind m^{-2}	g WM m^{-2}	mg C m^{-2}	ind m^{-2}	g WM m^{-2}	ind m^{-2}	g WM m^{-2}	ind m^{-2}	g WM m^{-2}	
Sep 2002 (2)	11.4	0.5	82.5	4.9	0.5	4.2	0	2.3	0	Arndt and Pavlova, 2005
Sep 2002 (3)	23.6	0.4	66.0	15.6	0.3	7.8	0	0.2	0	Arndt and Pavlova, 2005
Sep 2002 (4)	2.5	0.5	82.5	2.3	0.5	0.0	0	0.2	0	Arndt and Pavlova, 2005
Beaufort Sea (MYI)										
1952		1[f]	165.0	p		p		p		Barnard, 1959
1977–80	12.0			12.0		p		p		Melnikov, 1997
Canada (FY-F)										
May-Jul 1979	30.7	0.115	18.9	0.0	0	23.7	0.083	7.0	0.031	Cross, 1982
Feb 1981								60.0	0.12	Grainger *et al.*, 1985[d]
Mar 1981								110.0	0.22	Grainger *et al.*, 1985[d]
May 1981								120.0	0.23	Grainger *et al.*, 1985[d]
Jun 1981								0.0	0	Grainger *et al.*, 1985[d]
Feb 1982								19.0		Grainger *et al.*, 1985[d]
Mar 1982								140.0		Grainger *et al.*, 1985[d]
May 1982								9.0		Grainger *et al.*, 1985[d]
Russia/Franz-Josef-Land (MYI/FYI)										
Aug 1994	417.0	10.929	1803.3	368.0	10.12	34.0	0.099	15.0	0.71	Poltermann, 1998
Feb 1982	16.0	0.46	75.9	2.0	0.135	14.0	0.325	0.0	0	Averintzev, 1993

Mar-May 1982	16.8	2.086	344.2	10.5	1.656	6.3	0.13	0.0	0	Averintzev, 1993
Jun-Aug 1982	32.8	1.860	306.9	18.0	1.581	14.8	0.279	0.0	0	Averintzev, 1993
Aug 1981	490.0	4.46	735.9	0.0	0	490.0	4.46	0.0	0	Averintzev, 1993
Dec-Feb 1981-82	206.5	4.448	733.9	6.6	1.56	199.9	3.292	0.0	0	Averintzev, 1993
Mar-May 1982	80.7	21.095	3480.7	6.5	1.428	74.2	19.667	0.0	0	Averintzev, 1993
Jun-Aug 1982	47.9	2.546	420.1	20.8	1.87	27.0	0.676	0.0	0	Averintzev, 1993
Laptev Sea (FYI)										
Aug 1995	17.1	2.635	434.8	3.9	1.971	2.2	0.027	11.0	0.638	Werner, 1997a[e]
Greenland Sea (MYI)										
Aug-Sep 1994 and 95	171.0	14.720	2428.8	21.0	10.614	100.0	1.208	50.0	2.899	Werner, 1997a[e]
Alaska/Narwhal Island (FY-F)										
Mar-May 1979	268[g]			0.0		268[g]		0.0	0	Carey, 1992

[a] *Onisimus nanseni* and *O. glacialis*.

[b] Under-ice structure proportionally weighted and data recalculated.

[c] Calculated from unpublished data.

[d] Data estimated from qualitative samples.

[e] Means calculated from ranges by assuming the same asymmetric distribution as found by Lønne and Gulliksen (1991a,b).

[f] Biomass value based on rough estimation.

[g] Abundance value given per volume (ind m^{-3}).

Note: FYI: first year (pack) ice, FY-F: first year fast ice, MYI: multiyear ice.

"in great numbers adjacent to the ice but not in it at any size" (Grainger *et al.*, 1985).

Beneath land-fast ice around Franz Josef Land (Russia) (Averintzev, 1993; Poltermann, 1998) and Narwhal Island (Carey, 1992), *G. wilkitzkii, A. glacialis* and *Onisimus* spp. were sampled in numbers up to several hundred individuals per square metre. The maximum recorded biomass was found at Franz Josef Land: 21 g WM m^{-2}, which is equivalent to 3.5 g C m^{-2} (Averintzev, 1993). *O. glacialis* and *O. nanseni* are generally scarce in sea ice but may be numerically very important constituents in first-year ice, reaching abundances of up to 140 individuals m^{-2} (Grainger and Hsiao, 1990).

Ice amphipods have been sampled in the water column off the retreating ice border (Steele and Steele, 1974; Arndt and Pavlova, 2005) and apparently survive ice-free periods in the benthos (Carey, 1992; Weslawski, 1994; Arndt *et al.*, 2005a). There are a few observations of the pelagic occurrence of *A. glacialis* in the deep sea (e.g., Siferd *et al.*, 1997) and *O. glacialis* from >130 m water depth in the midst of the Arctic Ocean (Sars, 1900).

The distributions of the planktonic congeners *Parathemisto libellula* (Lichtenstein) and *Parathemisto abyssorum* (Boeck) (Hyperiidae) generally coincide with the extent of sea ice. These species are very common at the ice–water interface and may occur in swarms of up to 430 and 270 individuals m^{-2}, respectively (Dalpadado *et al.*, 2001). Most of the amphipod species found in association with sea ice are, however, of benthic origin and are, therefore, occasionally abundant beneath land-fast ice and ice that originates from shallow coastal areas. *G. setosus, Gammaracanthus loricatus* (Steele and Steele), *Ischyrocerus anguipes* Krøyer, *Weyprechtia pinguis* Krøyer, *Anonyx nugax* Phipps, *Anonyx sarsi* Steele and Brunel and *Onisimus litoralis* Krøyer are the most frequently recorded benthic amphipod species that occur as single individuals in the perennial ice pack and at times in numbers of several tens of individuals per square metre in annual fast ice.

4.2. Antarctic species

Copepods are the dominant crustaceans found within the sea ice interstices, while amphipods, copepods and euphausiids are important at the ice–water interface (Table 2, Figure 6). Much less commonly, mysids, ostracods and cumaceans have been observed. Interestingly, early work focused on amphipods (e.g., Gruzov *et al.*, 1967; Andriashev, 1968; Rakusa-Suszczewski, 1972), probably as a result of their direct observation by divers. Subsequently, amphipods have been under-represented in Antarctic sea ice studies, while copepods and euphausiids have become the dominant taxa for research. The early lack of recognition of the importance of copepods, in particular, probably derives from the sampling methods, as they are far less visible to the naked eye.

4.2.1. Copepods

Calanoid and harpacticoid copepods dominate the Antarctic sea ice, sometimes occurring in numbers up to hundreds of thousands per square metre of ice (Hoshiai and Tanimura, 1986; Swadling *et al.*, 1997a, 2000a; Schnack-Schiel *et al.*, 2001b). Highest abundances have been recorded from Prydz Bay (Swadling *et al.*, 1997a, 2000a) (Tables 8 and 9), while maximum carbon values, up to 1236 mg C m^{-2}, have also been recorded from that region. In general, the highest abundances for the interstitial ice crustacean fauna occur quite close to the coast, although this tends to reflect where most of the sampling has occurred. Smaller floes further out into the Southern Ocean have rarely been sampled, although it appears that crustacea that inhabit the sea ice interstices are generally rare or absent in the Indian Ocean sector (K. Swadling, unpublished data).

The three most frequently observed copepod species, the calanoids *Stephos longipes* Giesbrecht and *Paralabidocera antarctica* (I. C. Thompson) and the harpacticoid *Drescheriella glacialis* Dahms and Dieckmann, are thought to be circum-Antarctic in distribution, although there are clear differences in abundance. *Stephos longipes* is the dominant calanoid in the Weddell, Amundsen and Bellingshausen Seas, where it reaches up to 200,000 individuals m^{-2} (Schnack-Schiel *et al.*, 1995), while in Prydz Bay and Lutzow-Holm Bay (69°00′S, 39°35′E), it has not been observed to exceed 20,000 individuals m^{-2} (Hoshiai and Tanimura, 1986; Swadling *et al.*, 2000a). *S. longipes* has been recorded from the ice matrix, as well as slushy ice in summer and platelet ice in late spring (Schnack-Schiel *et al.*, 1998; Thomas *et al.*, 1998; Günther *et al.*, 1999), although in general its abundance is higher in the ice matrix. The highest abundance of *S. longipes* (up to 163 individuals l^{-1}) in the Amundsen Sea coincided with the high chlorophyll concentrations (up to 377 $\mu g\ l^{-1}$) in the more saline slush layer in the middle of the ice (Thomas *et al.*, 1998). High numbers of harpacticoids (up to 200 individuals l^{-1}) were also noted, although no identifications were given. The slush ice-filled gaps were >1 m from the surface, distinguishing them clearly from freeboard layers closer to the surface. This is one of few observations of high numbers of meiofauna inhabiting second or multiyear ice in Antarctica. Nauplii of *S. longipes* have also been noted in high abundance (up to 510 individuals L^{-1}) in sea ice of Terra Nova Bay (74°41′S, 164°07′E) (Costanzo *et al.*, 2002). A congener of *S. longipes, S. antarcticum* Wolfenden, has also been listed for Antarctic waters (Razouls *et al.*, 2000), but little is known about this species.

Table 8 Abundances (ind. m^{-2}) of interstitial copepods and nauplii and Carbon Biomass (mg C m^{-2}) in Antarctic sea ice; mean ±SD (min-max); mean carbon content for copepods: 3.5 μg C *Paralabidocera*$^{-1}$, 1.0 μg C harpacticoid^{-1}, 0.78 μg C *Oithona*$^{-1}$, 5.0 μg C *Stephos*$^{-1}$; and 0.26 μg C nauplius^{-1}

Location (ice type)/date	Copepodites		Nauplii		Reference
	ind m^{-2}	mg C m^{-2}	ind m^{-2}	mg C m^{-2}	
Weddell Sea (MYI)					
Sep/Oct 1989	3180		2820		Gradinger, 1999 (Fig. 3-4)
Jan–Mar 1991	5382	4.6	318	0.81	Gradinger, 1999 (Fig. 3-4)
Mar–May 1992	46,400		46,400		Gradinger, 1999 (Fig. 3-4)
Weddell Sea (FYI)					
Jan/Feb 1985	5640 (0–67,200)[a]	3.39 (0–40.2)[a,b]			Schnack-Schiel *et al.*, 2001a (Table 3-4)
Oct/Dec 1986	2340 (0–23,000)[a]	4.18 (0–43)[a,b]			Schnack-Schiel *et al.*, 2001a (Table 3-4)
Sep/Oct 1989	5260 (0–56,000)[a]	2.34 (0–23)[a,b]			Schnack-Schiel *et al.*, 2001a (Table 3-4)
Jan/Mar 1991	54,130 (200–221,000)[a]	10.89 (.01–68)[a,b]			Schnack-Schiel *et al.*, 2001a (Table 3-4)
Apr/May 1992	28,810 (0–246,000)[a]	11.23 (0–86)[a,b]			Schnack-Schiel *et al.*, 2001a (Table 3-4)
Jan/Mar 1997	3840 (0–31,000)[a]	4.13 (0–42)[a,b]			Schnack-Schiel *et al.*, 2001a (Table 3-4)
Lutzow-Holm Bay (FY-F)					
Mar–May 1970	1600 (0–2000)	1.6 (0–2.0)[c]	27,000 (0–44,000)	7 (0–11.4)[c]	Hoshiai and Tanimura, 1986 (Table 1-2)
Jun–Aug 1970	2000 (1000–3000)	2.0 (1.0–3.0)[c]	68,300 (49,000–79,000)	17.8 (12.7–20.5)[c]	Hoshiai and Tanimura, 1986 (Table 1-2)

Sep–Nov 1970	9830 (1000–22,000)	16.0 (1.0–32.0)[c]	21,000 (0–42,000)	5.5 (0–10.9)[c]	Hoshiai and Tanimura, 1986 (Table 1-2)
Dec 1970	0	0	0	0	Hoshiai and Tanimura, 1986 (Table 1-2)
Jul/Aug 1975	830 (0–1000)	0.8 (0–1.0)[c]	99,500 (52,000–131,000)	25.9 (13.5–34.0)[c]	Hoshiai and Tanimura, 1986 (Table 1-2)
Sep–Nov 1975	8830 (1000–23,000)	19.9 (2.3–50.7)[c]	100,600 (7000–215,000)	26.2 (1.8–55.9)[c]	Hoshiai and Tanimura, 1986 (Table 1-2)
Sep–Nov 1982	20,000 (4000–29,000)	37.5 (7.2–53.2)[c]	77,300 (2000–151,000)	20.1 (0.52–39.3)[c]	Hoshiai and Tanimura, 1986 (Table 1-2)
Dec 1982	0	0	0	0	Hoshiai and Tanimura, 1986 (Table 1-2)
Prydz Bay (FY-F)					
Mar 1994	330 (260–400)	1.2 (0.9–1.4)	44,110 (25,660–62,560)	11.5 (6.7–16.3)	Swadling, 1998 (Fig. 5.13)
Apr 1994	1010 (0–3610)	3.5 (0–12.6)	90,820 (64,420–151,330)	23.6 (16.7–39.3)	Swadling, 1998 (Fig. 5.13)
May 1994	3460 (1260–7820)	12.1 (4.4–27.4)	269,320 (162,650–324,230)	70.0 (42.3–84.3)	Swadling, 1998 (Fig. 5.13)
Jun 1994	820 (670–960)	2.9 (2.3–3.3)	107,180 (13,330–235,300)	27.9 (3.5–61.2)	Swadling, 1998 (Fig. 5.13)
Jul 1994	2500 (0–8060)	8.8 (0–28.2)	114,200 (29,800–206,470)	29.7 (7.7–53.7)	Swadling, 1998 (Fig. 5.13)
Aug 1994	820 (350–1310)	2.9 (1.2–4.6)	46,120 (2940–97,660)	12.0 (0.8–25.4)	Swadling, 1998 (Fig. 5.13)
Sep 1994	8760	30.6	157,580	41	Swadling, 1998 (Fig. 5.13)

(Continued)

Table 8 (Continued)

Location (ice type)/date	Copepodites		Nauplii		Reference
	ind m^{-2}	mg C m^{-2}	ind m^{-2}	mg C m^{-2}	
Oct 1994	68,250 (2480–219,250)	238.9 (8.7–767.4)	57,760 (2970–167,860)	15.0 (0.8–43.6)	Swadling, 1998 (Fig. 5.13)
Nov 1994	71,070 (0–353,050)	248.8 (0–1235.7)	5,340 (0–22,250)	1.4 (0.0–5.8)	Swadling, 1998 (Fig. 5.13)
Dec 1994	4490 (3700–55,100)	15.7 (13.0–19.3)	690 (340–950)	0.2 (0.1–0.2)	Swadling, 1998 (Fig. 5.13)
Bellingshausen Sea (SYI/MYI)					
Feb/Mar 1994	30,920 (18,540–59,340)	47.2 (26.8–80.0)	26,120 (5200–55,800)	6.8 (1.3–14.5)	Schnack-Schiel *et al.*, 1998 (Table 1, Fig. 2-3)
Amundsen Sea (SYI/MYI)					
Feb 1994	55,260 (0–12,0900)	94.9 (0–183.0)	25,560 (5200–59,400)	7.2 (1.3–15.4)	Schnack-Schiel *et al.*, 1998 (Table 1, Fig. 2-3)
Terra Nova (FY-F)					
Nov 1997	0–95,200	0–140	0–71,000	0–185.6	Costanzo *et al.*, 2002 (Fig. 2-3)

[a]Not separated into copepodites and nauplii.
[b]Blomass estimated according to Gradinger (1999).
[c]Carbon weights calculated after Swadling *et al.* (1997) and Swadling (unpublished, 2005).
Note: FYI: first year (pack) ice, SYI: second year (pack) ice, FY-F: first year fast ice, MYI: multiyear ice.

Table 9 Abundances (ind m^{-3}) of interstitial copepods in Antarctic sea ice, sorted by genera; mean (min-max)

Location (ice type)/date	*Stephos longipes* ind m^{-3}	*Harpacticus furcifer* ind m^{-3}	*Paralabidocera* spp. ind m^{-3}	*Drescheriella glacialis* ind m^{-3}	Reference
White Island (PL)			500,000[a]		Waghorn and Knox, 1988 (Figure 1)
Drescher Inlet (PL)	2500–75,000	0–200	200–26,000[b]	6300–825,000	Günther *et al.*, 1999 (Table 2)
Feb-95 Prydz Bay (FY-F)	1100		0–2300[b]		Swadling, 1998 (Figure 5.12)
Lutzow-Holm Bay (FY-F)			0–24,000[b]		Tanimura *et al.*, 1996
Weddell Sea (SI)	26,300 (0–185,000)			33,100 (0–154,000)	Schnack-Schiel *et al.*, 2001b (Table 1a-1b)
Weddell Sea (GW)	22,900 (0–194,000)			16,100 (0–267,000)	Schnack-Schiel *et al.*, 2001b (Table 1a-1b)
Weddell Sea (RG)	43,000 (0–240,000)			20,800 (0–111,000)	Schnack-Schiel *et al.*, 2001b (Table 1a-1b)
Weddell Sea (NI)	1500 (0–8000)			0	Schnack-Schiel *et al.*, 2001b (Table 1a-1b)

[a] *Paralabidocera grandispina.*
[b] *Paralabidocera antarctica.*
Note: FY-F: under first year ice, PL: platelet ice, SI: surface ice, GW: gap water, NI: new ice, RG: refrozen gap.

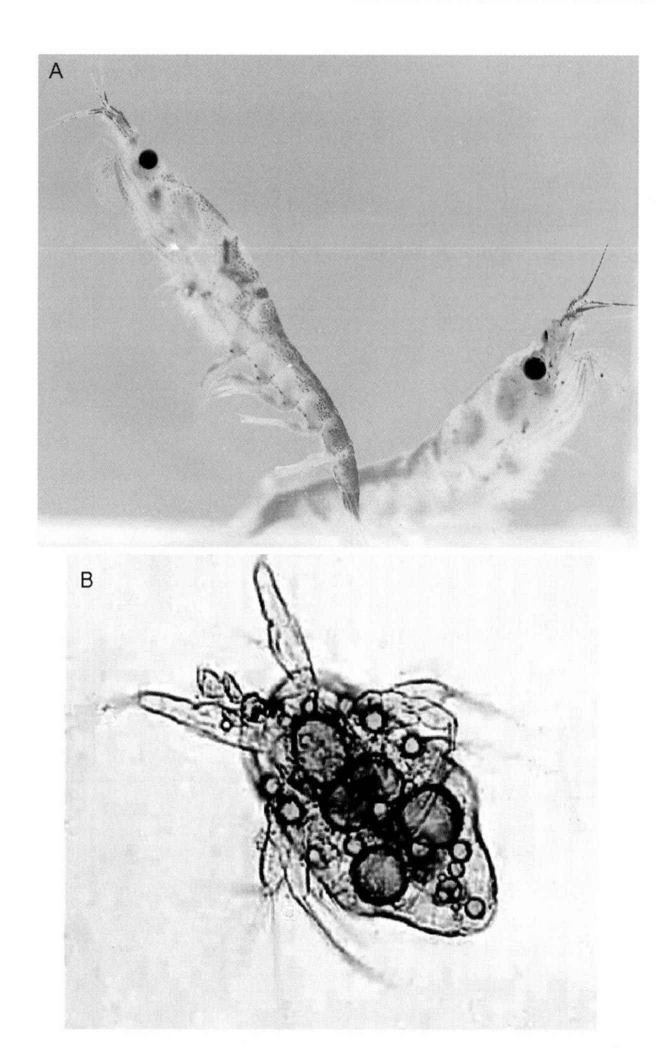

Figure 9 Examples of Antarctic sea ice crustacea. (A) Krill, *Euphausia superba* (maximal body length BL = 65 mm), is the key player in the ice covered Southern Ocean. (B) The copepod *Paralabidocera antarctica* is abundant in coastal ice of Antarctica. The nauplius, as shown here (BL = 0.4 mm) overwinters in the ice for approximately 5 mon. (Photos courtesy T. Hayashi, with permission, K. Swadling and M. Craven.)

Paralabidocera antarctica (Figure 9) is the dominant copepod in fast ice of eastern Antarctica (Hoshiai and Tanimura, 1986; Tanimura *et al.*, 1996; Swadling *et al.*, 1997a, 2000a). Nauplii can reach up to 900,000 individuals m^{-2} (corresponding to ~200 mg C m^{-2}) in autumn and winter, whereas young copepodid stages reach up to 200,000 individuals m^{-2} (equivalent to ~700 mg C m^{-2}). Older stages leave the sea ice matrix but still remain in close contact with the ice–water interface, where adults have been recorded up to 200,000 individuals m^{-3} (Tanimura *et al.*, 1984a). This species was observed in much lower numbers in the Weddell Sea; up to 26,000 individuals m^{-3} in platelet ice in Drescher Inlet (72°50′S, 19°33′W) (Günther *et al.*, 1999).

A congener of *Paralabidocera antarctica, Paralabidocera grandispina* Waghorn, occurred in very large numbers (>500,000 individuals m^{-3}) in platelet ice collected from a summer tide crack on the Ross Ice Shelf (Waghorn and Knox, 1988). This species was separated from *Paralabidocera antarctica* on the basis of an asymmetrical genital segment in the female, as well as a long terminal spine and well-developed spinose knob on the inner proximal border of the male fifth leg (Waghorn, 1979). *P. grandispina* has not been recorded from any other sea ice environment and almost nothing is known about its ecology apart from the work of Waghorn and Knox (1988). A third species of *Paralabidocera, Paralabidocera separabilis* Brodsky and Zvereva, has been described from open water in a fjord near the Bunger Oasis (66°15′S, 100°36′E), east Antarctica (Brodsky and Zvereva, 1976). As numerous adults were sampled from the surface waters in the summer, it is possible that the younger stages had been present in the ice cover during earlier months, although no further observations of this species have been made.

Other calanoid copepods have been sampled from the sea ice, although it is likely that they are occasional visitors that graze in the ice (see Section 6, later in this chapter) rather than undergo development there. The most frequently recorded species include *Ctenocalanus citer* Heron and Bowman, *Calanus propinquus* Brady, *Metridia gerlachei* Giesbrecht and *Microcalanus pygmaeus. C. citer* (originally cited as *Ctenocalanus vanus* Giesbrecht) occurred in numbers up to 1000 individuals m^{-2} (Hoshiai and Tanimura, 1986). *C. propinquus* is a biomass dominant species in the Southern Ocean. Its association with sea ice is equivocal and may have been overemphasised in the past. It was known that a proportion of the population remained trophically active in surface waters during winter (Burghart *et al.*, 1999), which led to the conclusion that *C. propinquus* might change its feeding mode to include sea ice algae. However, the species has generally been found to be more abundant in open water than in the ice and marginal ice zones and was shown to have a higher proportion of empty guts when it was found under sea ice (Burghart *et al.*, 1999). *M. gerlachei* was included in an early review by Bradford (1978) as a species common in the sub-ice fauna. However, this species is unlikely to be strongly associated with the ice at any time, as evidenced by the following: It can be abundant in the water column down to ~1,000 m (Boysen-Ennen and Piatkowski, 1988); it has been noted as being one of the least specialised species for high latitudes (e.g., Schnack-Schiel and Hagen, 1994; Atkinson,

1998); and its stable isotopic profile is not at all indicative of a sea ice algal diet (Schmidt *et al.*, 2003). For these reasons, *M. gerlachei* will not be considered further in this review. Finally, *Microcalanus pygmaeus* was recorded in small numbers under ice in the Weddell Sea (Menshinina and Melnikov, 1995), but there are no firm indications that this species has an association with the ice and it will not be considered further.

The most common and widely occurring harpacticoid copepod in Antarctic sea ice is *D. glacialis*. This species has a circum-Antarctic distribution and all stages are found in the ice (Dahms *et al.*, 1990; Swadling, 2001). It is common in both the Bellingshausen Sea and the Amundsen Sea, reaching high abundances (168 individuals L^{-1}) in the middle slush layer (Schnack-Schiel *et al.*, 1998; Thomas *et al.*, 1998) and up to 190,000 individuals m^{-2} in Prydz Bay fast ice (Swadling, 2001). *Harpacticus furcatus* Lang is probably the second most common harpacticoid species observed in sea ice, yet few measurements of abundance have been made: Costanzo *et al.* (2002) recorded it in numbers up to 60 individuals L^{-1} in sea ice of Terra Nova Bay. Other harpacticoids have been observed far less frequently. This is probably due, in part, to inadequate taxonomy, as individuals are often merely lumped together in the category "harpacticoids." One exception is *Tisbe prolata* Waghorn, described from platelet ice in a tide crack on the Ross Ice Shelf (Waghorn, 1979), which reached abundances up to 75,000 individuals m^{-3} (Waghorn and Knox, 1988). Other species listed in Table 2 have generally only been recorded as single observations.

To date, observations of cyclopoid copepods suggest that few are strongly associated with sea ice, yet there have been regular collections of three species from ice cores or at the ice–water interface, namely *O. similis, Oncaea curvata* Giesbrecht and *Pseudocyclopina belgicae* (Giesbrecht). *O. similis* and *O. curvata* are the dominant small copepods in Antarctic waters, where they reach abundances up to 5000 individuals m^{-3} (Kirkwood, 1993; Metz, 1995; Swadling *et al.*, 1997b). In fast ice cores from Prydz Bay, both *O. similis* and *O. curvata* have been recorded in numbers up to 600 individuals m^{-2}, with highest abundances occurring in May and June (Swadling, 1998). Up to 400 individuals m^{-2} of unidentified cyclopoid nauplii were recorded from the same region in August and November (Swadling, 1998). All three cyclopoid species were present in low numbers in platelet ice in Drescher Inlet: 0.2 l^{-1} for *O. curvata*, 0.3 l^{-1} for *O. similis* and 0.5 l^{-1} for *P. belgicae* (Günther *et al.*, 1999). At Ice Station Weddell 1 (ISW-1), located on drifting perennial ice at the western rim of the Weddell Sea Gyre (Figure 3), *O. similis* and *P. belgicae* were recorded in samples scraped from the bottom of the ice (Menshinina and Melnikov, 1995). Although quantitative estimates were not provided, the authors describe that a plankton net with mouth dimension 40 × 20 cm was scraped along the ice for 30 m. The mean numbers of cyclopoid copepods collected in this way from February to May were 23 and 14 animals per sample for *O. similis* and *P. belgicae*, respectively. *P. belgicae* has also been recorded from a summer tide crack on the Ross Ice Shelf near White Island, where its abundance in late November reached 225,000 individuals m^{-3} (Waghorn and Knox, 1988).

The status of *O. similis* as an ice-associated species requires some interpretation. Tucker and Burton (1988) suggested that *O. similis* is part of the "true" ice fauna, as both adults and juveniles were found at the ice-water interface, while only juveniles were present in the plankton in October and December. Gruzov *et al.* (1967) and El-Sayed (1971) also listed *O. similis* under the true ice fauna. Further, the species was found to be the most abundant prey item in the guts of fry of the ice fish *Pagothenia borchgrevinki* Boulenger, accounting for 30–70 % of identifiable gut contents (Hoshiai *et al.*, 1989). Clearly this is a species that is often found close to the ice if not generally living within the brine channel system. However, based on the relative abundances of planktonic *O. similis* versus those found in Antarctic ice cores, it seems more likely that this species is an occasional visitor (see Section 6). *O. similis* has also been sampled in Arctic sea ice and is believed to be cosmopolitan in the world's oceans (Gallienne and Robins, 2001), so it is demonstrably not an obligate ice dweller, but, given its numerical dominance in the Southern Ocean and its frequent occurrence in sea ice, we have elected to include this species in the remainder of this review. Similarly, the status of *O. curvata* is not clear. It was far less abundant in the stomach contents of the fry of *Pagothenia borchgrevinki* (~2%; Hoshiai *et al.*, 1989), but it is often recorded from ice cores. Therefore, we will continue to discuss this species also, with the caveat that it is likely to be only moderately ice associated, at best.

4.2.2. Amphipods

As in the Arctic, amphipods rarely if ever contribute to the Antarctic interstitial ice fauna, as their large size excludes them from the network of brine pockets and channels. Most of the taxa recorded near the sea ice are from the suborder Gammaridea, with only one species, *Hyperia macrocephala* Dana, coming from the suborder Hyperiidea. The gammarids are all primarily benthic species, while the hyperiid is pelagic, and none of the species observed can be regarded as obligate ice dwellers. In areas where they are common, amphipods are likely to have a considerable grazing impact on sea ice primary production.

There is very little information available on the abundance of amphipods associated with Antarctic sea ice (Table 10). Nevertheless, there are some species, in particular *Paramoera walkeri, Pontogeneia antarctica* Chevreux and *Orchomene* cf. *plebs*, that have been recorded in substantial numbers at the ice–water interface, where they probably feed on the underice algae.

Table 10 Abundances (ind m^{-2}) of sympagic amphipods and euphausiids, wet mass (mg WM m^{-2}) and estimated carbon content (mg C m^{-2}) in Antarctic ice-covered seas; mean values (min-max)

Location (ice type)	Date	Species	ind m^{-2}	mg WM m^{-2}	mg C m^{-2}	Reference
Amphipods						
Prydz Bay (MYI)						
	1987	*Paramoera walkeri*	1	8.0	1.1	Kirkwood, 1993
Signy Island (FYI)						
	June–Nov 1973	*Pontogeneia antarctica*	12,730	8300	465	Richardson and Whitaker, 1979
Mirny (FYI)						
	Dec 1965–Jan 1966	*Orchomenopsis* sp.	4500	130,000	7280	Gruzov *et al.*, 1967
	Dec 1965–Jan 1966	*Orchomenopsis* sp.	3040	86,000	4816	Andriashev, 1968
Euphausiids						
Prydz Bay (MYI)						
	1987	*Euphausia crystallorophias*	2.7	10.7	1.1	Kirkwood, 1993
Gerlache Strait (FYI)	Jul–Aug 1992	*Euphausia crystallorophias*	6 (0–83)			Nordhausen, 1994
Gerlache Strait (FYI)	Jul–Aug 1992	*Euphausia superba*	89 (0–810)			Nordhausen, 1994
Elephant Island	1990–1997	*Euphausia superba*		9600–100,401		Brierley *et al.*, 1999
Elephant Island	1978–1997	*Euphausia superba*	1.1–48.8	980–31,200		Siegel *et al.*, 1998
north Weddell Sea	1988–1989	*Euphausia superba*		30–2060		Siegel and Harm, 1996
Bellingshausen Sea	1993–1994	*Euphausia superba*		30–300		Siegel and Harm, 1996
Prydz Bay (FYI)	Jan–Mar 1996	*Euphausia superba*		0–909		Nicol *et al.*, 2000b
Weddell Sea	Jan–Feb 2001	*Euphausia superba*		61,600		Brierley *et al.*, 2002

Note: FYI: first year (pack) ice, MYI: multiyear ice.

Divers scraping nets along the bottom of the sea ice in nearshore waters of Prydz Bay collected *Paramoera walkeri* and *Orchomene* cf. *plebs* at the ice–water interface from April to December, and May to October, respectively (Tucker and Burton, 1988). At times *P. walkeri* and *Orchomene* cf. *plebs* were so numerous that they almost covered the undersurface of the ice (Tucker and Burton, 1988). Other amphipods recorded from the ice–water interface in Prydz Bay include *Orchomene* sp., *Hyperia macrocephala* and *Eusirus* cf. *fragilis* (Tucker and Burton, 1988). These latter taxa were represented by single catches of single individuals only. In Ellis Fjord, situated on the coast of Prydz Bay, *P. walkeri* was recorded at an abundance of 1 individual m^2 in a collection made by scraping a small quantitative net along the underside of the ice (Kirkwood, 1993); this equated to 1,058 μg C m^{-2} (K. Swadling, unpublished data). Elsewhere along east Antarctica, Rakusa-Suszczewski (1972) collected *P. walkeri* from the ice–water interface offshore from Molodeznaya Station (67°40′S, 45°51′E) in Enderby Land, between May and December, using a modified scraping/net device. These samples were not quantitative; however, numbers collected per sample ranged from <50 from May to July to >500 from August to December. While it can be common in the under-ice habitat of nearshore waters, *P. walkeri* was also found associated with shallow water macrophytes during most months of the year (Dhargalkar *et al.*, 1988).

Four species, including *P. walkeri, Pontogeneia antarctica, Cheirimedon femoratus* (Pfeffer) and *Lepidepecreum cingulatum* K.H. Barnard, were observed in fast ice near Signy Island (60°43′S, 045°36′W), although only the abundance of *Pontogeneia antarctica* was documented (Richardson and Whitaker, 1979). Up to 12,730 individuals m^{-3} (equivalent to 8.3 gWM m^{-3}) of this species were recorded at the ice–water interface, with juveniles accounting for 97% of the total. Richardson and Whitaker (1979) also noted that *C. femoratus* occurred sporadically in small numbers throughout the winter.

Other observations of ice-associated amphipods in Antarctica have been sparse. For example, De Broyer *et al.* (2001) noted the presence of three species in ice in the eastern Weddell Sea but gave no further information, while Günther *et al.* (1999) sampled *Oradarea* sp., *Probolisca* sp. and a member of the family Lysianassidae from Drescher Inlet (Weddell Sea). *Probolisca* sp. has been recorded in numbers up to 20,000 individuals m^{-3} in early December in the platelet ice of a tide crack on the Ross Ice Shelf (Waghorn and Knox, 1988), but it is unknown whether this is the same species listed by Günther *et al.* (1999). Finally Gruzov *et al.* (1967) recorded up to 4500 *Orchomenopsis* sp. m^{-2} (130 g WM m^{-2}) at the under-ice surface near Mirny Station, while Gruzov (1977) noted *Orchomene cavimanus* Pirlot in pockets at the undersurface of the ice throughout the winter.

4.2.3. Euphausiids

A great deal of information exists about *Euphausia superba* Dana (Figure 9) and it is not the purpose of this review to examine this species in depth. Compilations that cover research on *E. superba*, including Everson (2000), Mangel and Nicol (2000) and Kawaguchi and Nicol (2003), provide a wealth of background about this species. More recent information, some of which is already published (Hofmann *et al.*, 2004), comes from a Southern Ocean Global Ecosystem Dynamics program that had as one of its specific aims to study regional differences in the overwintering strategy of Antarctic krill in relation to the physical environment. Antarctic krill are found throughout the Southern Ocean, although they are not uniformly distributed (Table 10). Over 50% of the biomass has been estimated to be in the southwest Atlantic sector (Atkinson *et al.*, 2004).

One early study that examined the overwintering strategy of *E. superba* in the Weddell Sea used an ROV to film individuals from the undersurface of the ice down to 50 m depth between July and December (Marschall, 1988). These surveys indicated that *E. superba* occurred almost exclusively in close association with the ice, where krill density was estimated to be 40–400 m^{-2}. During the same time only one to two individuals were caught in the water column. These observations certainly suggested that sea ice was important for krill biology, and the first clear evidence of this came from a study that linked a recruitment index determined for *E. superba* with sea ice extent in the Elephant Island (61°01′S, 054°54′W) region of the Antarctic Peninsula (Siegel and Loeb, 1995). The study found that recruitment was highly variable from year to year, ranging from 2.2 to 511 individuals 1000 m^{-3}. One outcome of this work was that ice concentration and the duration of the ice cover during winter were found to influence krill recruitment. Early krill maturation was favoured in years when ice cover was extensive and of long duration, as these conditions promoted female gonadal development and spawning. This, in turn, resulted in a prolonged period of larval development and growth, which placed larvae at a more advantaged stage and in better physical condition to survive the winter. Converse ice conditions—low ice concentration and short duration of cover—favoured the development of extensive blooms of salps (Loeb *et al.*, 1997). These relationships between krill, salps and sea ice were further explored during a large-scale study off east Antarctica (Nicol *et al.*, 2000a). It was found that salps were located where sea ice extent was minimal, while the bulk of primary productivity and krill and their predators were found where sea ice extent was at its greatest. Abundances of *E. superba* in the region of study (80–150°E) reached 7 per 1000 m^3 (Nicol *et al.*, 2000b).

The Long-Term Ecological Research (LTER) program at Palmer Station (64°46′S, 064°03′W) has contributed significantly to our understanding of

Euphausia superba. By combining regular seasonal sampling transects with diver surveys and experimental studies, a wealth of information on the association of krill with sea ice has been gathered. This research has confirmed that recruitment success is highly variable and that greatest recruitment is associated with mean spring sea ice conditions and high total primary production in the ice (Quetin and Ross, 2003). However, these authors did not find an association with heavy winter sea ice, as had been indicated previously (Siegel and Loeb, 1995).

Juvenile Antarctic krill appear to be very strongly associated with ice, reaching concentrations up to 3000 individuals m^{-2} in under-ice crevices in spring (Daly and Macaulay, 1991). Schools of furciliae have been found with pieces of ice as small as 1.6 m diameter. In dense aggregations beneath bigger floes (10–20 m diameter), these schools reached 10^6 individuals m^{-3} (Hamner *et al.*, 1989).

Detailed studies of larval krill during winter in the Palmer LTER revealed that larvae and juveniles occupied both the sea ice and the water column, but that the size structure was different in each environment, with animals captured by divers being larger than those captured with nets (Frazer *et al.*, 2002). Comparisons of net-collected and diver-collected data from June and September 1991 and 1993 showed that estimates based on nets were often higher than those from diver observations, and that abundances reached 260 m^{-2}. Furciliae as young as stage 2 were observed by divers under the ice, although the bulk of the larval aggregations associated with the ice consisted of stages F5 and F6. It became clear from this study that larval krill require structurally complex habitats; upward facing surfaces and regions of over-rafting and/or erosion were a more attractive habitat than downward facing and smooth surfaces.

Daly (2004) examined the distribution of larval stages west of the Antarctic Peninsula and found that their abundance and depth distribution varied between locations and years. In autumn 2001, densities reached 5627 individuals m^{-2} at stations furthest offshore and young stages (calyptopsis 2 to furcilia 2) were most abundant. The younger stages were concentrated between 50 and 200 m, while older stages (furcilia 3 to furcilia 6) were found between 25 and 100 m. Closer inshore, older larvae were more common and often concentrated in a narrow band near the surface. Furcilia were observed feeding on the undersurface of the ice at all the dive sites visited in 2001 (Daly, 2004). Larval densities were generally lower at the same sites when revisited in 2002, and, while the dominant overwintering stage was again F6, the individuals were not strongly associated with the undersurface of the sea ice. Daly (2004) concluded that larvae of *E. superba* have the flexibility to exploit habitats other than sea ice and that larvae might trade the refuge potential of sea ice for the grazing potential of the water column in years when sea ice biota is not abundant or is inaccessible to the young krill. It was further suggested that a rigorous evaluation of the role of sea ice in the winter survival of larval krill is needed, and that this evaluation should include detailed information on the depth distribution of larvae relative to that of their food sources.

Euphausia crystallorophias often replaces *E. superba* as the dominant euphausiid in shelf waters (Pakhomov and Perissinotto, 1996) and high Antarctic ecosystems (Boysen-Ennen *et al.*, 1991). Its distribution does not extend northward beyond the Antarctic Divergence. Perhaps surprisingly, the association of *E. crystallorophias* with sea ice appears to be tenuous. The species was originally named *E. crystallorophias*, which means "exclusively a dweller beneath the roof of ice" because specimens were first known only from those caught through holes in the ice (Holt and Tattersall, 1906, in Brinton and Townsend, 1991). However, it appears that these "ice krill" have not been recorded directly in or very close to sea ice (Schnack-Schiel, 2002).

Gruzov *et al.* (1967) observed numerous whales feeding along the coastline near Mirny Station and attributed their presence to an abundance of *Euphausia crystallorophias*. However, these authors did not catch, or observe, any ice krill during their extensive diving work in the area; this lack of krill was attributed to inadequate sampling equipment. Later work by Foster (1987) consistently found more *E. crystallorophias* in hauls from 300 m than in hauls from 100 m in McMurdo Sound. Similarly, Kirkwood and Burton (1987) did not collect any of this species with a diver-operated push net that was scraped along the bottom of the ice, even though both calyptopes and furciliae 1 to 6 were sampled lower in the water column consistently by other nets. Nordhausen (1994) did measure up to 83 individuals m^{-2} west of the Antarctic Peninsula in winter 1992 and found that this species did not show a clear preference for any depth. In general, the distribution of *E. crystallorophias* appears to be highly patchy and related to the distribution of polynyas. It can reach high abundances but does not appear to attain the biomass reached by *E. superba*. Williams *et al.* (1986) recorded a mean density of 418 individuals 1000 m^{-3} (comprising 64% of total biomass) in Prydz Bay, within 60 km of the Antarctic coastline, and Brinton and Townsend (1991) recorded $\sim$100 larvae m^{-2} in January in Bransfield Strait. Pakhomov and Perrisinotto (1996) noted that the spatial distribution of larvae and adults is very patchy, and that sharp fluctuations in abundance and biomass of all larval stages are typical features of *E. crystallorophias*.

4.3. Polar comparison

When comparing the crustacean assemblages in Arctic and Antarctic sea ice, it becomes clear that while species richness is comparable at both poles, taxonomic dominance and biomass differ considerably. In the Arctic, harpacticoid and cyclopoid copepods are abundant in the interstitial spaces of the ice cover and calanoids are found at the ice–water interface in loose association with sea ice, whereas amphipods appear to be ubiquitous and high in biomass all over the region. In the Antarctic, calanoid and harpacticoid copepods inhabit the brine channels of the ice in considerable abundance, although three species contribute overwhelmingly to these numbers. Cyclopoid copepods, often numerically dominant in the water column, have a weaker association with the ice. Based on limited data, it appears that amphipods can be locally very abundant at the ice–water interface, whereas several stages of *E. superba* have developed a highly complex association with the ice.

Large differences between the polar ecosystems are apparent when comparing abundance and production values. Abundances are on average 10 times higher in the Antarctic. In both primary and crustacean production, Antarctic sea ice is also at least one order of magnitude higher (Tables 3 and 4, Table 8). The biometric values of Arctic ice copepods differ substantially from those of Antarctic copepods. *Halectinosoma* sp. and *T. furcata* males and females range from 0.6 to 0.8 mm in body length (Friedrich, 1997). These size dimensions form the basis for an often-used carbon conversion factor, which for ice copepods is 0.6 μg per specimen (Friedrich, 1997). In the Antarctic ice, calanoid copepods are larger (up to 2 mm in length) and mean carbon contents generally range between 0.8 and 5.0 μg per specimen (Swadling *et al.*, 1997b, 2004; K. Swadling, unpublished data).

Interestingly, in the Arctic, there is no crustacean that reaches a comparable biomass or fills a similar key role to that played by *S. superba* in the Southern Ocean; in the Arctic, marine vertebrates mainly graze on the lipid-rich pelagic *Calanus* species (Falk-Petersen *et al.*, 1990, 2000). Arctic euphausiids do not penetrate ice-covered waters because of their physiological requirements and feeding ecology. *Thysanoessa inermis* (Krøyer) and *T. raschii* (M. Sars) are important constituents of the macroplankton community in the southern fjords of Spitsbergen (Weslawski *et al.*, 2000), but only within the range of the Atlantic water inflow (Drobysheva, 1982; Søreide *et al.*, 2003). The distributions of *Thysanoessa* spp. rarely overlap with the extent of sea ice. As herbivores, they graze on detrital material on bottom sediments and *Phaeocystis pouchettii* (Hariot) rather than ice-associated diatoms (Falk-Petersen *et al.*, 2000). Also, the distribution of *Meganyctiphanes norvegica* (M. Sars) is at its northern extent limited by the 5°C isotherm (Einarsson, 1945). This species only occurs as an expatriate carnivore in Arctic shelf waters (Falk-Petersen *et al.*, 2000), but never in association with sea ice.

The question arises why the secondary production in Antarctic sea ice is 10 times higher than in the Arctic, where the ice is much thicker and older, and where about one-third extends over shallow water enabling exchange with the productive benthic systems. This is especially intriguing given that much of the very high biomass in Antarctic ice occurs in the narrow belt of coastal fast ice (i.e., in ice that also lies over shallow water). Antarctic pack ice, while supporting high amounts of primary production, lies over very deep water and is often depleted in the smaller ice-associated grazers, particularly in the Indian Ocean sector (K. Swadling, unpublished data). The biggest difference between Arctic ice and Antarctic fast ice might be the age. Antarctic fast ice generally breaks out and re-forms every year, so only a small proportion supports multiyear ice. Antarctic sea ice is generally more porous and, therefore, has more habitat space than Arctic ice (H. Eicken, personal communication).

It has been speculated that nitrogen resources are limited in Arctic sea ice, which in turn limits the production of ice algae (Demers *et al.*, 1989; Cota *et al.*, 1991; Gradinger *et al.*, 1999). On the other hand, ice crustaceans do not seem to control the standing stock of algae by grazing and there is no apparent food shortage during the year (Gradinger *et al.*, 1999; Poltermann, 2001). Because of the relatively low grazing pressure, up to 75% of the bottom ice carbon is exported to the pelagic and benthic realms (Michel *et al.*, 2002). Similar estimates are not yet available for much of the Southern Ocean, although for the Weddell Sea, Gradinger (1999) calculated that copepods could have accounted for between 12 and 50% of integrated potential ingestion. Grazing was estimated to range between 0 and 58 mg C m^{-2} d^{-1}, while primary production was between 0.5 and 240 mg C m^{-2} d^{-1}. From these estimates, it seems that sea ice meiofauna are not limited and their grazing probably does not control ice algal accumulation during summer, although selective feeding might affect algal diversity. It remains an open question why the production of ice crustaceans does not increase to exploit the available algal biomass.

5. REPRODUCTION AND LIFE HISTORY STRATEGIES

Compared to many of their temperate and tropical congeners, the life history strategies of polar crustaceans are characterised by longer life cycles, slower growth rates and fewer generations per year. The seasonal peak in primary production can be a strong driving force for the timing of reproduction and growth.

5.1. Arctic

For some harpacticoid and cyclopoid copepod species, ovigerous females have been collected from Arctic sea ice, indicating that these species actively reproduce in the ice. Many studies, however, did not assign the copepodites and nauplii to species, so the information about the reproductive cycle of these copepods is incomplete. Since amphipods brood their young, the developmental sequence is somewhat clearer. However, in general, a lack of comprehensive information on the life cycles of ice-associated species is apparent (Table 11).

5.1.1. Copepods

The life cycles of those harpacticoid and cyclopoid copepod species that are most commonly found in sea ice are not well understood, but knowledge can be inferred from the population structure *in situ* and from studies on related species that live in temperate waters.

For *H. superflexus*, only one cohort of juveniles has been observed in the fast ice near Narwhal Island (Montagna and Carey, 1978; Kern and Carey, 1983). To our knowledge, no ovigerous females of the genera *Harpacticus* and *Halectinosoma* have been reported from sea ice. Carey (1992), therefore, concluded that these two genera do "grow and develop but not actively reproduce in sea ice." Somewhat contradictory observations are that populations of *H. finmarchicum* sampled in (and beneath) the ice in the Laptev Sea and Greenland Sea comprised "various developmental stages" (Friedrich, 1997; Werner, 1997a), and the male-to-female ratio in the population of *Halectinosoma* sp. was 1:1.7 for all ice cores examined from the Barents, Laptev and Greenland Seas (Friedrich, 1997). Furthermore, the occurrence of high densities of harpacticoid nauplii in ice from the Arctic Ocean (Gradinger, 1999) and Baffin Bay (Nozais *et al.*, 2001) indicates that reproduction is continually performed in the ice environment. Continual reproduction has been observed for harpacticoid copepods both in sub-Arctic waters (Schizas and Shirley, 1996) and at lower latitudes (Hall and Bell, 1993; Steinarsdóttir *et al.*, 2003).

Harpacticoid copepods are among the most common emergers from sediments and, hence, couplers between the water column and sea floor, and they are believed to do so for mating and reproduction (Thistle, 2003). Since harpacticoids generally lack planktonic larval stages but develop without habitat transition in the benthos (Hicks and Coull, 1983; Dahms and Quian, 2004), it is not the nauplii that colonise new habitats but the egg-carrying females. Indeed, ovigerous females of *Harpacticus* sp. have been observed to be very good swimmers that colonise drifting objects such as seaweed

Table 11 Life cycle strategies of Arctic crustacea sampled within or beneath sea ice

Species	Generations yr^{-1}	Life span (yr)	Location	Reference
Arctic				
Copepods				
Cyclopoids				
Cyclopina sp.	3		Narwhal Island	Kern and Carey, 1983
Oithona similis		1	Baffin Island	McLaren, 1969
Calanoids				
Metridia longa	1		Svalbard	Grønvik and Hopkins, 1984
			Balsfjord/Norway	Grønvik and Hopkins, 1984
		1.5?	Resolute Passage	Conover and Huntley, 1991
		2?	W Spitsbergen	Diel, 1991
Acartia longiremis	4		W North Sea	Evans, 1977
Calanus glacialis	1		W Greenland	MacLellan, 1967
		1.5	Svalbard	Weslawski *et al.*, 1991
		2–3	Resolute Bay	Conover *et al.*, 1991
C. hyperboreus	1		Korsfjord/Norway	Mathews *et al.*, 1978
		2	Fram Strait	Smith, 1990; Diel, 1991
		2–3	Resolute Bay	Conover *et al.*, 1991
			W Barents Sea	Hirche, 1997
		3–4	Canada	Conover and Siferd, 1993
			Greenland Sea	Hirche, 1997
Microcalanus pygmaeus	1		Ellesmere Island	Cairns, 1967
Pseudocalanus minutus	1		Baffin Island	McLaren, 1969
Pseudocalanus acuspes		1	Svalbard	Weslawski *et al.*, 1991

Pseudocalanus sp.	2		N Labrador	Corkett and McLaren, 1978
			Ellesmere Island	Cairns, 1967
		1	Svalbard	Weslawski *et al.*, 1991
			Resolute Passage	Conover and Huntley, 1991
Amphipods				
Apherusa glacialis		2	Pond Inlet	Cross, 1982
			Franz Josef Land	Poltermann *et al.*, 2000
			Svalbard	Beuchel and Lønne, 2002
Gammarus wilkitzkii		6.5	Siberia	Gurjanova, 1951
			Franz Josef Land	Poltermann, 1997
			Svalbard	Beuchel and Lønne, 2002
G. setosus		3	Svalbard	Weslawski and Legezynka, 2002
Gammaracanthus loricatus		3	Arctic Ocean	Melnikov, 1997
Gammarellus homari		3–4	Svalbard	Weslawski and Legezynka, 2002
Ischyrocerus anguipes		1	Pond Inlet	Cross, 1982
			Svalbard	Weslawski and Legezynka, 2002
Anonyx nugax		>4	Svalbard	Weslawski and Legezynka, 2002
Onisimus nanseni		2.5	Franz Josef Land	Poltermann, 1997
			Svalbard	Arndt and Beuchel (2006)
O. glacialis		3.5	Svalbard	Arndt and Beuchel (2006)
O. litoralis		2.5	Svalbard	Weslawski *et al.*, 1991
O. edwarsi		2.5	Svalbard	Weslawski *et al.*, 1991
Weyprechtia pinguis		2.5	Svalbard	Weslawski *et al.*, 1991
Parathemisto libellula		2–3	N Pacific	Wing, 1976
			Svalbard	Weslawski *et al.*, 1991

clumps, where they release their eggs. These drifting platforms reportedly act as a means of long-distance dispersal for harpacticoid copepods (Ólafsson *et al.*, 2001). Ovigerous females and copepodid stages of *Harpacticus* sp. were present all year round in the intertidal benthos near Auke Bay (southeast coast of Alaska), although densities peaked between November and March (Schizas and Shirley, 1996). Given the strong seasonal signal in Arctic waters, it is likely that spawning of *Harpacticus* sp. and *Halectinosoma* spp. is timed with the onset of primary production in early spring.

In contrast to the aforementioned species, another harpacticoid, *Tisbe furcata*, was always present in sea ice as various stages including ovigerous females (Grainger *et al.*, 1985; Grainger and Hsiao, 1990; Melnikov, 1997). This might indicate that *T. furcata* spawns and completes its life cycle in the ice. *T. furcata* generally is the most abundant harpacticoid copepod in (and beneath) sea ice with peak densities in March (Grainger *et al.*, 1985). In contrast to what is known about *Harpacticus* and *Halectinosoma*, it is the numerical dominance of *T. furcata* in sea ice that helps to elucidate the population dynamics and life cycle of this species. In temperate waters, *T. furcata* is iteroparous with 7–12 broods yr^{-1}, each producing ~50 eggs. The lifetime potential fecundity of this rather short-lived species (40–50 d) is very high: >500 eggs are produced per female life span and the generation cycle is only 20 d (Johnson and Olson, 1948). In the ice cores taken from the annual ice pack in the Laptev Sea, stage CII were numerically dominant followed by stages CIII and CI (Friedrich, 1997).

Gravid females of the benthic harpacticoid species *Dactylopodia signata* and *D. vulgaris*, which are occasionally very abundant in sea ice (Kern and Carey, 1983; Grainger and Hsiao, 1990), appear in the ice. However, there is nothing else known about their biology.

It is generally accepted that harpacticoids have an extended to continuous breeding season. Whether the offspring are released within the ice or the eggs, nauplii and copepodid stages are advected or actively migrate to the ice to seek shelter and/or food is not known. Given the good swimming (and dispersal) capacities of adult harpacticoids and the rich food supply in sea ice throughout the year (see Section 6), it makes the ice a likely spawning and/or nursery ground. Further field observations will help to elucidate the role of sea ice in the reproduction of the different harpacticoid species.

There is less doubt that cyclopoid copepods use sea ice as a platform for reproduction and development. The large *Cyclopina* populations found in sea ice generally comprise ovigerous females (Grainger *et al.*, 1985; Carey, 1992; Melnikov, 1997; Werner *et al.*, 2002a) and up to three cohorts (Carey and Montagna, 1982; Kern and Carey, 1983). The generation cycle is 31 d, indicating an iteroparous multivoltine brood rhythm.

Other copepod species that have been sampled beneath sea ice but are considered to be pelagic follow different life strategies. All calanoid species, except *Pseudocalanus* sp., sampled from the ice–water interface have perennial life histories (Table 11). The life spans of *Calanus* spp. are prolonged in higher latitudes. *C. glacialis* sampled from sub-Arctic waters lives to 1.5 yr (Weslawski *et al.*, 1991), but those specimens sampled from north of Canada live for 2–3 yr (Conover *et al.*, 1991). *C. hyperboreus*, which is restricted to Arctic waters, may live up to 4 yr (Cairns, 1967; Conover and Siferd, 1993; Hirche, 1997). Their slower growth and development in colder water is associated with ontogenetic resting stages that spend the winter diapausing in deeper water (Smith, 1990). *C. glacialis* stages CIII and CIV descend to deep waters in autumn, where maturation takes place. The development of the gonads is fueled by ambient food, whereas during the second winter, the mature females produce their eggs by utilising stored lipid depots (Hirche and Kattner, 1993). Pre-bloom spawning by *C. hyperboreus* in October to March is also fuelled by lipid stores (Hirche, 1997). In contrast to *C. glacialis*, the large eggs of *C. hyperboreus* are positively buoyant and develop while they rise to the surface (Conover and Huntley, 1991). The arrival of nauplii in the surface waters and underneath sea ice is timed with the springtime peak of primary producers, which has led to the suggestion that both *C. glacialis* and *C. hyperboreus* use the underside of sea ice as a nursery ground (Conover and Huntley, 1991). The nauplii of *C. hyperboreus* develop to stage CIII and ontogenetically descend below 500 m in mid-August. The second winter is spent as stage CIV and the third as stage CV. Maturation and reproduction, therefore, occur after 3 yr (Hirche, 1997).

Under temperature-controlled conditions, a single *C. hyperboreus* female may spawn up to seven times within 2 mo at 0°C, producing >1000 eggs (Hirche and Niehoff, 1996). *C. glacialis* females produce egg clutches at 3- to 12-d intervals in May and at 3- to 4-d intervals in June under the ice in Hudson Bay (Tourangeau and Runge, 1991). *Calanus* spp. and *Acartia* spp. are both free spawners, in contrast to *Pseudocalanus* spp., which carry an egg sac. The fecundity of egg-carrying species is, on average, lower than that of free spawners. For example, in *Pseudocalanus* sp., an average of 180 eggs is produced per female life span, whereas in *C. glacialis*, the average is 340 (maximum of 1270), in *C. hyperboreus* 450 (maximum of 3800) and in *Acartia* spp. 1280 eggs (see Table 45 in Mauchline, 1998).

Diapause has not been observed in other calanoid copepods sampled from beneath the ice, and the shorter living species tend to employ different strategies. *Pseudocalanus acuspes* overwinters as stage CIII or CIV (Conover and Huntley, 1991; Fortier *et al.*, 1995), and then maturation and gonad development occur between January and March and are fueled by sympagic production (Conover and Siferd, 1993) and lipid depots (Norrbin *et al.*, 1990; Fortier *et al.*, 1995). In contrast, *A. longiremis* and *M. longa* continually feed during winter so that growth and reproduction is supported by ambient food (Hopkins *et al.*, 1984; Norrbin *et al.*, 1990). While *Pseudocalanus* sp. spawns before the spring bloom, *A. longiremis* and *M. longa* rely on constant food supply for reproduction and, therefore, spawn later in spring.

5.1.2. Amphipods

Ice-inhabiting amphipods in the Arctic have perennial life cycles and presumably spawn once a year when sexually mature. *A. glacialis* has the shortest life span (Cross, 1982; Poltermann *et al.*, 2000; Beuchel and Lønne, 2002), reaching up to 2 yr, but its fecundity is extraordinary: >550 eggs develop for 6 mo (Melnikov, 1997) in a female's brood pouch during reproduction in the second year (Poltermann *et al.*, 2000). The offspring are released between March and August (Klekowski and Weslawski, 1991; Poltermann *et al.*, 2000). The high fecundity and relatively short life span characterise *A. glacialis* as an *r*-strategist.

The reproductive cycles of the *Onisimus* congeners are temporally offset in that *O. glacialis* spawns a few months earlier in the year but reaches sexual maturity 1 yr later and grows 1 yr older (3.5 yr) than *O. nanseni* (Arndt and Beuchel, in press). Both species are believed to be univoltine iteroparous and produce a maximum of 100 eggs per female life span (Arndt, 2004). *Gammarus wilkitzkii* reaches 6 yr in life span and is the longest living species among ice amphipods and probably within the entire ice ecosystem in the Arctic (Gurjanova, 1951; Poltermann *et al.*, 2000; Beuchel and Lønne, 2002). This species matures sexually after 2 yr and reproduces once a year (Poltermann *et al.*, 2000; Beuchel and Lønne, 2002). Depending on the age of the female, between 90 and 250 eggs are produced annually (Steele and Steele, 1975), resulting in fecundity rates of 300–500 eggs per female life span (Poltermann *et al.*, 2000). Embryo development takes 6–7 mo (Barnard, 1959). The timing of the release of the annual brood appears to not be directly linked to the vegetation period, since newly hatched individuals have been observed in April and May (Melnikov, 1997; Poltermann, 1997), in July (Steele and Steele, 1975) and in September (Werner, 1997a). Given its longevity and low reproductive potential, *G. wilkitzkii* has been described as a *K*-strategist (Poltermann *et al.*, 2000).

Apart from the relatively short-living *Ischyrocerus anguipes* (Weslawski and Legezynska, 2002), other amphipod species found in loose association with sea ice reach sexual maturity relatively late and have perennial life cycles. Examples are the pelagic hyperiid *Parathemisto libellula* (life span: 2–3 yr) and the benthic gammarids *G. setosus* (3 yr), *Gammaracanthus loricatus* (3 yr), *Gamarellus homari* (Fabricius) (3–4 yr), *Weyprechtia pinguis* (2.5 yr), and *Onisimus edwarsi* and *O. litoralis* (2.5 yr) (Wing, 1976; Melnikov, 1997; Weslawski and Legezynska, 2002). The use of sea ice as a nursery ground, however, has only been suggested for *Gammaracanthus loricatus* (Melnikov,

1997). *Gamarellus homari* performs ontogenetic migrations to secure the release of the hatchlings in shallow waters near shore, where they may become part of the sympagic ecosystem (Steele, 1972). Benthic amphipod species that live in association with sea ice are potentially able to complete their perennial life cycle in the sympagic, particularly if they inhabit multiyear ice. In all the aforementioned species, the eggs incubate during winter and hatch in spring/early summer at peak times of primary production in the euphotic zone (e.g., Carey, 1985; Weslawski *et al.*, 1991; Weslawski and Legezynska, 2002).

5.2. Antarctic species

There is sufficient information available to clarify at least part of the life histories of the three dominant ice-associated copepods. Further information exists on the life cycles of those primarily pelagic copepods that associate with ice. Far less is known about the life cycles of the amphipods, although some general trends are apparent. The euphausiids undergo complex life cycles that span several years (Table 12).

5.2.1. Copepods

Although several harpacticoid copepods have been recorded from the sea ice (see Table 2), in most cases these were single observations of a few individuals. Therefore, there is presently insufficient information available to comment on the life cycles or strength of association with the ice for these species. *Drescheriella glacialis* is the harpacticoid copepod that exhibits the strongest association with sea ice, and it is the one species where the life cycle has been reasonably well elucidated. What is known is summarised briefly here. Most developmental stages of *D. glacialis*, including nauplii, copepodids and females with egg sacs, have been found concurrently, both in the ice in Prydz Bay and in the Weddell Sea (Dahms *et al.*, 1990; Swadling *et al.*, 2000a; Swadling, 2001; Schnack-Schiel *et al.*, 2001a,b), suggesting that this species is capable of reproducing year round. Bergmans *et al.* (1991) maintained cohorts in the laboratory and discovered that *D. glacialis* has two to three generations per year and is capable of reproducing in winter. As a result of its relatively fast growth, shorter longevity, high fecundity and small size, this species was described as an *r*-strategist by Bergmans *et al.* (1991). Such traits presumably facilitate *D. glacialis* to colonise patches of sea ice that are unpredictable in formation and breakout. Observations suggest that nauplii of *D. glacialis* are not strong swimmers (Dahms *et al.*, 1990), and it is likely that dispersal in the ice occurs via the copepodids or the egg-carrying adults. The presence of nauplii in winter ice (Swadling, 2001)

Table 12 Life cycle strategies of Antarctic crustacea sampled within or beneath sea ice

Species	Generations yr^{-1}	Life span (yr)	Location	Reference
Antarctic				
Copepods				
Harpacticoids				
Drescheriella glacialis	>3	<1	Weddell Sea	Dahms *et al.*, 1990; Swadling, 2001
			Prydz Bay	
			Weddell Sea	Bergmans *et al.*, 1991
Tisbe prolata	1?	1?	McMurdo Sound	Waghorn and Knox, 1988
Microsetella norvegica	5–6	<1	Inland Sea of Japan	Uye *et al.*, 2002
Cyclopoids				
Oncaea curvata	>1	<1	Weddell Sea	Metz, 1995
Oithona similis		1	Baffin Island	McLaren, 1969
Pseudocyclopina belgicae	1?	1?	McMurdo Sound	Waghorn and Knox, 1988
Calanoids				
Stephos longipes	>1?	1	Weddell Sea	Schnack-Schiel *et al.*, 1995
			Weddell Sea	Kurbjeweit *et al.*, 1993
Paralabidocera antarctica	1	1	Lutzow-Holm Bay	Tanimura *et al.*, 1996; Swadling *et al.*, 2004
			Prydz Bay	
Paralabidocera grandispii	1	1	McMurdo Sound	Waghorn and Knox, 1988
Metridia gerlachei	1	1	Weddell Sea	Hagen and Schnack-Schiel, 1996
			Weddell Sea	Niehoff *et al.*, 2002
Ctenocalanus citer	1	1	Weddell Sea	Mizdalski and Schnack-Schiel, 1994
			Weddell Sea	Niehoff *et al.*, 2002
Amphipods				
Paramoera walkeri	1	>2	Molodeznaya	Rakusa-Suszczewski, 1972
Pontogenia antarctica	1	>2	Signy Island	Richardson and Whitaker, 1979
Orchomene plebs	1	>2	McMurdo Sound	Rakusa-Suszczewski, 1982
Euphausiids				
Euphausia superba	1	4–6	Antarctic Peninsula	Siegel, 1987
Euphausia crystallorophia	1	M: 4, F: 5	Antarctic Peninsula	Siegel, 1987

Note: M: males, F: females.

indicates that *D. glacialis* is an opportunistic species, with a reproductive strategy designed to take advantage of patches that open up in the sea ice habitat.

Other, less detailed, observations of harpacticoids provide some insights into their reproductive biology. The population of *Tisbe prolata* in the Ross Sea consisted primarily of females in summer (Waghorn and Knox, 1988), of which a large proportion was ovigerous. It was suggested that this species overwintered as late juvenile or adult stages and used the summer food maximum to induce gonad maturity. *Microsetella norvegica* (Boeck), in the Inland Sea of Japan, has larval stages that are confined to the warmer months. Adults reach a peak in early October and overwinter and remain in the plankton until the reappearance of nauplii in May. The occurrence of ovigerous females is confined to the period from August to October (Uye *et al.*, 2002). It is not known to what extent this species associates with ice during its reproductive cycle. Finally, Costanzo *et al.* (2002) observed *Harpacticus furcifer* in sea ice in early spring in Terra Nova Bay. Only copepodids and adults were observed and the presence of high numbers of exuviae suggested that the population was in a period of growth.

Of the calanoid copepods, the life cycles of *Paralabidocera antarctica* and *Stephos longipes* are quite well understood. *Paralabidocera antarctica* hatches from eggs in the sediments in late summer, has a brief pelagic phase during stage NI and undergoes most of its development (NII to CIII) within the brine channel system of the sea ice cover (Tanimura *et al.*, 1996; Swadling *et al.*, 2004). It is a fecund species, producing up to 69 eggs per female per day over its short (<1-mo) reproductive span (Swadling *et al.*, 2004). The life cycle of *Paralabidocera antarctica* is approximately 1 yr long and includes a long overwintering period (>5 mo) in the sea ice by the naupliar stages, which is an unusual strategy in calanoids, as generally the overwintering stage is a late-stage copepodid. The shift from the sea ice to the pelagic habitat occurs at stage CIII, although adults remain in close proximity to the under-ice surface, where they graze on ice algae (see Section 6). Tanimura *et al.* (2002) found stages CIII to adults in large numbers just below the ice in November, and they suggested that sea ice is essential for the development of *Paralabidocera antarctica*. However, in several marine-derived populations of *Paralabidocera antarctica* that are currently isolated in saline lakes in the Vestfold Hills, this species undergoes complete development in the pelagic and seldom, if ever, enters the lake ice cover (Swadling *et al.*, 2004), indicating that it is not an obligate ice dweller. An intriguing record of *Paralabidocera antarctica* was the presence of adults under Weddell Sea ice in March and April (Menshinina and Melnikov, 1995). Whether this population was taking longer than 1 yr to complete its life cycle or whether there was a shift in the timing of reproduction and development is unknown. Finally, a congener of *Paralabidocera antarctica, P. grandispina*, was present as young copepodids in early December and had developed to adults by late January (Waghorn and Knox, 1988), suggesting this species might follow a broadly similar reproductive cycle to that deduced for *Paralabidocera antarctica.*

A life cycle was constructed for *Stephos longipes* by Schnack-Schiel *et al.* (1995) from collections in the Weddell Sea. Briefly, rapid development from CI to CIII occurred during summer in the upper 50 m of the water column. In autumn, CIV were found in mid-water, whereas nauplii inhabited the sea ice. Part of the Weddell Sea population overwintered as CIV and CV in deeper water, while a younger generation overwintered in the sea ice. Reproduction appeared to take place over an extended period from late winter to autumn (Schnack-Schiel *et al.*, 1995), and egg production is up to one egg per female per day (Schnack-Schiel, 2001).

As discussed earlier, other calanoid copepods have far weaker associations with sea ice and it is likely that they do not use the ice at all as a breeding ground. Therefore, only brief mention will be made of what is known about the life cycles of *Calanus citer* and *C. propinquus. C. citer* undergoes an ontogenetic migration to deeper waters and possibly overwinters as mid-stage copepodids. The species first reproduces in springtime, before the onset of the phytoplankton bloom (Fransz, 1988; Schnack-Schiel and Mizdalski, 1994; Niehoff *et al.*, 2002); however, its main period of reproduction occurs during times of high phytoplankton biomass (Swadling, 1998; Niehoff *et al.*, 2002). The egg production rate is around one egg per female per day (Schnack-Schiel, 2001). Reproductive activity extends well into autumn (Niehoff *et al.*, 2002), as high proportions of mature females are found in summer and autumn (Schnack-Schiel and Mizdalski, 1994).

Calanus propinquus has a life cycle that is different to other *Calanus* species from both poles (Hagen and Schnack-Schiel, 1996). While *Calanus* spp. generally undergo a substantial diapause at depth, *C. propinquus* shows a less pronounced seasonal vertical migration (Schnack-Schiel *et al.*, 1991). The proposed model for the life cycle of this species is that while a portion of the population overwinters at depth as stages CIV and CV, a second group remains active in surface waters, possibly feeding on sea ice algae (but see Section 6) (Bathmann *et al.*, 1993; Delgado *et al.*, 1998). Gonad maturation starts around August, gravid females are present in mid-October and egg spawning occurs in November and December. Egg production rate is highly dependent on food supply and can reach up to 112 eggs per female per day (summarised in Schnack-Schiel, 2001), and these eggs take up to 5 d to hatch (Ward and Shreeve, 1998).

Of the cyclopoid copepods, only *Pseudocyclopina belgicae* appears to have any strong association with sea ice during its reproductive cycle. At the ice–water interface in the perennial ice of ISW-1, *P. belgicae* was represented by stages CIV to CVI from late February to late May (Menshinina and

Melnikov, 1995), with the younger stages dominating over that period. In the summer tide crack on the Ross Ice Shelf, the dominant stages were CI and CII in late November, but the population structure changed to a dominance of CIII–CV in early December (Waghorn and Knox, 1988). Adults were only represented by the presence of two males. From mid-December, *P. belgicae* was then absent from both the tide crack and the nearby water column. Based on these two data sets, it seems possible that *P. belgicae* overwinters as late stage copepodids, but it is uncertain exactly when breeding takes place.

There are several similarities in the life history strategies of *Oithona curvata* and *O. similis*, including prolonged periods of reproduction, low rates of egg production throughout the year, short development times for the nauplii, the lack of clear cohorts over a year and copepodite development times of 14–28 d per stage (Fransz, 1988; Metz, 1995, 1996; Tanimura *et al.*, 1997). From these studies, there is no evidence that *O. curvata* or *O. similis* uses sea ice as a platform for reproduction.

5.2.2. Amphipods

The sea ice-associated amphipod that has received the most attention in the literature is *Paramoera walkeri*, which was studied in some detail in the early 1970s by Rakusa-Suszczewski and colleagues. Ovigerous females of *P. walkeri* are first observed in early winter, and development of the embryos occurs for ~4.5 mo before hatching begins in October. Hatching continues throughout November and December, and brooding females can still be seen in December (Tucker and Burton, 1988). Young possibly remain in the brood pouch for up to 1 mo (Sagar, 1980). For its size, *P. walkeri* produces a large number of eggs: up to 200 eggs per clutch per female (Rakusa-Suszczewski, 1972). This species has slow growth and lives for at least 3 yr (Sagar, 1980).

Orchomene cf. *plebs* also broods its young over winter and no newly released juveniles are observed after October (Rakusa-Suszczewski, 1982; Tucker and Burton, 1988). *Orchomene* cf. *plebs* spawns in shallow waters, the eggs develop during winter and hatching occurs in spring. Up to 47 eggs can be carried in a single clutch. In summer, juveniles occur on the shelf, whereas adults can be as deep as 760 m (Rakusa-Suszczewski, 1982).

Pontogeneia antarctica has extended embryonic development, lasting 4.5–5.0 mo (Richardson and Whitaker, 1979). Ovigerous females are found throughout most of the year, although most hatchlings are released in late August. Fecundity is ~65 young per female (Richardson and Whitaker, 1979). The young are positively phototrophic and congregate at the ice just under tide cracks, where illumination is strongest. At night, they disperse across the bottom of the sea ice, while larger animals move out into the main body of the water column.

5.2.3. Euphausiids

The life cycles of most euphausiids are complex, occurring over several years (Siegel, 2000). In *Euphausia superba*, the main spawning period occurs for ~2.5 mo between late December and March, with females laying up to 10,000 eggs at one time (Ross and Quetin, 1986; Nicol, 1995). Embryos are released in the surface waters and sink to deep waters, where they hatch into Nauplius stage I (NI). Laboratory experiments have shown that the embryos can sink up to 195 m d^{-1} in the first 24 hr, and up to 95 m d^{-1} after 36 hr (Ross and Quetin, 1982). Based on these estimates, it was calculated that krill should hatch between 700 and 1300 m; however, deep net tows in the Scotia Sea found abundant krill eggs in the 1000–2000 m layer, with few eggs above 1000 m or below 2000 m (Hempel and Hempel, 1986). During development from the NI through NII and Metanauplius, the larvae ascend to the surface; this phenomenon was originally described by Marr (1962) as an ontogenetic ascent. The nauplii do not feed during this phase of their development and the ascent is fuelled by rich lipid stores laid down in the egg (Nicol, 1995).

Metanauplii metamorphose to Calyptopsis stage 1 (C1) at the surface and C1 is the first feeding stage (Ross and Quetin, 1989). Phytoplankton is a critical resource for C1 and they are usually found in surface waters between January and March, when phytoplankton are most abundant. However, if insufficient food is available, the C1 can reach a point of no return (~10–14 d), beyond which starved larvae cannot recover even if they are fed (Ross and Quetin, 1989). The larvae continue to develop through the winter and by early spring have developed to an advanced furcilia (F5–6) stage. These late-stage furciliae are routinely captured at or very near the surface and, as mentioned previously, often form dense schools just below ice floes (Hamner *et al.*, 1989). As furciliae develop into juvenile krill, they retain their association with the sea ice as they move into their second winter (Daly and Macauly, 1991). Females begin to mature by their third summer, and all are mature after 3 yr, while males mature 1 yr later than females (Siegel and Loeb, 1994).

Euphausia crystallorophias spawning occurs from October to December and may extend into January. This early spawning might reduce food competition between their larvae and those of *Euphausia superba*. Calyptopsis I appears 34 d after spawning and furcilia stage I 55 d later. Larval development from the egg to furcilia stage VI occurs during winter and takes ~240 d (Ikeda, 1986; Kirkwood, 1996). Juveniles are found in the second summer, males mature in April and females in July. All have reached maturity by their third summer. It appears that *E. crystallorophias* exhibits broadly similar spawning and developmental times at widely separated geographical locations: Gravid females collected from Enderby Land pack ice (Ikeda, 1986) spawned around the same time as those in the western Bransfield Strait (Brinton and Townsend, 1991). Further, in Admiralty Bay in the South Shetland Islands, juveniles were observed in summer, males began maturing in April and females in July (Stepnik, 1982).

The eggs of *Euphausia crystallorophias* are neutrally buoyant, a result of the presence of a large perivitelline space (Harrington and Thomas, 1987), a feature that is atypical of euphausiids (Ikeda, 1986). This neutral buoyancy might enable *E. crystallorophias* to spawn on the continental shelf. In contrast, *E. superba* has negatively buoyant eggs and spawns offshore in deep water, as discussed earlier. High wax ester concentrations have been reported for *E. crystallorophias* and only a small proportion of body lipid appears to be lost during spawning season, compared with up to 60% lost by *E. superba*.

The beginning of spawning in *Euphausia crystallorophias* shows a good correlation with the time of ice breakout, and it appears that the best conditions for spawning and growth coincide with regions where stationary polynyas are frequently observed (Pakhomov and Perrisinotto, 1996). It is possible that this species is attracted to enhanced productivity at ice edges.

5.3. Polar comparison

Broadly speaking, the life history strategies of crustaceans are similar at both poles. Their life cycles are longer than those of congeners in temperate and tropical waters, they often enter a period of arrested development and there is usually only one and rarely more than three generations per year (Tables 11 and 12). Among the copepods, there are some differences between the calanoids and the harpacticoids and cyclopoids, the main one being that the calanoids almost all have only one generation per year, while the smaller harpacticoids and cyclopoids may have three or four. Diapause is common in the large Arctic *Calanus* species, whereas in the Antarctic few species appear to undergo a true diapause at depth (Niehoff *et al.*, 2002). One important difference between the calanoid copepods found at the poles is that, in the Arctic, no calanoids appear to tie their reproduction tightly to the ice habitat.

Most harpacticoids and cyclopoids seem to have a multivoltine brood rhythm (Tables 11 and 12), and their life cycles are decoupled from the cycle of algal growth and development. Many of these smaller copepods carry egg sacs, which is a strategy that enables them to take advantage of grazing patches that develop in the ice. For instance, decreased snow cover can lead to an increase in light penetration, which then creates conditions favourable for algal growth. The copepods can colonise these patches and, by releasing eggs, undertake effective population expansion. The presence of ovigerous females in sea ice was observed for *Tisbe furcata, Dactylopodia* spp., *Cyclopina* sp. (Arctic), and *Tisbe prolata* and *Drescheriella glacialis* (Antarctic). It appears to be a rule for both hemispheres that the adult phase of harpacticoids and cyclopoids, rather than the reproductive stages (eggs, nauplii, copepodids), colonises the ice and transfers the recruits to the ice habitat (e.g., Ólafsson *et al.*, 2001; Dahms and Quian, 2004); conversely, for calanoids in the Antarctic, it is the egg and naupliar stages that are the colonisers. The major difference between the Arctic and the Antarctic sea ice ecosystems is that a large pool of perennial ice in the Arctic serves as a seeding platform for ice organisms to colonise new ice through lateral dispersion (Melnikov, 1997; Arndt and Pavlova, 2005), whereas in the Antarctic 80% of the sea ice is annual and needs to be inhabited anew every year. Antarctic species, therefore, need a mechanism whereby they can either survive for short periods in the open water or can recolonise the ice from the benthos. As much of the Antarctic ice is over very deep water, the secondary strategy will only be effective in shallow water habitats.

The life history strategies of amphipods appear to be similar at both poles. In general they live for at least 2 yr, carry their broods for 5 (Antarctic) to 7 mo (Arctic) and rarely, if ever, enter the ice for the purposes of reproduction. Many species time the release of their broods to coincide with the spring-summer pulse of algal production, although there is often extended breeding over several months and the coupling between their life cycles and the spring bloom is not as strong as that seen for many copepods. The lifetime fecundity appears to be higher in the Arctic (~500 eggs female^{-1}), but knowledge is insufficient to allow a thorough comparison.

The life cycles of euphausiids are more complex than those of copepods and amphipods, in that they require longer times to reach maturity, their fecundity is substantially higher and their greater mobility means that their life cycles are completed at a different scale. While it is clear that *Euphausia superba* has a strong association with ice during part of its development, the fundamental associations between this species and sea ice still require clarification. Finally, it appears that the life cycle of *E. crystallorophias* is primarily coupled to the development of ice edges (i.e., those associated with ice breakout or with polynyas).

6. DIET

Along with ice algae, particulate organic matter incorporated during ice formation and detritus-lumps that enter crevices and holes with the under-ice currents and small-scale turbulence provide food throughout the year. In the central Arctic ice pack, average values for particulate organic carbon range between 0.09 mg L^{-1} for young ice and 0.4 mg l^{-1} for perennial ice. Compared to the underlying water column the concentration of organic matter can be up to five times higher in sea ice (Melnikov, 1997; Swadling, 1998). Previous studies have suggested that pelagic forms might rely substantially upon ice-derived material that is released during ice melt (Conover *et al.*, 1986, 1988), while sympagic forms loosen their preferred food items from the ice proper (Grainger and Hsiao, 1990). Sea ice crustaceans feed on a wide range of food items. Most taxa show a high degree of plasticity in their diets and exhibit both ontogenetic and seasonal shifts.

6.1. Arctic

There is only a single study that has investigated the feeding ecology of Arctic ice-associated copepods using classical gut content analysis and *in situ* observations (Grainger and Hsiao, 1990), so there is still uncertainty about their diets. In contrast, our understanding for ice amphipods is more advanced; a variety of methods including lipid analysis (but not stable isotope analysis) (Table 13) have been applied to describe the ingested food items and define the trophic status of amphipods.

6.1.1. Copepods

Harpacticoid and cyclopoid copepods that have been sampled from the interstitial spaces of Arctic sea ice have been described as "entirely herbivorous" and showing "no evidence of selective feeding" (Grainger and Hsiao, 1990). These authors studied mouthpart morphology and setation of harpacticoids and cyclopoids from the ice and found fewer and simpler setae than on the pelagic herbivores, and furthermore, "some degree of modification of mouth-area appendages for grasping or holding." The authors concluded that sympagic copepods were feeding by seizing individual particles rather than by suspension feeding. While marine harpacticoids generally feed on algae, bacteria, fungi and detritus (Hicks and Coull, 1983), little is known about the feeding ecology of sympagic harpacticoid copepods. Major components of their diet appear to be ice-derived material such as single-celled algae, autotrophic flagellates and heterotrophic microbes (Grainger *et al.*, 1985; Grainger and Hsiao, 1990) (Table 13). *Harpacticus superflexus, Halectinosoma* sp., *Dactylopodia signata* and *T. furcata* are thought to be primarily herbivores that supplement their diatom-based diets with detritus and lumped organic matter during periods of low primary production. In particular, *Halectinosoma* sp. appears to be very unselective

Table 13 Feeding ecology of Arctic crustacea sampled within or beneath sea ice

Species	Diet	Method	Reference
Arctic			
Copepods			
Harpacticoids			
Harpacticus superflexus	(Herbivorous-omnivorous)	Gut content analysis	Grainger and Hsiao, 1990
	-diatoms, bacteria	Gut content analysis	Grainger *et al.*, 1985 (references herein, Table 6); Grainger and Hsiao, 1990
Halectinosoma sp.	(Herbivorous-omnivorous)	Gut content analysis	Grainger and Hsiao, 1990
	-diatoms (unselective)	Gut content analysis	Grainger and Hsiao, 1990
Tisbe furcata	(Herbivorous-omnivorous)	Gut content analysis	Grainger *et al.*, 1985
	-Algae, bacteria, fish larvae	Gut content analysis	Grainger *et al.*, 1985
	-Detritus	?	Johnson and Olsen, 1948
	-Flagellates, diatoms	Feeding experiment	Abu-Rezq *et al.*, 1997
Dactylopodia signata	(Herbivorous-omnivorous)	Gut content analysis	Grainger *et al.*, 1985
	-Ice diatoms	Gut content analysis	Grainger *et al.*, 1985
Cyclopoids			
Cyclopina schneideri	(Omnivorous)	Gut content analysis	Grainger and Hsiao, 1990
	-Diatoms (unselective)	Gut content analysis	Grainger and Hsiao, 1990
Oithona similis	(Omnivorous-carnivorous)	Gut content/lipid analysis	Grainger *et al.*, 1985; Kattner *et al.*, 2003
	-Diatoms; phytoplankton	Gut content analysis	Runge and Ingram, 1988; Hopkins, 1985
	-Diatoms, flagellates	Lipid analysis	Kattner *et al.*, 2003
	-Copepod faeces	Lab observation	Gonzalez and Smetacek, 1994
Oncaea borealis	(Omnivorous-carnivorous)	Gut content/lipid analysis	Grainger *et al.*, 1985; Kattner *et al.*, 2003

Calanoids			
Metridia longa	(Omnivorous-carnivorous)	Lipid analysis	Falk-Petersen *et al.*, 1987; Sargent and Falk-Petersen, 1988; Stevens *et al.*, 2004
	-*Calanus*-eggs	Feeding experiment	Conover and Siferd, 1993
		Lipid analysis	Stevens *et al.*, 2004
	-Copepods	Lipid analysis	Sargent and Falk-Petersen, 1988; Albers *et al.*, 1996; Stevens *et al.*, 2004
	-Ice diatoms	Gut content/pigment	Runge and Ingram, 1988
		Lipid analysis	Stevens *et al.*, 2004
	-Ciliates, (dino)flagellates	Lipid analysis	Stevens *et al.*, 2004
Acartia longiremis	(Herbivorous-omnivorous)	Lipid analysis	Norrbin *et al.*, 1990
	-Algae, bacteria, crustacea	(Various)	Grainger *et al.*, 1985 (references herein, Tab. 6)
Calanus glacialis	(Herbivorous-omnivorous)	Lipid analysis	Stevens *et al.*, 2004
	-Ice diatoms	Gut content/pigment	Runge and Ingram, 1988
		Lipid analysis	Stevens *et al.*, 2004
	-*Calanus*-eggs	Feeding experiment	Conover and Siferd, 1993
C. hyperboreus	(Herbivorous-omnivorous)	Lipid analysis	Stevens *et al.*, 2004
	-Ice diatoms	Lipid analysis	Sargent and Falk-Petersen, 1988; Stevens *et al.*, 2004
Jaschnovia brevis	(Herbivorous-omnivorous)	Lipid analysis	Scott *et al.*, 2002
	-Diatoms, flagellates	Lipid analysis	Scott *et al.*, 2002
	-Particulate organic matter	Lipid analysis	Scott *et al.*, 2002
	-Calanoid copepods	Lipid analysis	Scott *et al.*, 2002
Pseudocalanus acuspes	(Herbivorous-omnivorous)	Lipid analysis	Norrbin *et al.*, 1990
	-Ice diatoms, flagellates	Gut content/pigment	Runge and Ingram, 1988

(Continued)

Table 13 (Continued)

Species	Diet	Method	Reference
Amphipods			
Apherusa glacialis	(Herbivorous-omnivorous)		
	-Diatoms, phytodetritus	Lipid analysis	Scott *et al.*, 1999
	-Diatoms, flagellates, detritus, crustacea	Gut content analysis	Bradstreet and Cross, 1982; Poltermann, 2001
	-Ice diatoms	Feeding experiment	Werner, 1997b
Gammarus wilkitzkii	(Carnivorous-omnivorous)		
	-Crustacea/live prey	Lipid analysis	Scott *et al.*, 1999
		Feeding experiment	Arndt, 2002; Werner *et al.*, 2002a
		Gut content analysis	Bradstreet and Cross, 1982; Poltermann, 2001
		In situ observation	Gulliksen and Lønne, 1989
	-Carcasses	Baited traps	Poltermann, 1997; Arndt, 2002
	-Diatoms, flagellates, detritus	Gut content analysis	Bradstreet and Cross, 1982; Poltermann, 2001
	-Particulate organic matter	Behaviour/morphology	Poltermann, 1997; Arndt, 2002
G. setosus	(Carnivorous-omnivorous)		
	-Copepods	Gut content analysis	Grainger *et al.*, 1985
Onisimus nanseni	(Omnivorous)		
	-Carcasses	Baited traps	Barnard, 1959; George and Paul, 1970; Poltermann, 1997; Arndt, 2002
	-Phytodetritus, crustacea	Gut content analysis	Poltermann 2001
		Lab observation	George and Paul, 1970
		Behaviour/morphology	Arndt *et al.*, 2005b

O. glacialis	(Herbivorous-omnivorous)		
	-Diatoms, phytodetritus	Gut content analysis	Bradstreet and Cross, 1982; Poltermann, 2001
		Behaviour/morphology	Arndt *et al.*, 2005b
	-Crustacea/carcasses	Baited traps	Poltermann, 1997
		Gut content analysis	Poltermann, 2001
O. litoralis	(Omnivorous-carnivorous)		
	-Crustacea/live prey	Gut content analysis	Grainger and Hsiao, 1990; Carey, 1992
	-Ice diatoms	Gut content analysis	Grainger and Hsiao, 1990; Carey, 1992
Anonyx nugax	(Omnivorous-carnivorous)	Gut content analysis	Weslawski, 1991
	-Omnivory	Lipid analysis	Graeve *et al.*, 1997
	-Carrion	Baited traps	Melnikov and Kulikov, 1980
A. sarsi	(Omnivorous-carnivorous)		
	-Macroalgae, fish eggs, carrion	Gut content analysis	Werner *et al.*, 2004
	-Copepods	Lipid analysis	Werner *et al.*, 2004
Weyprechtia pinguis	-Carrion	Baited traps	Melnikov and Kulikov, 1980
Parathemisto libellula	(Omnivorous-carnivorous)		
	-Diatoms	Lipid analysis	Scott *et al.*, 1999; Auel *et al.*, 2002
	-Crustacea/live prey	Gut content analysis	Bradstreet and Cross, 1982
Decapod			
Eualus gaimardii	(Omnivorous-carnivorous)		
	-Small organisms	?	Feder and Lewett, 1981 cited in Graeve *et al.*, 1997
	-Carrion	Baited traps	Melnikov and Kulikov, 1980
		Gut content analysis	Weslawski, 1991

when foraging for diatoms in the ice. Gut content analysis of sympagic harpacticoids revealed their omnivorous feeding patterns during certain times of the year (Grainger *et al.*, 1985; Grainger and Hsiao, 1990). Fish larvae could be identified in the guts of *T. furcata* collected from the ice (Grainger *et al.*, 1985), and it is likely that *Tisbe* sp. also preys on the eggs and young stages of copepods.

Omnivory supplemented by carnivory is a characteristic feature of marine cyclopoids (Grainger and Hsiao, 1990). The guts of sympagic *Cyclopina schneideri* contained a wide range of diatom frustules, along with amorphous material (Grainger and Hsiao, 1990). It is well known that soft animal tissue and detrital material are generally underestimated in studies based on gut content analysis, since plant material, particularly silicious diatom frustules, are very robust and usually the last to break down in the digestive tract. Gut content analysis showed that *Oithona similis* sampled from beneath sea ice ingested ice diatoms and phytoplankton (Hopkins, 1985; Runge and Ingram, 1988). Lipid biomarkers also revealed a carnivorous dietary input (Kattner *et al.*, 2003), and coprophagy was observed during laboratory experiments (Gonzalez and Smetacek, 1994). The cyclopoid *O. borealis* supplements its diet of diatoms and phytodetritus with copepod eggs and nauplii that float in the water column (Kattner *et al.*, 2003).

The calanoid species *C. glacialis* and *C. hyperboreus* ingest a large variety of food, but they are predominantly herbivorous (Mauchline, 1998). These species base their diet on ice-derived diatoms that are then converted into large lipid stores (Sargent and Falk-Petersen, 1988; Runge and Ingram, 1991; Stevens *et al.*, 2004) (Tables 13, 14 and 15). Wax esters represent the highest energy reserve in copepods. The highest levels of wax esters are generally found in species that undergo diapause and do not feed during the winter (Sargent and Falk-Petersen, 1988, and references therein). In addition, these large lipid depots are effective at providing buoyancy, particularly in species such as *C. hyperboreus* that overwinter in dormancy in deeper water layers (Auel *et al.*, 2003). In feeding experiments, *C. glacialis* also preyed on *Calanus* spp. eggs (Conover and Siferd, 1993).

Pseudocalanus sp., which is occasionally very abundant beneath sea ice, is predominantly herbivorous (Runge and Ingram, 1988; Mauchline, 1998). However, Grainger *et al.* (1985), who identified *Pseudocalanus* sp. as a key species for ice-based trophic interactions in the near-shore Canadian Arctic, saw no evidence that this copepod forages on ice diatoms that are attached to the ice surface but removes single cells from the ice–water boundary layer. Lipid biomarkers in *P. acuspes* indicate a predatory feeding mode (Norrbin *et al.*, 1990).

The omnivorous calanoid copepod *Metridia longa* preys on *Calanus* eggs and small crustaceans (Sargent and Falk-Petersen, 1988; Conover and Siferd, 1993; Albers *et al.*, 1996; Stevens *et al.*, 2004). Gut pigment and lipid analysis

Table 14 Individual dry mass (DM) of Arctic crustacea sampled within or beneath sea ice

Species	Stage/sex	DM (mg)	Reference
Arctic			
Copepods			
Calanoids			
Metridia longa	C6/F	0.34 ± 0.066	Conover and Huntley, 1991
	C6/F	(0.4–0.5)	Stevens *et al.*, 2004
	C6/M	0.15 ± 0.036	Conover and Huntley, 1991
	(stage?)	0.35	Lee, 1975
Acartia longiremis	C6/F	0.008	Norrbin *et al.*, 1990
Calanus glacialis	C6/F	0.85 ± 0.30	Conover and Huntley, 1991
	C6/F	0.96	Tande and Henderson, 1988
	C6/M	0.71 ± 0.29	Conover and Huntley, 1991
	(stage?)	0.7	Lee, 1975
	C5	(1.0–1.4)	Stevens *et al.*, 2004
	C5	1.58 ± 0.46	Scott *et al.*, 2000
C. hyperboreus	(stage?)	2.0	Lee, 1975
	C6/F	3.77 ± 0.83	Conover and Huntley, 1991
	C6/M	1.5	Conover and Huntley, 1991
	C5	(2.4–3.6)	Stevens *et al.*, 2004
	C5	1.58 ± 0.46	Scott *et al.*, 2000
Jaschnovia brevis	C5/F	0.1	Scott *et al.*, 2002
	C5/M	0.2	Scott *et al.*, 2002
Microcalanus sp.			
Pseudocalanus spp.	C6/F	0.017 ± 0.0054	Conover and Huntley 1991
	C6/M	0.016 ± 0.0078	Conover and Huntley 1991
P. acuspes	C4	(0.006–0.0075)	Norrbin *et al.*, 1990
	C5	(0.011–0.014)	Norrbin *et al.*, 1990
Amphipods			
Gammarus wilkitzkii		104.6 ± 50.8	Werner, 1997a
Apherusa glacialis		2.5 ± 2.0	Werner, 1997a
Onisimus spp.		12.1 ± 5.1	Werner, 1997a
Onisimus nanseni			
Onisimus glacialis			
Parathemisto libellula F		76.3 ± 6.6	Auel *et al.*, 2002
	M	93.4 ± 24.6	Auel *et al.*, 2002
P. abyssorum		(6.9–10.8)	Auel *et al.*, 2002
Anonyx sarsi		(90–114)	Werner *et al.*, 2004
A. nugax			

Note F: female, M: male.

Table 15 Total lipid (% DM, see Table 14) and lipid class composition (% total lipid) of Arctic crustacea sampled within or beneath sea ice; mean ± SD (min, max)

Species	Stage/ sex	Total lipid (% DM)	Reference	Lipid class (% total lipid) WE	TAG	PL or:	PolarL	Reference
Arctic Copepods Calanoids								
Metridia longa	C6/F							
	C6/F	(11–35)	Stevens *et al.*, 2004	71.9 ± 3.2	12.2 ± 2.1	10.0 ± 1.3		Stevens *et al.*, 2004
	(stage?)	57	Lee, 1975					
Acartia longiremis	C6/F	(not determined)	Norrbin *et al.*, 1990	(4–12)	max. 55			Norrbin *et al.*, 1990
Calanus glacialis	C6/F	24	Tande and Henderson, 1988					
	(stage?)	56	Lee, 1975					
	C5	(32–58)	Stevens *et al.*, 2004	82.8 ± 3.4	8.3 ± 3.2	3.4 ± 0.8		Stevens *et al.*, 2004
	C5	60.6	Scott *et al.*, 2000	71.5 ± 6.7	5.8 ± 5.2		3.7 ± 2.1	Scott *et al.*, 2000
C. hyperboreus	(stage?)	64	Lee, 1975					
	C5	(46–70)	Stevens *et al.*, 2004	84.2 ± 2.8	5.5 ± 2.9	4.5 ± 1.3		Stevens *et al.*, 2004
	C5	65	Scott *et al.*, 2000	74.6 ± 4.5	7.6 ± 5.5		1.6 ± 2.9	Scott *et al.*, 2000
		40.9 ± 8.2	Scott *et al.*, 1999	53.4 ± 8.0	9.7 ± 7.1	15.0 ± 4.7		Scott *et al.*, 1999

Jaschnovia brevis	C5/F	70	Scott *et al.*, 2002	4.1 ± 0.8	81.9 ± 2.9		8.4 ± 0.6	Scott *et al.*, 2002
	C5/M	40	Scott *et al.*, 2002	5.8 ± 5.0	80.2 ± 7.7		9.8 ± 2.3	Scott *et al.*, 2002
P. acuspes	C4	(not determined)	Norrbin *et al.*, 1990	(55–72)	max. 14			Norrbin *et al.*, 1990
Amphipods								
Gammarus wilkitzkii		24.6 ± 3.3	Werner, 1997a					
		21	Lee, 1975	6	13	72		Lee, 1975
		27.6 ± 5.4	Scott *et al.*, 1999	18.6 ± 6.2	43.5 ± 10.8	14.5 ± 5.1		Scott *et al.*, 1999
Apherusa glacialis		51.1 ± 4.4	Werner, 1997a					
		50.9 ± 6.2	Scott *et al.*, 1999	20.1 ± 10.1	58.4 ± 5.8	8.1 ± 2.7		Scott *et al.*, 1999
				24				Lee, 1975
Onisimus spp.		12.1 ± 5.1	Werner, 1997a	42.2 ± 7.6				Werner 1997a
				35.4 ± 4.9				Scott *et al.*, 1999
Onisimus nanseni		34	Lee, 1975	16	7	56		Lee, 1975
Onisimus glacialis		38.6 ± 10.9	Scott *et al.*, 1999	30.0 ± 13.1	32.0 ± 8.2	11.9 ± 8.0		Scott *et al.*, 1999
Parathemisto libellula	F	38.5 ± 3.3	Auel *et al.*, 2002	42.6 ± 14.4	29.8 ± 6.9	12.8 ± 9.5		Auel *et al.*, 2002
	M	22.3 ± 7.0	Auel *et al.*, 2002					
		38.9 ± 7.9	Scott *et al.*, 1999	17.1 ± 1.6	10.5 ± 2.1	39.9 ± 26.4		Scott *et al.*, 1999
P. abyssorum		(19.4–29.6)	Auel *et al.*, 2002	40.8 ± 9.9	31.9 ± 12.1	11.4 ± 7.5		Auel *et al.*, 2002
Anonyx sarsi		(2.6–17.2)	Werner *et al.*, 2004					
A. nugax					69	31		Graeve *et al.*, 1997

WE: wax ester, TAG: triacylglycerol, PL: phospholipid, PolarL: polar lipid.
Note: F: female, M: male.

indicate that diatoms form an important component of the diet of *M. longa* (Runge and Ingram, 1988; Kattner *et al.*, 2003; Stevens *et al.*, 2004), and lipid biomarkers of ciliates, flagellates and dinoflagellates have been recorded in its tissues (Stevens *et al.*, 2004). This opportunist shifts its diet with season; during the primary production period in spring *M. longa* has an omnivorous, possibly diatom-based, diet but changes to a more carnivorous diet of calanoid copepods in winter (Sargent and Falk-Petersen, 1988). *Acartia* sp. is also known as a herbivorous-omnivorous species (Mauchline, 1998). The suggested predatory feeding mode for *A. longiremis* (Grainger *et al.*, 1985) was confirmed by lipid analysis (Norrbin *et al.*, 1990). The diet of this species shifts ontogenetically; young stages are herbivorous, whereas the adult stage feeds opportunistically (Grainger *et al.*, 1985, and references therein). The benthopelagic calanoid *Jaschnovia brevis* (Farran) feeds on diatoms and flagellates at the ice–water interface but may also ingest remnants of other crustacea as part of detrital material (Scott *et al.*, 2002).

The weight-specific ingestion rates of "ice taxa" (namely harpacticoids and cyclopoids) have been estimated to be on the order of 0.5 mg C m^{-2} d^{-1} in thick multiyear ice in the Greenland Sea, assuming a density of grazers of 2400 individuals m^{-2} (Gradinger *et al.*, 1999). Compared to other infauna, the grazing impact of ice-associated copepods was $<10\%$ (Gradinger *et al.*, 1999). In the annual ice near the North Water Polynya (Baffin Bay), the maximum potential daily ingestion rate by copepods was 8 mg C m^{-2} d^{-1}, and densities were high (11,080 individuals m^{-2} [Nozais *et al.*, 2001]).

6.1.2. Amphipods

The first studies on the feeding ecology of the primary ice amphipods date back to the 1950s when Barnard (1959) deployed baited traps under the drift ice station Fletcher T-3 (for a detailed review, see Arndt *et al.*, 2005b).

Ice diatoms and (phyto)detritus represent a major food source for sympagic amphipods. *Apherusa glacialis* is primarily herbivorous (Arndt *et al.*, 2005b). This species tends to concentrate along ice edges and beneath more translucent (new) ice, where the onset of primary production takes place (Poltermann, 1998; Hop *et al.*, 2000). Its diet is composed of ice algae, flagellates and detrital lumps (Bradstreet, 1982; Scott *et al.*, 1999; Poltermann, 2001) (Table 13). It is likely that the diet of this species not only shifts with ontogenetic stage (from herbivorous juveniles to more detritivorous adults) but also with season as "fresh" algae become scarce during winter.

Phytodetritus is the major food item for *Onisimus nanseni* and *O. glacialis*. These closely related species are very much alike morphologically but differ in their second gnathopod, suggesting more predatory food acquisition for

O. nanseni, while *O. glacialis* selectively picks algal cells (Arndt *et al.*, 2005b). *O. nanseni* has been repeatedly collected by means of baited traps (Barnard, 1959; George and Paul, 1970; Poltermann, 1997; Arndt, 2002). This species scavenges on carcasses and live prey, whereas its congener, *O. glacialis*, is facultatively herbivorous. Remnants of crustaceans found in its gut may be ingested as part of detrital material rather than captured prey (Poltermann, 2001). *O. litoralis*, which may occur in high densities beneath ice in shallow waters, preys on diatoms in the ice interior as juveniles but grazes on the lower ice surface as adults (Grainger and Hsiao, 1990). In land-fast ice, *O. litoralis* has been identified to be a major trophic link between the ice and the pelagic communities (Grainger and Hsiao, 1990), most probably because of its numerical dominance in some locations. Apart from feeding on algae, this species is known to be necrophagous and predatory and hunts meiofaunal crustacea (Carey, 1992). Other Lysianassid amphipod species occasionally found in sea ice, such as *Anonyx nugax* (e.g., Green and Steele, 1977; Melnikov and Kulikov, 1980; Lønne and Gulliksen, 1991a) and *A. sarsi* (e.g., Werner *et al.*, 2004), are scavengers that feed not only on "large carrion" but also on fish eggs and copepods, as well as macroalgae (Weslawski *et al.*, 1991; Sainte-Marie, 1992; Graeve *et al.*, 1997; Werner *et al.*, 2004).

Gammarus wilkitzkii is primarily a predator (e.g., Bradstreet and Cross, 1982; Scott *et al.*, 1999; Poltermann, 2001; Arndt, 2002). This large species is an opportunistic omnivore that catches copepods, chaetognaths and other live prey (including conspecifics) and feeds on carrion as well as amorphous organic debris, diatoms and microflagellates. The amphipods' pronounced grooming behaviour and the dense and long setae on both pairs of antennae and the mouthparts suggest they are capable of filtering suspended microparticles (Poltermann, 1997; Arndt *et al.*, 2005b). Its congener *G. setosus* is a higher trophic level predator that represents an energetic link to vertebrates feeding at the underside of sea ice (Bradstreet and Cross, 1982; Grainger and Hsiao, 1990).

In the pelagic, the hyperiid amphipod *Parathemisto libellula* is a large visual predator that performs vertical migration in response to the movements of its prey (Fortier *et al.*, 2001). Lipid biomarkers have shown that this species feeds extensively on large calanoid copepods (Scott *et al.*, 1999; Auel and Werner, 2003). *P. libellula* also ingested ice-derived algal cells that sloughed off during ice melt and dispersed in the water column (Bradstreet and Cross, 1982). *Parathemisto* sp. populations show geographic differences in their diet between ice-covered and ice-free waters (Auel and Werner, 2003). Fatty acid analysis of its wax esters indicated an omnivorous diet as diatom, dinoflagellate and copepod markers were present (Scott *et al.*, 1999). Being generally opportunistic feeders, *Parathemisto* sp. shows an ontogenetic shift in its diet from algae-based as juveniles towards carnivory as adults. This species also shifts diet with season depending on the availability of certain food items (Wing, 1976). Because of its large individual size and its tendency to form dense swarms in the water column (Vinogradov, 1999; Eiane and Daase, 2002), and in particular beneath ice (Melnikov, 1997), *P. libellula* represents an important food source for marine vertebrates in ice-covered waters.

The only decapod that might associate with sea ice above shallow waters is *Eualus gaimardii* (H. Milne Edwards) (Melnikov and Kulikov, 1980; C. Arndt, unpublished observations). From gut content analysis, we know that this species feeds on carrion and small organisms (Feder and Lewett, 1981, cited in Graeve *et al.*, 1997; Weslawski *et al.*, 1991).

In feeding experiments with ice-encrusted diatoms, Werner (1997b) estimated the ingestion rates of the main sympagic amphipods. The herbivore *Apherusa glacialis* consumed >50 ng algal biomass (mg DM)$^{-1}$ d^{-1}, which was 4% more and 20% more than in the more mixotrophic species *Gammarus wilkitzkii* and *Onisimus* spp., respectively. At locations with maximum amphipod abundances, up to 800 individuals m^{-2}, the ice-algal biomass might then be reduced by 30% d^{-1} (Werner, 1997b), and by as much as 63% in shallow coastal areas (Siferd *et al.*, 1997). A comparison between the two carnivores *G. wilkitzkii* and *Parathemisto libellula* showed 10 times higher ingestion rates for *G. wilkitzkii* (<7 mg dried copepods d^{-1}) (Werner *et al.*, 2002b; Auel and Werner, 2003). In contrast to holoplankters, ice-associated amphipod species have a quasi-benthic and, hence, energy-saving lifestyle. For example, *G. wilkitzkii* spends only 8% of the time swimming (Poltermann, 1997). The metabolic rate (measured as oxygen consumption) is, therefore, approximately four times higher in *P. libellula* than in *G. wilkitzkii O. glacialis* (recalculated after Aarset and Aunaas, 1990: 0.13 and 0.12 μl O_2 (mg DM)$^{-1}$ hr^{-1} for the two latter species, respectively).

6.2. Antarctic

Few studies have addressed the diets of ice-associated Antarctic species in detail. However, information can be gleaned from a combination of gut content observations and biochemical and morphological analyses (Tables 16, 17 and 18).

6.2.1. Copepods

The diets of the primary ice-associated species appear to be dominated by protists, in particular sea ice diatoms. The gut contents of all stages of *Paralabidocera antarctica* collected from or near sea ice were examined by Hoshiai *et al.* (1987) using SEM. They found no remains of prey in the first

Table 16 Feeding ecology of Antarctic crustacea sampled within or beneath sea ice

Species	Diet	Method	Reference
Antarctic Copepods Calanoids			
Stephos longipes	(Herbivorous-omnivorous)		
	-Ice algae	Faecal pellet analysis	Kurbjeweit *et al.*, 1993; Schnack-Schiel *et al.*, 1995
	-Zooplankton and detritus	Mandibular gnathobase Analysis	Michels and Schnack-Schiel, 2005
Paralabidocera antarctica	(Herbivorous)		
	-Sea ice diatoms	SEM gut content analysis	Hoshlai *et al.*, 1987
	-Diatoms and cryptomonads	Lipid analysis	Swadling *et al.*, 2000b
Calanus propinquus	(Herbivorous-omnivorous)	Lipid analysis	Hagen *et al.*, 1993
	-Phytoplankton detritus, diatoms, protozoa	Gut content analysis	Hopkins, 1985, 1987
	-Diatom frustules	Mandibular gnathobase Analysis	Michels and Schnack-Schiel, 2005
	-Phytoplankton	Stable isotope analysis	Schmidt *et al.*, 2003; Swadling, unpublished
	-Copepods	Predation experiments	Metz and Schnack-Schiel, 1995
Ctenocalanus citer	(Herbivorous-omnivorous)		
	-Phytoplankton detritus, diatoms	Gut content analysis	Hopkins, 1985, 1987
	-Diatom frustules	Mandibular gnathobase Analysis	Michels and Schnack-Schiel, 2005

(Continued)

Table 16 (Continued)

Species	Diet	Method	Reference
Cyclopoids			
Oncaea curvata	(Omnivorous)	Lipid analysis	Kattner *et al.*, 2003
	-Algal aggregations, Phaeocystis Colonies	Feeding experiment	Metz, 1998
	-Diatoms	Lipid analysis	Kattner *et al.*, 2003
	-Detritus (phytoplankton?)	Lipid analysis	Kattner *et al.*, 2003
	-Copepod eggs/nauplii	Lipid analysis	Kattner *et al.*, 2003
	Diatoms, phytoplankton detritus	Gut content analysis	Hopkins, 1985, 1987
Oithona similis	(Omnivorous-carnivorous)	Gut content/lipid analysis	Kattner *et al.*, 2003
	-Diatoms; pytoplankton	Gut content analysis	Hopkins, 1985
	-Diatoms, flagellates	Lipid analysis	Kattner *et al.*, 2003
	-Copepod faeces	Lab observation	Gonzalez and Smetacek, 1994
Harpacticoids			
Drescheriella glacialis	(Herbivorous)		
	-Sea ice algae	Stable isotope analysis	Schmidt *et al.*, 2003
Amphipods			
Paramoera walkeri	(Omnivorous)		
	-Diatoms, copepods, sponge spicules	Gut content analysis	Rakusa-Suszczewski, 1972
	-Ice algae	*In situ* observation	Tucker and Burton, 1988
	-Juvenile amphipods	*In situ* observation	Whitaker, 1977

Pontogenia antarctica	(Herbivorous)		
	-Ice algae, crustacea, Phaeocystis	Gut content analysis	Richardson and Whitaker, 1979
Orchomene sp.	(Omnivorous-carnivorous)	Feeding experiment	
	-Carcasses, bacterial aggregations	Baited traps	Stockton, 1982
	-Ice algae	Gut content analysis	Stockton, 1982
Eusirus cf. *antarcticus*	(Carnivorous)		
	-Carrion, amphipods	Feeding experiments	Dauby *et al.*, 2001
	-Krill larvae	Observation	Hamner *et al.*, 1989
Euphausiids			
Euphausia superba	(Herbivorous-omnivorous)		
	-Sea ice algae	*In situ* observation	Hamner *et al.*, 1989
	-Sea ice algae	Biochemical analysis	Ikeda and Kirkwood, 1989
	-Ice algal assemblages	Laboratory observation	Hamner *et al.*, 1983
	-Phytoplankton	DNA analysis	A. J. Passmore, pers. comm.
	-Phytoplankton/POM	Stable isotope analysis	Schmidt *et al.*, 2003
	-Phytoplankton, copepods	Biochemical analysis	Perissinotto *et al.*, 2000
	-Heterotrophic material	Grazing experiments	Atkinson and Snÿder, 1997
Euphausia crystallorophias	(Omnivorous-carnivorous)		
	-Animal matter	Lipid analysis	Kattner and Hagen, 1998; Nicol *et al.*, 2004
	-Diatoms, benthic algae	Gut content analysis	Kittel and Ligowski, 1980
	-Detritus (ice algae)	Faecal pellet analysis	O'Brien, 1987

Table 17 Individual dry mass (DM) of Antarctic crustacea sampled within or beneath sea ice

Species	Stage/sex[a]	DM (μg)	Reference
Antarctic			
Copepods			
Cyclopoids			
Oithona similis	C1-C4	1.56 ± 3.24	Swadling *et al.*, 1997a
Oncaea curvata	C1-C4	1.68 ± 3.48	Swadling *et al.*, 1997a
Calanoids			
Stephos longipes	C2-C5	4.78 ± 3.36	Swadling *et al.*, 1997a
Paralabidocera antarctica	N1-N6	2.1 (1.5–2.7)	Swadling *et al.*, 2000a
	C1	3.2	Swadling *et al.*, 2000a
	C2	3.4	Swadling *et al.*, 2000a
	C3	4.2 (3.9–5.6)	Swadling *et al.*, 2000a
	C4	5.1	Swadling *et al.*, 2000a
	C5/F	19.7	Swadling *et al.*, 2000a
	C6/F	29.2 (19.6–35.8)	Swadling *et al.*, 2000a
	C6/M	16.4 (12.6–20.1)	Swadling *et al.*, 2000a
Calanus propinquus	C5	1108 ± 270[b], 884 ± 323[c]	Hagen *et al.*, 1993
	C6/F	1762 ± 271[b], 1264 ± 97[c]	Hagen *et al.*, 1993
	C2	79[b]	Kattner *et al.*, 1994
	C3	239 ± 29[b], 54 ± 9[c]	Kattner *et al.*, 1994
	C4	149 ± 13[c], 134[d]	Kattner *et al.*, 1994
	C5	190[d]	Kattner *et al.*, 1994
	C6/F	941[d]	Kattner *et al.*, 1994
	C6/M	921[c]	Kattner *et al.*, 1994

Amphipods			
Eusirus antarcticus		47,000[e]	Torres *et al.*, 1994a
Hyperia macrocephala		104,000	Torres *et al.*, 1994b
Orchomene plebs	F	180,000	Rakusa-Suszczewski, 1982
Orchomene plebs	M	15,300–18,000	Rakusa-Suszczewski, 1982
Euphausiids			
Euphausia superba	F (gravid)[f]	200,000–590,000[e]	Mayzaud *et al.*, 1998
	Males	400,000–1,210,000[e]	Mayzaud *et al.*, 1998
	Juvenile	47,000,000[e]	Mayzaud *et al.*, 1998
	Calyptopsis	69–235	Hagen *et al.*, 2001
	Furcilia	850–2100	Hagen *et al.*, 2001
	Juvenile	20,600–48,300	Hagen *et al.*, 2001
	Subadult	30,700–75,900	Hagen *et al.*, 2001
	Female	37,400–204,700	Hagen *et al.*, 2001
	Male	63,700–222,200	Hagen *et al.*, 2001
Euphausia crystallorophias	Calyptopes I-III	64–136	Kattner and Hagen, 1998
	Furciliae I-VI	260–3270	Kattner and Hagen, 1998
	Immatures	2400–49,100	Kattner and Hagen, 1998
	Females	13,700–101,500	Kattner and Hagen, 1998
	Males	8500–67,300	Kattner and Hagen, 1998

[a]F: female, M: male.
[b]Summer stages.
[c]Lipid-rich winter stages.
[d]Lipid-poor winter stages.
[e]Wet mass.
[f]Large range of dry mass, gravid females represent the largest individuals.

Table 18 Total lipid (% DM, see Table 17) and lipid class composition (% total lipid) of Antarctic crustacea sampled within or beneath sea ice

Species	Stage/sex[a]	Total lipid (% DM)	Reference	Lipid class (% total lipid) WE	TAG	PolarL	Reference
Antarctic							
Copepods							
Cyclopoids							
Oithona similis	C6/F			15.2 ± 6.4			Kattner *et al.*, 2003
	C6/M			68.8 ± 8.7			Kattner *et al.*, 2003
	C1-C4	16	Swadling, 1998	<1–70		25–70	Swadling, 1998
Oncaea curvata	C6			86			Kattner *et al.*, 2003
	C1-C4	5–18	Swadling, 1998	2–80		20–80	Swadling, 1998
Calanoids							
Stephos longipes		45	Schnack-Schiel, 2001	7	>80		Schnack-Schiel, 2001
Paralabidocera antarctica	N1–N6	23.5 (16.0–36.4)	Swadling *et al.*, 2000a		54 (23–76)	41 (24–70)	Swadling *et al.*, 2000a
	C1	30.1	Swadling *et al.*, 2000a		5	87	Swadling *et al.*, 2000a
	C2	38.2	Swadling *et al.*, 2000a		2	88	Swadling *et al.*, 2000a
	C3	38.0 (30.1–39.0)	Swadling *et al.*, 2000a		9 (5–12)	86 (80–89)	Swadling *et al.*, 2000a
	C4	27.4	Swadling *et al.*, 2000a		0	97	Swadling *et al.*, 2000a
	C5/F	30.4	Swadling *et al.*, 2000a		20	75	Swadling *et al.*, 2000a
	C6/F	18.2 (6.0–40.9)	Swadling *et al.*, 2000a		44 (0–69)	50 (24–100)	Swadling *et al.*, 2000a
	C6/M	9.5 (6.0–11.9)	Swadling *et al.*, 2000a		21 (2–55)	77 (49–96)	Swadling *et al.*, 2000a
Calanus propinquus	C5	36 ± 13, 35 ± 10	Hagen *et al.*, 1993		89 ± 6		Hagen *et al.*, 1993
	C6/F	40 ± 11, 30 ± 4	Hagen *et al.*, 1993		91 ± 6		Hagen *et al.*, 1993
	C2	18	Kattner *et al.*, 1994		42		Kattner *et al.*, 1994
	C3	19 ± 2, 21 ± 5	Kattner *et al.*, 1994		68 ± 1		Kattner *et al.*, 1994
	C4	16 ± 4	Kattner *et al.*, 1994				
	C5	15	Kattner *et al.*, 1994				
	C6/F	8	Kattner *et al.*, 1994				
	C6/M	29	Kattner *et al.*, 1994				

Microcalanus pygmaeus				52			Schnack-Schiel, 2001
Ctenocalanus citer		22	Schnack-Schiel, 2001				
Amphipods							
Paramoera walkeri		4.8–14.8	Rakusa-Suszczewski and Dominas, 1974				
Euphausiids							
Euphausia superba	F (gravid)[b]	5–47	Hagen, 1988		33–58		Hagen, 1988
	Calyptopsis	12.3–15	Hagen *et al.*, 2001				
	Furcilia	7.3–18.4	Hagen *et al.*, 2001		55	44	Ju and Harvey, 2004[c]
	Juvenile	9.8–42.9	Hagen *et al.*, 2001				
	Subadult	8.6–40.4	Hagen *et al.*, 2001		48	48	Ju and Harvey, 2004[c]
	Female	9.6–37.9	Hagen *et al.*, 2001				
	Male	7.1–39.4	Hagen *et al.*, 2001				
	Adult				46	53	Ju and Harvey 2004[c]
Euphausia crystallorophias	Calyptopes I-III	10.3–21.8	Kattner and Hagen, 1998	13.7 ± 2.7			Kattner and Hagen, 1998
	Furciliae I-VI	14.7–29.4	Kattner and Hagen, 1998	16.4–34.4			Kattner and Hagen, 1998
	Immatures	11.3–47.4	Kattner and Hagen, 1998	31.9–53.5			Kattner and Hagen, 1998
	Females	11.7–51.5	Kattner and Hagen, 1998	47.4 ± 8.2			Kattner and Hagen, 1998
	Males	6.5–47.3	Kattner and Hagen, 1998	24.3 ± 15.3			Kattner and Hagen, 1998

[a]F: female, M: male.
[b]Large range of dry mass, gravid females represent the largest individuals.
[c]Estimated from their Figure 2.
Note: WE: wax ester, TAG: triacylglycerol, PolarL: Polar lipid. mean ± SD (min-max).

three naupliar stages and suggested that this species does not start feeding until stage NIV. The gut contents of NIV to NVI and those of CI to III closely resembled the composition of diatoms in the sea ice cores, in particular species in the genera *Amphipora, Nitzschia* and *Navicula.* Copepodid stage CIV to adults were collected from the ice–water interface. Their gut contents also resembled the suite of species found in the sea ice.

Fatty acid profiles of *Paralabidocera antarctica* provide further support for a diatom-based diet. This species stores large amounts of triacylglycerides at certain stages of development (Tables 17 and 18), suggesting that it does not undergo long periods without grazing. The fatty acid 16:1(n−7), which is a diatom biomarker, was present in high proportions in adult males (15% of total fatty acids) and females (25%) collected from the under-ice surface in Prydz Bay (Swadling *et al.*, 2000b). The ratio of fatty acids 16:1 to 16:0 in diatoms is usually >1 (Skerratt *et al.*, 1995) and tends to be higher in herbivorous copepods than in omnivorous species (Ward *et al.*, 1996). The ratio of these fatty acids in *Paralabidocera antarctica* females (1.4) was higher than that of males (0.6). Together with the high proportions of 20:5 (n−3) and 14:0, these findings indicated that diatoms were the dominant component of the diet of the females but were less important for the males (Swadling *et al.*, 2000b). 18:4(n−3), a biomarker for cryptomonads when considered in combination with other profile features of the fatty acids (Volkman *et al.*, 1989; Skerratt *et al.*, 1995), was also detected in the *Paralabidocera antarctica* adults (females 7%, males 4%). Cryptomonads were present at 2 m in the water column at the sampling site (Gibson *et al.*, 1997), and were possibly ingested by the copepods. Further, while colonies of *Phaeocystis antarctica* (Karsten) were common in the water column during the same period (Gibson *et al.*, 1997), there was no evidence that this alga made a significant contribution to the diet of *Paralabidocera antarctica.*

There is less direct evidence for the diets of *Stephos longipes* and *Drescheriella glacialis*, although high abundances of both species often correspond with high levels of chlorophyll *a* in the sea ice (Schnack-Schiel *et al.*, 1998; Thomas *et al.*, 1998). Some preliminary feeding experiments and examination of faecal pellets have shown that *S. longipes* feeds on ice algae (Kurbjeweit *et al.*, 1993; Schnack-Schiel *et al.*, 1995). *S. longipes* also stores triacylglycerides (Kurbjeweit, 1993), indicating that it has frequent access to a food supply. Recent morphological examination of several Antarctic copepods, including *S. longipes*, suggests that animal prey contributes substantially to the diet of this species (Michels and Schnack-Schiel, 2005), as the structure of its mandibular gnathobases has similarities with those of known carnivores. *S. longipes* spends part of its life cycle in deep water layers, where it presumably overwinters as stage CIV, and Michels and Schnack-Schiel (2005) suggested that this stage, at least, might prey on zooplankton and detritus. The same study suggested that *Ctenocalanus citer* and *Calanus propinquus* use their short and compact teeth to crack silicious diatom frustules. There is good evidence that *C. citer* has a mainly herbivorous diet (Hopkins, 1987), whereas *C. propinquus* is hypothesised to undergo dietary switching from mainly herbivorous in summer to omnivorous in winter. *C. propinquus* was found to be feeding actively in early spring (October–early November), with at least half of stage CV and many females with food in their guts (Pasternak and Schnack-Schiel, 2001). Gut content analysis revealed the presence of both diatoms and metazoan material. Stable isotope ratios did not indicate that sea ice algae form a major part of the diet of this species in spring (K. Swadling, unpublished data) or autumn (Schmidt *et al.*, 2003).

Evidence for a sea ice–based diet for *D. glacialis* comes from stable isotopic ratios that showed this species to have a signature ($\delta^{15}N$ of 5.5‰, $\delta^{13}C$ of −25‰) resembling that of particulate matter in sea ice ($\delta^{15}N$ of 3.7‰, $\delta^{13}C$ of −21.9‰) (Schmidt *et al.*, 2003). There is even less information available for other harpacticoids. *Tisbe prolata* found on the Ross Ice Shelf were believed to feed on ice algal production (Waghorn and Knox, 1988), but there was no direct evidence for this suggestion. *Microsetella norvegica* might use floating substrates (marine snow aggregates) as a source of food in oligotrophic waters (Uye *et al.*, 2002).

6.2.2. Amphipods

Paramoera walkeri appears to be opportunistic and omnivorous. Its gut contents were examined during winter and spring by Rakusa-Suszczewski (1972), who observed that, in winter, plant remnants plus diatom frustules of species found on the bottom were common, along with copepod remains. In spring, sponge spicules were discovered in the guts, along with some remains of *Oithona* spp. However, the bulk of food in spring consisted of diatoms found at the under-ice surface (no species given). These observations indicate that *P. walkeri* was migrating between the benthic and the under-ice habitats. Tucker and Burton (1988) recorded that divers noted that *P. walkeri* congregated on algal patches on the under surface of ice in spring, where they fed on filiform algae, changing colour from their characteristic red and white bands to brown and green. It is also possible that both *P. walkeri* and *Cheirimedon femoratus* actively prey on newly-released young of *Pontogeneia antarctica* (Whitaker, 1977).

Pontogeneia antarctica is the predominant herbivore of ice-associated microalgae around Signy Island (Richardson and Whitaker, 1979). Gut content analysis of mainly ovigerous females revealed the presence of crustacean and microalgal remains, and there was some indication that they ingested *Phaeocystis antarctica*. This species is likely to be opportunistic, as their gut contents mainly paralleled the dominance of different prey groups in their environment. There is also some evidence of ontogenetic dietary shifts in *Pontogeneia antarctica.* Hatchlings fed actively on *Nitzschia curta* (Van Heurck) Hasle and *Navicula glacei* (Van Heurck), small diatoms that are important in the ice algal mat in spring. The adults revealed a wider diet, as outlined earlier, which also included cannibalism. Feeding by *Pontogeneia antarctica* is substantially reduced in winter, and ovigerous females do not feed as intensively as non-ovigerous females (Richardson and Whitaker, 1979).

Little else is known about the dietary habits of ice-associated amphipods in Antarctica. Briefly, *Orchomene* sp. appears to be a facultative scavenger (Stockton, 1982), as some individuals taken from beneath the Ross Ice Shelf (caught with a baited trap on the sea floor) had copepod hard parts in the guts, while stomach contents of hatchlings indicated the presence of bacterial aggregations. In laboratory experiments when hatchlings were fed on larger diatom species, such as *Cocconeis imperatrix* Schmidt, there was high mortality. A change in the gut contents of *Orchomene* sp. from 60% detritus in May to 100% algal material in October and November reflected the growth and sloughing off of the under-ice algal mat. *Eusirus* cf. *antarcticus* is predominantly a benthopelagic species that was shown to eat amphipods in laboratory experiments. It has been characterised as a macropredator that is able to feed partially on carrion (Dauby *et al.*, 2001) and has also been known to prey on furciliae of *Euphausia superba* (Hamner *et al.*, 1989). Finally, examination of the stomach contents of *Orchomenopsis* sp. revealed the presence of very yellow-green amorphous masses and oil droplets, but no diatom frustules (Andriashev, 1968).

6.2.3. Euphausiids

The mouthparts of *Euphausia superba* appear to be well adapted to grazing in the sea ice habitat. Both the furciliae and the adults scrape algae from the undersurface of the ice with their thoracic endopods, a method thought to be an efficient mechanism for concentrating large quantities of algae (Marschall, 1988; Hamner *et al.*, 1989). Adult *E. superba* have been observed feeding on concentrated ice algal assemblages that were frozen into ice blocks in the laboratory (Hamner *et al.*, 1983). As the algae melted out of the ice in concentrated streams, the krill could detect them and responded with active feeding behaviour; they increased their activity with more turning, somersaulting and rapid swimming. This was accompanied by rapid opening and closing of the feeding baskets. Hamner *et al.* (1983) noted that this initial frenzied behaviour was then often followed by a period of grazing on the ice surface itself.

While their food-gathering techniques may be similar, it is becoming apparent that furciliae adopt quite different overwintering strategies to both juveniles and adults. At the beginning of autumn, postlarval krill were found to have accumulated large lipid stores, reduced their excretion and respiration rates up to 60–80% of summer values and reduced feeding rates compared to summer (Atkinson *et al.*, 2002). In contrast, furciliae examined during the same study had high feeding and metabolic rates, suggesting that these stages had switched to under-ice feeding (Meyer *et al.*, 2002). These results are intriguing, as they indicate fundamental differences in the strategies of larval and postlarval krill for surviving the winter and raise the question of why juveniles and adults do not exhibit high feeding rates on ice algae. Perhaps, as was suggested, it is a mechanism for reducing competition between the stages, a way of lowering predation risks, or simply arises because postlarvae cannot fit into narrow crevices and gaps in the ice (Atkinson *et al.*, 2002). Whatever the case, it does appear certain that larval krill cannot tolerate the long periods of starvation (Meyer and Oettl, 2005) that adult krill can endure (Ikeda and Dixon, 1982).

In a comparison of the pigment contents of field-caught versus laboratory-fed furciliae, Ross *et al.* (2004) found those of field animals feeding on under-ice algae were only ~20% of those fed phytoplankton in the laboratory. Thus, it appears that field-caught larvae were not reaching their maximum ingestion rates when feeding on sea ice algae. One explanation might be that while larval krill are primarily herbivorous in winter, they might also graze on heterotrophic protists that are associated with sea ice surfaces; these protists would not contribute to measured pigment contents (Ross *et al.*, 2004). Microscopic analysis of gut and faecal pellets has certainly highlighted the ability of furciliae to ingest a range of small particles found in sea ice and water column, including phytoplankton, sea ice biota, exuviae, detritus and microzooplankton (Daly, 2004). Although furciliae appear capable of ingesting a range of food, there is substantial geographic and seasonal variation in their survival; changes in body size and condition suggest furciliae are often food limited in winter (Daly, 2004).

Although adult *Euphausia superba* exhibit quite complex feeding behaviour in the presence of sea ice algae, there are biochemical indications that this food source is not of critical nutritional importance. The early work of Ikeda and Kirkwood (1989) found that feeding on ice algae contributed to the activation of carbon and phosphorus catabolism and to small increases in carbon, nitrogen and phosphorus content. However, these authors observed no appreciable improvement in body condition of *E. superba* and they suggested that ice algae might not be as nutritious as phytoplankton for this species. Stübing and Hagen (2003) found that while the fatty acid and lipid class profiles of adult *E. superba* were not altered by being fed different diets, including sea ice algae, those of the furciliae were affected. Clearly *E. superba* is an opportunistic species, and, while primary producers might form the bulk of their diet during the year, at certain times heterotrophic material can make a substantial contribution (e. g., Perissinotto *et al.*, 2000). As mentioned earlier, there is also strong evidence for a substantially reduced metabolism in winter (Atkinson *et al.*, 2002), so a limited amount of heterotrophic grazing might be sufficient to ensure adult krill survive that period.

There is evidence to suggest that *Euphausia crystallorophias* utilises many food sources and that ice algae play at best a minor role in their diet. Ikeda and Kirkwood (1989) observed this species in pack ice off Enderby Land, where they aggregated in swarms ~1–5 m below the sea ice bottom. They were not seen to feed on sea ice algae, and at that time, chlorophyll in the water column was very low, leading these authors to suggest that *E. crystallorophias* was feeding on zooplankton and/or faecal pellets. Fatty acid profiles support the idea of carnivorous feeding, as *E. crystallorophias* were found to contain large quantities of 18:1(n−9), a biomarker fatty acid associated with ingestion of animal matter (Kattner and Hagen, 1998).

The gut contents of animals caught in the shelf waters of Ezcurra Inlet in Admiralty Bay, King George Island, indicated that they were filter feeding on a wide range of diatoms, especially *Chaetoceros criophilus* Castracane, *Nitzschia curta* and *Navicula* spp. Neritic, oceanic and benthic algae were all recorded in the stomachs, yet the great quantity of benthic algae suggested they were spending considerable time feeding on the bottom sediments (Kittel and Ligowski, 1980). It has been calculated that dense swarms of *Euphausia crystallorophias* might remove 14–97% of local primary production (Pakhomov and Perissinotto, 1996).

Both herbivory and carnivory are important feeding strategies for *Euphausia crystallorophias* and there are strong indications that this species undergoes both ontogenetic and seasonal shifts in diet. In winter, the digestive gland of *E. crystallorophias* was 33% of carapace length, compared with 63% in spring, suggesting feeding intensity had been reduced considerably in the winter. The colour of the gland was also different, being pale yellow-green in winter and dark green in spring (Nicol *et al.*, 2004). Lipids comprised between 3 and 5% of dry mass in winter, and up to 15% in summer, with wax esters as the main storage compound. Feeding in winter was depressed relative to spring and summer, and significantly higher levels of 20:1 and 22:1 fatty acids in winter suggested a shift to a carnivorous diet. C16 polyunsaturated fatty acids, which are markers of the phytoplankton diet common in summer krill, were lacking in winter krill. It appears the winter diet is probably more typically omnivorous, consisting of small metazoans and faecal pellets, along with detrital material (Kattner and Hagen, 1998; Nicol *et al.*, 2004). Detritus can include sinking flakes of ice algae, and these have been observed in the faecal pellets of *Euphausia crystallorophias* (O'Brien, 1987).

Finally, Boysen-Ennen and Piatkowski (1988) observed that larvae of *Euphausia crystallorophias* fed on phytoplankton, while the older stages ingested both detritus and benthic diatoms. They further found that the distribution of the stages reflected this shift in feeding type; larvae were found in the uppermost stratum of the water column, while juveniles and adults were located in deeper waters.

6.3. Polar comparison

Both classical gut content and mouthpart analysis and biochemical assays reveal that crustaceans at both poles tend towards omnivory and opportunism. This is not surprising given the strong seasonality of the food supply. While the ice cover clearly provides a rich source of food, this supply can be both patchy and ephemeral. Being able to switch between varieties of prey is an important adaptive strategy for the polar environment. For those species (mainly copepods) that are small enough to inhabit the interstitial spaces of sea ice, there is probably enough foraging potential on ice protists that they do not need to switch diets too markedly. However, the larger species, predominantly amphipods and euphausiids, which mainly rely on accessing algae that grow on ice surfaces, often need to employ other strategies. These strategies appear to range along a continuum from no feeding at all to a switch from herbivorous to carnivorous grazing.

As discussed earlier, only large copepods belonging to the genus *Calanus* undergo true diapause, generally at depth, where metabolism is severely reduced and other bodily processes reach a minimum. These species rely on large stores of wax esters and link their emergence and reproduction tightly to the spring bloom. Strict herbivores perform strong diel vertical migration, in contrast to facultatively herbivorous organisms that are often the numerically dominant species in ice-covered waters. In particular, algae-detritus aggregates appear to be a year-round food source for many taxa. Further along the continuum are species such as the copepods *C. propinquus* and *Metridia longa*, which graze heavily on diatoms during the productive summer months, tie their reproduction to this production, but then switch to a carnivory-based diet to survive the winter. In most crustaceans, it has been shown that the diet switches not only with seasonal availability of different food types but also with ontogenetic stage and even sexual differences, and differences between mature and non-mature females have been found. Many of the polar amphipods are similar opportunists, taking advantage of a range of food from diatoms to metazons and carrion. The amphipods are generally restricted to grazing only on the bottom of the ice and in the shallow waters at both poles they readily migrate between the ice and the benthos. Amphipods brood their young over the long winter months, and many of them time the release of their young to coincide with the growth of the underice algal mat. In contrast to their pelagic counterparts, ice-associated amphipods lead a low-cost life, as reflected by their lower metabolic rates and, hence, lower dietary requirements. Finally, amongst the Antarctic euphausiids, it appears that only *Euphausia superba* has a strong association with sea ice. This species undergoes a complex life cycle, with many stages showing varying dependency on the ice; however, it seems that late furciliae and juvenile stages generally do graze on sea ice as a means of surviving the winter.

In Section 8, we discuss the question of why only a few species have discovered the rich feeding ground of sea ice and why these species do not fully exploit the ice-associated annual primary production.

7. BEHAVIOURAL AND PHYSIOLOGICAL STRATEGIES

It is apparent that the sea ice habitat is available only to those species that combine behavioural and physiological adaptations to exploit this resource. Behavioural adaptations include both reproductive and trophic strategies, while important physiological traits encompass salinity and temperature tolerance, storage of lipid reserves and metabolic flexibility. Finally, anatomical traits, such as specialised mouthparts and feeding appendages, might assist in the exploitation of the ice habitat. In this brief section, we consider what is known about the strategies adopted by species at both poles.

There have been many studies on physiological and behavioural adaptations in cold-water species, but knowledge is sparse on those crustaceans that associate with sea ice. Clearly, an important feature for success in the sea ice environment is the ability to cope with the variable salinities and lower temperatures that occur in the brine channel system, compared to those of the underlying water column. The Arctic amphipods *Gammarus wilkitzkii Onisimus glacialis* and the Antarctic amphipod *Eusirus antarcticus* are "cold tolerant" but "freeze sensitive" and do not tolerate being frozen into solid sea ice (Aarset and Torres, 1989; Aarset, 1991). In contrast, the Arctic littoral species *Gammarus oceanicus* Segerstråle and the ice amphipods *A. glacialis* and *Gammaracanthus loriatus* can survive short-term exposure to temperatures as low as −7°C (Green and Steele, 1977; Aarset and Zachariassen, 1988), while *Euphausia superba* can tolerate temperatures down to −4°C in solid ice (Aarset, 1987). Arctic ice amphipods show relatively high supercooling points (−4°C) compared to Antarctic species (−11.4°C for *Eusirus antarcticus*, −8°C for *Euphausia superba* [Aarset, 1987]), which suggests that Antarctic fauna have better supercooling capacity than Arctic species (Aarset and Aunaas, 1987). Neither "thermal hysteresis agents" nor anti-freeze proteins were found in the body fluids of any of these species, but inorganic ions and free amino acids lowered the freezing point and reduced ice nucleation (Aarset and Aunaas, 1987, 1990).

Low-temperature adaptation promotes greater tolerance to changing external salinity. The Arctic ice amphipods *Gammarus wilkitzkii* and *Onisimus glacialis* are hyperosmotic regulators at low salinity values (3–4), both regulating extracellular concentrations of sodium and chloride in the haemolymph. At higher salinities, they respond by being osmoconformers over the tolerated salinity range (34–60 [Aarset and Aunaas, 1987]). In contrast, the pelagic hyperiid *Parathemisto libellula* is stenohaline and needs to avoid osmotic stress (Aarset and Aunaas, 1987). *Eusirus antarcticus* and *Euphausia superba* become osmoconformers at salinities ~40–50, when temperatures are ~−1.5°C. These species might tolerate staying in the vicinity of the ice by conforming to ambient brine salinity, through the lowering of the freezing points of their body fluids (Aarset, 1987). Arctic species are able to tolerate salinities down to those close to freshwater, while the two Antarctic species mentioned could survive in salinities down to 25. These differences between the two poles are probably a function of the melt processes of the ice. An interesting adaptation to the problem of variable salinities was observed in *Paramoera walkeri* eggs, which underwent deformation during development when the egg case swelled up with diluted seawater. This was possibly a mechanism that protected the embryos against extensive variation in salinity (Rakusa-Suszczewski, 1972).

To our knowledge, the cold and freezing resistances of ice-associated copepods have not been established. However, there are some studies on the cryobiology of harpacticoid copepods sampled from supralittoral habitats that demonstrate their tolerance to freezing (e.g., McAllen and Block, 1997). As discussed earlier for amphipods, this tolerance is a result of depressing the supercooling point with increased internal salinity. In laboratory experiments comparing sympagic with pelagic copepods, Grainger and Mohammed (1990) demonstrated that survival was longest at salinities of 40 in the sympagic animals, but at salinities of 20 in the pelagic animals. Ice copepods were the only survivors at salinities of 60 and 70. *Tisbe furcata*, *Cyclopina* spp. and *Arctocyclopina pagonasta* Mohammed and Neuhof are also obvious osmoconformers that are well adapted to temperature and salinity stresses in sea ice. In the Antarctic, *D. glacialis* can survive in salinities from 18 to 90, although it shows lethargic behaviour at both extremes of this range (Dahms *et al.*, 1990). *Paralabidocera antarctica* prospers in three hyposaline lakes (10–20) in the Vestfold Hills (Bayly, 1978), and 100% survival rates were observed for nauplii maintained for 12 d at salinities from 35 to 50 (K. Swadling, unpublished data).

Other adaptations to the ice environment are evident in the trophodynamic and reproductive strategies of ice crustaceans, which have been addressed earlier in detail. As an adaptation to the seasonal variability in food supply, polar crustaceans often deposit large lipid stores (e.g., Kattner and Hagen, 1995). Opportunistic feeders, including amphipods and copepods that stay trophically active throughout the year, do not store wax esters but triacylglycerols, which are short-term buffers against the seasonally oscillating food supply (e.g., Scott *et al.*, 1999; Swadling *et al.*, 2000b). Arctic ice crustaceans do not appear to suffer from food shortage at any time of the year (Gradinger *et al.*, 1999; Poltermann, 2001).

The reproductive strategies of many of the successful groups among ice crustaceans, particularly the harpacticoid copepods and amphipods, involve brooding and direct development. The advantages are obvious: By omitting the (planktonic) larval stages, there is no risky habitat transition between larvae and adults, and the survival rate is higher for the next generation. However, the two highly successful Antarctic calanoids, *Paralabidocera antarctica* and *S. longipes*, are broadcast spawners, releasing their eggs into the water column. These species adopt different strategies to ensure that their young reach the sea ice at the appropriate time; *S. longipes* has sticky eggs that facilitate attachment to frazil ice crystals as they rise to the surface (Kurbjeweit *et al.*, 1993), while it is the NI stage of *Paralabidocera antarctica* that swims to the surface and is incorporated into new ice as it forms (Swadling *et al.*, 2004).

Tanimura *et al.* (2002) hypothesised a behavioural switch in *Paralabidocera antarctica* as they metamorphosed from the naupliar to the copepodid stages. Nauplii might exhibit a strong thigmotactic response, indicating they respond positively to contact with solid substrate. As they develop, this response is supplanted by a phototactic response, resulting in migration to areas where solar radiation is stronger and indicates the presence of patches of algae. This reaction is particularly strong in the adults, which form dense swarms at the underice surface in response to undersurface brightness (Tanimura *et al.*, 1984a). A similar phenomenon was observed in the Antarctic amphipod *Pontogeneia antarctica*; the young are positively phototactic and congregate at points where illumination is greatest. Thus, the hatchlings are attracted to sites where microalgal growth is potentially strongest. There is also some suggestion that this species undergoes a switch in both habitat and food size, which was hypothesised to be an adaptation to avoid cannibalism (Richardson and Whitaker, 1979).

Euphausia superba show a strong escape response when threatened by predators. When the furciliae were approached slowly by divers, they moved in closely to crevices in the ice, as opposed to deep diving, suggesting the ice is a superior refuge (Hamner *et al.*, 1989). Similarly, Marschall (1988) observed that, when approached by an ROV, krill usually tail-flipped until they found shelter in pockets and crevices in the ice, and only rarely did they try and escape downwards. As mentioned earlier, krill furciliae are preyed on by *Eusirus antarcticus*. The amphipods cling to the ice when not foraging and then dart out to capture individual furcilia, before returning to the ice for ingestion. Similar behaviour has been observed in amphipods foraging under Arctic sea ice. Disturbances caused by divers led the amphipods to seek shelter in holes that often appeared to resemble their body shapes exactly (C. Arndt, personal observation). In *Gammarus wilkitzkii*, the setose antennae are extended out of the hole and await floating particles that are then brushed off by the mouthparts and ingested.

8. FINAL REMARKS AND CONCLUSIONS

Crustaceans play a key role in the sea ice ecosystems in the Arctic and the Antarctic. Species richness is similar at both poles, and this richness is characterised by the predominance of copepods and amphipods over other crustacean groups. However, there are some fundamental differences in how the communities are structured and the magnitude of secondary production associated with the ice.

Taxonomic dominance differs considerably. In the Arctic, harpacticoid and cyclopoid copepods are abundant in the interstitial spaces of the ice cover, while amphipods appear to be ubiquitous and high in biomass all over the region. In the Antarctic, calanoid and harpacticoid copepods inhabit the brine channels of the ice in considerable numbers, and amphipods can be locally very abundant at the ice–water surfaces, as can several stages of *Euphausia superba*.

In Arctic sea ice, harpacticoid and cyclopoid copepods appear to fill a similar niche to calanoids in the Antarctic. If calanoids were the dominant copepod group in Arctic sea ice and appeared in comparable abundances to harpacticoids and cyclopoids, carbon biomasses would be as high as, or higher than, in Antarctic sea ice. The question then arises why different copepod taxa dominate in the sea ice at both poles. Vast areas of Arctic sea ice are above shallow water, where harpacticoid and cyclopoid copepods have a tight benthic connection and are good colonisers. Furthermore, the same taxa are often found in estuarine systems, such as in the large Siberian and Canadian river deltas, where there is a propensity for organisms to exhibit broad salinity tolerances.

In Antarctica, deep water under the pack ice restricts colonisation by small benthic dwellers, such as harpacticoid copepods, although larger species, including the amphipods and euphausiids, can migrate vertically over long distances. Much of the very high Antarctic interstitial biomass occurs in ice over shallow water (<100 m), and this is particularly true for the coast along the Indian Ocean sector. The most prolific ice-associated calanoid copepod in that region is *Paralabidocera antarctica*, which is restricted to surface waters south of the Antarctic Divergence. This species is a member of the family Acartiidae, which is ecologically very successful and is distributed widely throughout marine inshore waters and estuaries. The nauplii of *Paralabidocera antarctica* are reasonably strong swimmers and are capable of covering the distances needed to colonise the ice once they have hatched from their eggs in the sediments.

Secondary production is up to 10 times higher in the Antarctic interstitial ice. As mentioned previously, this is often attributed to the higher pore volume in Antarctic sea ice, which is related to the presence of platelet ice and other factors such as heavy snow load, warmer temperatures and snow-ice formation that produces more habitable space throughout the ice matrix (H. Eicken, personal communication). One of the most intriguing findings is that the sympagic crustacea do not appear to maximise their grazing potential but take only moderate advantage of the available ice related primary production. It was once believed that diatoms were universally high-quality food for herbivores. However, evidence has accumulated that certain diatoms contain anti-feedants and toxins that make them unpalatable or even harmful to their predators (Pohnert, 2005). For example, the pennate diatoms *Pseudonitzschia multiseries* (Hasle) and *P. seriata* (Cleve) both grow at the underside of sea ice in the Arctic Ocean and are known to produce phycotoxins (Bates, 2004). Toxicity, along with reduced nutritional value (e.g., low C/N), in diatoms may have negative effects on fecundity, egg viability and/or naupliar condition (Paffenhöfer, 2002, and references therein; Jones and Flynn, 2005). Copepods are highly selective and it is mainly calanoids that consume diatoms, while other major groups, such as many harpacticoid and cyclopoid species, do not rely heavily on diatoms. In general, our knowledge of the feeding ecology of ice-associated copepods is sparse. Moreover, little work has been done on the nutritional value of their preferred diet and their grazing rates.

To evaluate the diets of ice-associated crustaceans, we need to know more about their prey. Ciliates, acoel turbellarians and crustaceans are common in both systems, while nematodes are abundant in the Arctic (Gradinger, 1999) but have been recorded only once in the Antarctic (Blome and Riemann, 1999). Similarly, to the best of our knowledge, rotifers have not been recorded from Antarctic ice, although they are common in the Arctic. Given the limited number of research projects that have focused on sympagic microfauna and meiofauna and their trophic roles in both hemispheres, these taxa may have been simply overlooked so far. Crustaceans at both poles tend towards omnivory and opportunism. Whether the interstitial meiofauna substantially contributes to their diet is not yet clear. Future studies using DNA, stable isotopes and/or fatty acid analyses will provide valuable information about the food-web relationships of sympagic organisms and their role in trophic transfer.

8.1. Future climatic impacts

The information provided in this review may enable a baseline against which different species associated with sea ice can be assessed for their tolerance to impacts of climate change on the sea ice habitat. It is clear that only a few species, at best, are obligate ice dwellers, so it is feasible that many could continue to thrive under scenarios of thinner or less frequent ice cover. However, other changes associated with a reduction in sea ice, including increased water temperatures that might result in the northward (Arctic) or southward (Antarctic) expansion of warmer-water species, will, at the very least, upset the current ecological balance by increasing competition for space and other resources. Life history plasticity will be the key to a species' continual success and it is clear from this review that many high-latitude species are flexible, in terms of both their diets and their life cycles. Understanding the relative importance of factors that drive the organisms to colonise sea ice (e.g., response to predation vs reproductive needs) would enhance our ability to predict their responses to environmental change.

Because the Arctic and Antarctic regions underwent glaciation during different geological epochs (Pleiostocene and Miocene, respectively), the polar marine ecosystem is some 4 Ma old in the Antarctic, while Arctic polar biota are not older than 0.7 Ma (Herman and Hopkins, 1980). Whether evolutionary time scales help to explain why the majority of sea ice organisms originate from the benthos in the Arctic and from the pelagos in the Antarctic is largely a matter for speculation. Interestingly, it is in the Antarctic where the affiliation between crustaceans and sea ice appears to be strongest. *Euphausia superba* seems to be dependent on ice for a crucial part of its life cycle. While the possible effects of a large-scale reduction in sea ice on the enormous reproductive potential currently displayed by this species are still under much discussion (reviewed in Brierley and Thomas, 2002; Atkinson *et al.*, 2004; Smetacek and Nicol, 2005), it is certain that any sea ice–related changes to krill biomass will have a significant effect on the Southern Ocean ecosystem. Such effects could be manifested through decreased survival and recruitment of key krill predators, such as penguins (Reid and Croxall, 2001; Fraser and Hofmann, 2003).

8.2. Knowledge gaps/future research

Large knowledge gaps result from a geographic imbalance in polar research. In the Antarctic, much work has focused on the Weddell Sea, where predominantly German-led research cruises have undertaken extensive well-planned studies. The long-term studies initiated by the United States at Palmer Station have provided important insights into the complexities of Antarctic krill biology and ecology. By contrast, the Indian Ocean sector is characterised by geographically patchy station-based studies that have enabled a good understanding of the temporal dynamics of fast ice. Several transect-based cruises in the Indian sector have resulted in a wealth of data relating distribution of krill and pelagic zooplankton to large-scale variations in sea ice and water masses. However, the east Antarctic pack ice ecosystem is still understudied; there are large areas that have not been visited for the purposes of sampling the smaller ice-associated metazoans. In the Arctic, the focus has been more on land-fast ice in the Canadian and Alaskan Arctic, where access and logistics are easy. There have been only few—but in their findings spectacular—research operations from drift ice stations in the perennial ice pack. Research in the inner ice pack is logistically difficult and cost intensive, so most of the ship-based ice research has been undertaken along the marginal ice zone, which is highly disturbed by swell and decay.

Further gaps in our understanding exist for organism and ecosystem functioning because processes affecting life in the polar environment occur year round. There are, at best, limited winter data from most ice-covered regions.

Research foci have also changed during the last century of ice research. For many taxa, basic knowledge of life history strategies is missing. For instance, the momentum of early work on amphipods in the Antarctic has not been sustained, while in the Arctic copepods have received far less attention than amphipods. Although there has been a focus on the use of Antarctic amphipods, primarily *Paramoera walkeri*, as indicators of benthic contaminants and ultraviolet exposure (Duquense *et al.*, 2000; Liess *et al.*, 2001), there have been few studies designed to understand the basic biology of these animals. Yet it seems certain that amphipods could be locally very important grazers. What little we do know about the diets of these organisms comes mainly from gut content analyses and we know almost nothing about physiological properties such as lipid storage. In general, there has been very little application of new technologies to ice-associated animals, other than as a means for assessing abundance and distribution (i.e., acoustics). We understand the life cycles of only a few of the strongly ice-associated species and there has been almost no work done on their metabolism, diets or trophic roles. Finally, ecophysiological investigations have been rare; we know almost nothing about the physiological responses of ice crustacea to variations in the physical conditions of their habitat. It is 15 yr since Conover and Huntley's (1991) review of Arctic and Antarctic copepods, but it is still fair to say that "there is much to do both north and south."

ACKNOWLEDGEMENTS

We thank Sarah Jones for help with finding references and Dr. John Gibson for frequent discussions on the manuscript. Dr. Rob Massom provided information on large-scale sea ice movements in the Southern Ocean. An Australian Research Council Discovery grant (DP0209308) and Antarctic Science Advisory Committee grants (1328, 691, 875) provided funding to K. Swadling. C. Arndt thanks Dr. Ole Jørgen Lønne (IMR) and Prof. Bjørn Gulliksen (NFH) for inspiring discussions and great support, Prof. Alastair Richardson and Prof. Andrew McMinn for their hospitality at the University of Tasmania, Prof. Heinz Wanner and Prof. Brigitta Ammann for their hospitality at the University of Bern, Switzerland, and the University Centre on Svalbard and Total EandP Norway for logistics and funding. Two anonymous reviewers and Dr. So Kawaguchi are thanked for their critical reviews of the manuscript.

REFERENCES

Aarset, A. V. (1987). AMERIEZ 1986: Under-ice fauna from the Weddell Sea—responses to low temperature and osmotic stress. *Antarctic Journal of the United States* **22**, 170–171.

Aarset, A. V. (1991). The ecophysiology of under-ice fauna. *Polar Research* **10**, 309–324.

Aarset, A. V. and Aunaas, T. (1987). Osmotic responses to hyposmotic stress in the amphipods *Gammarus wilkitzkii, Onisimus glacialis* and *Parathemisto libellula* from Arctic waters. *Polar Biology* **7**, 189–193.

Aarset, A. V. and Aunaas, T. (1990). Metabolic responses of the sympagic amphipods *Gammarus wilkitzkii* and *Onisimus glacialis* to acute temperature variations. *Marine Biology* **107**, 433–438.

Aarset, A. V. and Torres, J. J. (1989). Cold resistance and metabolic responses to salinity variations in the amphipod *Eusirus antarcticus* and the krill *Euphausia superba*. *Polar Biology* **9**, 491–497.

Aarset, A. V. and Zachariassen, K. E. (1988). Low temperature tolerance and osmotic regulation of the amphipod *Gammarus oceanicus* from Spitsbergen waters. *Polar Research* **6**, 35–41.

Abu-Rezq, T. W., Yule, A. B. and Teng, S. K. (1997). Ingestion, fecundity, growth rates and culture of the harpacticoid copepod, *Tisbe furcata*, in the laboratory. *Hydrobiologia* **347**, 109–118.

Ackley, S. F. and Sullivan, C. W. (1994). Physical controls on the development and characteristics of Antarctic sea ice biological communities—a review and synthesis. *Deep-Sea Research I* **41**, 1583–1604.

Albers, C. S., Kattner, G. and Hagen, W. (1996). The composition of wax esters, triacylglycerols and phospholipids in Arctic and Antarctic copepods: Evidence of energetic adaptations. *Marine Chemistry* **55**, 347–358.

Andriashev, A. P. (1968). The problem of the life community associated with the Antarctic fast ice. *In* "Symposium on Antarctic Oceanography" (R. I. Currie, ed.), pp. 147–155. W Heffer & Son, Cambridge.

Arndt, C. E. (2002). Feeding ecology of the Arctic ice-amphipod *Gammarus wilkitzkii*—physiological, morphological and ecological studies. *Reports on Polar and Marine Research* **405**, 1–74.

Arndt, C. E. (2004). Ecosystem dynamics in arctic sea ice: The impact of physical and biological processes on the occurrence and distribution of sympagic amphipods. Ph.D. Thesis, University Centre on Svalbard, Norway.

Arndt, C. E. and Beuchel, F. (2006). Life cycle and population dynamics of the Arctic sympagic amphipods *Onisimus nanseni* SARS and *O. glacialis* SARS (Grammaridea: Lysianassidae). *Polar Biology* **29**, 239–248.

Arndt, C. E. and Pavlova, O. (2005). Origin and fate of ice fauna in the Fram Strait and Svalbard area. *Marine Ecology Progress Series* **301**, 55–66.

Arndt, C. E., Kanapathippillai, P., Kluge, R. and Krapp, R. (2000). Abundance of sympagic amphipods north of Svalbard considering the ice conditions. *In* "Report of AB-310 Course at UNIS 2000" (O. J. Lønne, ed.), pp. 1–23. Longyearbyen. Available at www.unis.no.

Arndt, C. E., Fernandez-Leborans, G., Seuthe, L., Berge, J. and Gulliksen, B. (2005a). Ciliated epibionts on the Arctic sympagic amphipod *Gammarus wilkitzkii* benthic coupling. *Marine Biology* **147**, 643–652.

Arndt, C. E., Berge, J. and Brandt, A. (2005b). Mouthpart-atlas of Arctic sympagic amphipods—trophic niche separation based on mouthpart morphology and feeding ecology. *Journal of Crustacean Biology* **25**, 401–412.

Arrigo, K. R. (2003). Primary production in sea ice. *In* "Sea Ice: An Introduction to Its Physics, Chemistry, Biology and Ecology" (D. N. Thomas and G. S. Dieckmann, eds), pp. 143–183. Blackwell Publishing, Oxford.

Atkinson, A. (1998). Life cycle strategies of epipelagic copepods in the Southern Ocean. *Journal of Marine Systems* **15**, 289–311.

Atkinson, A. and Snÿder, R. (1997). Krill-copepod interactions at South Georgia, Antarctica, I. Omnivory by *Euphausia superba*. *Marine Ecology Progress Series* **160**, 63–76.

Atkinson, A., Meyer, B., Stübing, D., Hagen, W., Schmidt, K. and Bathmann, U. V. (2002). Feeding and energy budgets of Antarctic krill *Euphausia superba* at the onset of winter—II. Juveniles and adults. *Limnology and Oceanography* **47**, 953–966.

Atkinson, A., Siegel, V., Pakhomov, E. and Rothery, P. (2004). Long-term decline in krill stock and increase in salps within the Southern Ocean. *Nature* **432**, 100–103.

Auel, H. and Werner, I. (2003). Feeding, respiration and life history of the hyperiid amphipod *Themisto libellula* in the Arctic marginal ice zone of the Greenland Sea. *Journal of Experimental Marine Biology and Ecology* **296**, 183–197.

Auel, H., Harjes, M., da Rocha, R., Stübing, D. and Hagen, W. (2002). Lipid biomarkers indicate different ecological niches and trophic relationships of the Arctic hyperiid amphipods *Themisto abyssorum* and *T. libellula*. *Polar Biology* **25**, 374–383.

Auel, H., Klages, M. and Werner, I. (2003). Respiration and lipid content of the Arctic copepod *Calanus hyperboreus* overwintering 1 m above seafloor at 2300 m water depth in the Fram Strait. *Polar Biology* **143**, 275–282.

Averintzev, V. G. (1993). Cryopelagic life at Franz Josef Land. *In* "Environment and Ecosystems of the Frans Josef Land (Archipelago and Shelf)" pp. 171–186. Apatity, Kola Scientific Center, Russian Academy of Science.

Barnard, J. L. (1959). Epipelagic and under-ice Amphipoda of the central Arctic Basin. Scientific Studies at Fletcher's Ice Island T-3, 1952–1955. *Geophysical Research Paper* **63**, 115–153.

Bates, S. S. (2004). Amnesic shellfish poisoning: Domoic acid production by *Pseudo-nitzschia* diatoms. *Aqua Info Aquaculture Notes* **16**, 1–4.

Bathmann, U. V., Makarov, R. R., Spiridonov, V. A. and Rohardt, G. (1993). Winter distribution and overwintering strategies of the Antarctic copepod species *Calanoides acutus, Rhincalanus gigas* and *Calanus propinquus* (Crustacea, Calanoida) in the Weddell Sea. *Polar Biology* **13**, 333–346.

Bayly, I. A. E. (1978). The occurrence of *Paralabidocera antarctica* (I. C. Thompson) (Copepoda: Calanoida: Acartiidae) in an Antarctic saline lake. *Australian Journal of Marine and Freshwater Research* **29**, 817–824.

Bergmans, M., Dahms, H.-U. and Schminke, H. K. (1991). An *r*-strategist in Antarctic pack ice. *Oecologia* **86**, 305–309.

Beuchel, F. and Lønne, O. J. (2002). Population dynamics of the sympagic amphipods *Gammarus wilkitzkii* and *Apherusa glacialis* in sea ice north of Svalbard. *Polar Biology* **25**, 241–250.

Beuchel, F., Borgå, K., Karlsson, S. and Lilleøkdal, G. (1998). Distribution of the sympagic fauna at three different locations north of Svalbard. *In* "Report of AB-310 course at UNIS 1998" (O. J. Lønne, ed.), pp. 1–31. Longyearbyen. Available at www.unis.no.

Birula, A. A. (1937). Beiträge zur Kenntnis der Crustaceen-Fauna des Kara-Busens und des Unterlaufes des Flusses Kara. *Travaux de l'Institute Zoologique de l'Académie des Sciences de l'URSS* **4**, 701–747.

Blome, D. and Riemann, F. (1999). Antarctic sea ice nematodes, with description of *Geomonhystera glaciei* sp. nov. (Monhysteridae). *Mitteilung des Hamburgischen Zoologischen Museum Instituts* **96**, 15–20.

Boysen-Ennen, E. and Piatkowski, U. (1988). Meso- and macrozooplankton communities in the Weddell Sea, Antarctica. *Polar Biology* **9**, 17–35.

Boysen-Ennen, E., Hagen, W., Hubold, G. and Piatkowski, U. (1991). Zooplankton biomass in the ice-covered Weddell Sea, Antarctica. *Marine Biology* **111**, 227–235.

Bradford, J. M. (1978). Sea ice organisms and their importance to the Antarctic ecosystem (Review). *Antarctic Record* **1**, 43–50.

Bradstreet, M. S. W. (1982). Occurrence, habitat use, and behavior of seabirds, marine mammals, and arctic cod at the Pond Inlet ice edge. *Arctic* **35**, 28–40.

Bradstreet, M. S. W. and Cross, W. E. (1982). Trophic relationships at high Arctic ice edges. *Arctic* **35**, 1–12.

Brierley, A. S. and Thomas, D. N. (2002). Ecology of Southern Ocean pack ice. *Advances in Marine Biology* **43**, 171–277.

Brierley, A. S., Demer, D. A., Watkins, J. L. and Hewitt, R. P. (1999). Concordance of interannual fluctuations in acoustically estimated densities of Antarctic krill around South Georgia and Elephant Island: Biological evidence of same-year teleconnections across the Scotia Sea. *Marine Biology* **134**, 675–681.

Brierley, A. S., Fernandez, P. G., Brandon, M. A., Armstrong, F., Millard, N. W., McPhail, S. D., Stevenson, P., Pebody, M., Perrett, J., Squires, M., Bone, D. G. and Griffiths, G. (2002). Antarctic krill under sea ice: Elevated abundance in a narrow band just south of ice edge. *Science* **295**, 1890–1892.

Brinton, E. and Townsend, A. W. (1991). Development rates and habitat shifts in the Antarctic neritic euphausiid *Euphausia crystallorophias*, 1986–87. *Deep Sea Research* **38**, 1195–1211.

Brodsky, K. A. and Zvereva, J. A. (1976). *Paralabidocera separabilis* sp. n. and *P. antarctica* (J. C. Thompson) (Copepoda, Calanoida) from Antarctica. *Crustaceana* **31**, 233–240.

Budd, W. F. and Wu, X. (1998). Modelling long term global and Antarctic changes resulting from increased greenhouse gases. *In* "Coupled Climate Modelling" (P. J. Meighen, ed.), Bureau of Meteorology, Canberra. Australia, Research Report **69,** 71–74.

Bunt, J. S. (1963). Diatoms of Antarctic sea ice as agents of primary production. *Nature* **199**, 1255–1257.

Bunt, J. S. and Wood, E. J. F. (1963). Microalgae and Antarctic sea-ice. *Nature* **199**, 1254–1255.

Burghart, S. E., Hopkins, T. L., Vargo, G. A. and Torres, J. J. (1999). Effects of a rapidly receding ice edge on the abundance, age structure and feeding of three dominant calanoid copepods in the Weddell Sea, Antarctica. *Polar Biology* **22**, 279–288.

Cairns, A. A. (1967). The zooplankton of Tanquary Fjord, Ellesmere Island, with special reference to the calanoid copepods. *Journal of the Fisheries Research Board of Canada* **24**, 555–568.

Carey, A. G., Jr. (1985). Marine ice fauna: Arctic. *In* "Sea Ice Biota" (R. A. Horner, ed.), pp. 173–190. CRC Press, Florida.

Carey, A. G., Jr. (1992). The ice fauna in the shallow southwestern Beaufort Sea, Arctic Ocean. *Journal of Marine Systems* **3**, 225–236.

Carey, A. G., Jr. and Montagna, P. A. (1982). Arctic sea ice faunal assemblage: First approach to description and source of the underice meiofauna. *Marine Ecology Progress Series* **8**, 1–8.

Comiso, J. C. (2002). A rapidly declining perennial sea ice cover in the Arctic. *Geophysical Research Letters* **29**, 17.1–17.4.

Comiso, J. C. (2003). Large-scale characteristics and variability of the global sea ice cover. *In* "Sea Ice: An Introduction to Its Physics, Chemistry, Biology and Ecology" (D. N. Thomas and G. S. Dieckmann, eds), pp. 112–142. Blackwell Publishing, Oxford.

Conover, R. J. and Huntley, M. (1991). Copepods in ice-covered seas—distribution, adaptations to seasonally limited food, metabolism, growth patterns and life cycle strategies in polar seas. *Journal of Marine Systems* **2**, 1–41.

Conover, R. J. and Siferd, T. D. (1993). Dark-season survival strategies of coastal zone zooplankton in the Canadian Arctic. *Arctic* **4**, 303–311.

Conover, R. J., Herman, A. W., Preusenberg, S. S. and Harris, L. R. (1986). Distribution of and feeding by the copepod *Pseudocalanus* under fast ice during the arctic spring. *Science* **232**, 1245–1247.

Conover, R. J., Bedo, A. W., Hermann, A. W., Head, E. J. H., Harris, L. R. and Horne, E. P. W. (1988). Never trust a copepod—some observations on their behaviour in the Canadian Arctic. *Bulletin of Marine Science* **43**, 650–662.

Conover, R. J., Harris, L. R. and Bedo, A. W. (1991). Copepods in cold oligotrophic waters—how do they cope? *Bulletin of the Plankton Society of Japan* **38**, 177–199.

Corkett, C. J. and McLaren, I. A. (1978). The biology of *Pseudocalanus*. *Advances in Marine Biology* **15**, 1–231.

Costanzo, G., Zagami, G., Crescenti, N. and Granata, A. I. (2002). Naupliar development of *Stephos longipes* (Copepoda: Calanoida) from the annual sea ice of Terra Nova Bay, Antarctica. *Journal of Crustacean Biology* **22**, 855–860.

Cota, G., Legendre, L., Gosselin, M. and Ingram, R. G. (1991). Ecology of bottom ice algae: I. Environmental controls and variability. *Journal of Marine Systems* **2**, 257–278.

Cross, W. E. (1982). Under-ice biota at the Pond Inlet ice edge and in adjacent fast ice areas during spring. *Arctic* **35**, 13–27.

Dahms, H.-U. and Schminke, H. K. (1992). Sea ice inhabiting Harpacticoida (Crustacea, Copepoda) of the Weddell Sea (Antarctica). *Bulletin de l'Institute Royal des Sciences naturelles de Belgique* **62**, 91–123.

Dahms, H.-U. and Quian, P.-Y. (2004). Life histories of the Harpacticoida (Copepoda, Crustacea): A comparison with meiofauna and macrofauna. *Journal of Natural History* **38**, 1725–1734.

Dahms, H.-U., Bergmans, M. and Schminke, H. K. (1990). Distribution and adaptations of sea ice inhabiting Harpacticoida (Crustacea, Copepoda) of the Weddell Sea (Antarctica). *Marine Ecology* **11**, 207–226.

Dalpadado, P., Borkner, N., Bogstad, B. and Mehl, S. (2001). Distribution of *Themisto* (Amphipoda) spp. in the Barents Sea and predator–prey interactions. *Journal of Marine Science* **58**, 876–895.

Daly, K. L. (1990). Overwintering development, growth, and feeding of larval *Euphausia superba* in the Antarctic marginal ice zone. *Limnology and Oceanography* **35**, 1564–1576.

Daly, K. L. (2004). Overwintering, growth and development of larval *Euphausia superba*: An interannual comparison under varying environmental conditions west of the Antarctic Peninsula. *Deep-Sea Research II* **51**, 2139–2168.

Daly, K. L. and Macaulay, M. C. (1991). Influence of physical and biological mesoscale dynamics on the seasonal distribution and behaviour of *Euphausia superba* in the Antarctic marginal ice zone. *Marine Ecology Progress Series* **79**, 37–66.

Dauby, P., Scailteur, Y. and De Broyer, C. (2001). Trophic diversity within the eastern Weddell Sea amphipod community. *Hydrobiologia* **443**, 69–86.

De Broyer, C., Scailteur, Y., Chapelle, G. and Rauschert, M. (2001). Diversity of epibenthic habitats of gammaridean amphipods in the eastern Weddell Sea. *Polar Biology* **24**, 744–753.

de la Mare, W. K. (1997). Abrupt mid-twentieth-century decline in Antarctic sea-ice extent from whaling records. *Nature* **389**, 57–60.

Delgado, J. R., Jaña, R. and Marin, V. (1998). Testing hypotheses on life-cycle models for Antarctic calanoid copepods, using qualitative, winter, zooplankton samples. *Polar Biology* **20**, 74–76.

Demers, S., Legendre, L., Maestrini, S., Rochet, M. and Ingram, R. G. (1989). Nitrogenous nutrition of sea-ice microalgae. *Polar Biology* **9**, 377–383.

Dhargalkar, V. K., Burton, H. R. and Kirkwood, J. M. (1988). Animal associations with the dominant species of shallow water macrophytes along the coastline of the Vestfold Hills, Antarctica. *Hydrobiologia* **165**, 141–150.

Dieckmann, G. S. and Hellmer, H. H. (2003). The importance of sea ice: An overview. *In* "Sea Ice: An introduction to Its Physics, Chemistry, Biology and

Ecology" (D. N. Thomas and G. S. Dieckmann, eds), pp. 1–22. Blackwell Publishing, Oxford.

Dieckmann, G. S., Arrigo, K. R. and Sullivan, C. W. (1992). A high-resolution sampler for nutrient and chlorophyll *a* profiles of the sea ice platelet layer and underlying water column below fast ice in polar oceans: Preliminary results. *Marine Ecology Progress Series* **80**, 291–300.

Diel, S. (1991). Zur Lebensgeschichte dominanter Copepodenarten (*Calanus finmarchicus, C. glacialis, C. hyperboreus, Metridia longa*) in der Framstraße. *Reports on Polar Research* **88**, 133.

Drobysheva, S. S. (1982). Degree of isolation of *Thysanoessa inermis* (Krøyer) and *Th. raschii* (M. Sars) (Crustacea, Euphausiacea) populations in the southern Barents Sea. *International Council for Exploration of the Sea* 21CM/1982 L:**19**, 21.

Dunbar, M. J. (1954). The amphipod crustacea of Ungava Bay, Canadian Eastern Arctic. *Journal of the Fisheries Research Board of Canada* **11**, 709–798.

Dunton, K. H., Reimnitz, E. and Schonberg, S. (1982). An arctic kelp community in the Alaskan Beaufort Sea. *Arctic* **35**, 464–484.

Duquense, S., Riddle, M. J., Schulz, K. L. and Liess, M. (2000). Effects of contaminants in the Antarctic environment—potential of the gammarid crustacean *Paramoera walkeri* as a biological indicator for Antarctic ecosystems based on toxicity and bioaccumulation of copper and cadmium. *Aquatic Toxicology* **49**, 131–143.

Eiane, K. and Daase, M. (2002). Observations of mass mortality of *Themisto libellula* (Amphipoda, Hyperidae). *Polar Biology* **25**, 396–398.

Eicken, H. and Lange, M. A. (1989). Development and properties of sea ice in the coastal regime of the southeastern Weddell Sea. *Journal of Geophysical Research* **94**, 8193–8206.

Einarsson, H. (1945). Euphausiacea. I. Northern Atlantic species. *Dana Report* No. 27.

El-Sayed, S. Z. (1971). Biological aspects of the pack ice ecosystem. *In* "Symposium on Antarctic Ice and Water Masses" (G. E. R. Deacon, ed.), pp. 35–54. Scientific Committee on Antarctic Research, Cambridge.

Evans, F. (1977). Seasonal density and production estimates of the commoner planktonic copepods of Northumberland coastal waters. *Estuarine and Coastal Marine Science* **5**, 223–241.

Everson, I. (2000). "Krill: Biology, Ecology and Fisheries" Blackwell, Oxford.

Falk-Petersen, S., Sargent, J. R. and Tande, K. (1987). Food pathways and life strategy in relation to the lipid composition of sub-Arctic zooplankton. *Polar Biology* **8**, 115–120.

Falk-Petersen, S., Hopkins, C. and Sargent, J. R. (1990). Trophic relationships in the pelagic, Arctic food web. *In* "Trophic Relationships in the Marine Environment" (M. Barnes and R. N. Gibson, eds), "Proceedings of the 24th European Marine Biological Symposium", pp. 315–333. Aberdeen University Press, Aberdeen.

Falk-Petersen, S., Hagen, W., Kattner, G., Clarke, A. and Sargent, J. R. (2000). Lipids, trophic relationships, and biodiversity in Arctic and Antarctic krill. *Canadian Journal of Fisheries and Aquatic Science* **57**, 179–191.

Feder, H. M. and Lewett, S. C. (1981). Feeding interaction in the eastern Bering Sea with emphasis on the benthos. *In* "The eastern Bering Sea shelf: Oceanography and Resources, Vol. II" (D. W. Hood and J. Calder, eds), pp. 1229–1261. University of Washington Press, Seattle.

Fortier, L., Fortier, M. and Demers, S. (1995). Zooplankton and larval fish community development: Comparative study under first-year sea ice at low and high

latitudes in the northern hemisphere. *Proceedings of the NIPR Symposium on Polar Biology* **8**, 11–19.
Fortier, M., Fortier, L., Hattori, H., Saito, H. and Legendre, L. (2001). Visual predators and the diel vertical migration of copepods under Arctic sea ice during the midnight sun. *Journal of Plankton Research* **23**, 1263–1278.
Foster, B. A. (1987). Composition and abundance of zooplankton under the spring sea-ice of McMurdo Sound, Antarctica. *Polar Biology* **8**, 41–48.
Fransz, H. G. (1988). Vernal abundance, structure and development of epipelagic copepod populations of the eastern Weddell Sea (Antarctica). *Polar Biology* **9**, 107–114.
Fraser, W. R. and Hofmann, E. E. (2003). A predator's perspective on causal links between climate change, physical forcing and ecosystem response. *Marine Ecology Progress Series* **265**, 1–15.
Frazer, T. K., Quetin, L. B. and Ross, R. M. (2002). Abundance, sizes and development of larval krill, *Euphausia superba*, during winter in ice-covered seas west of the Antarctic Peninsula. *Journal of Plankton Research* **24**, 1067–1077.
Friedrich, C. (1997). Ökologische Untersuchungen zur Fauna des arktischen Meereises (Ecological investigations on the fauna of the Arctic sea-ice). *Berichte zur Polarforschung* **246**, 211.
Gallienne, C. P. and Robins, D. B. (2001). Is *Oithona* the most important copepod in the world's oceans. *Journal of Plankton Research* **23**, 1421–1432.
Garrison, D. L. (1991). Antarctic sea ice biota. *American Zoologist* **31**, 17–33.
Garrison, D. L. and Buck, K. R. (1989). The biota of Antarctic pack ice in the Weddell Sea and Antarctic Peninsula regions. *Polar Biology* **10**, 211–219.
George, R. Y. and Paul, A. Z. (1970). "USC-FSU-Biological Investigations from the Fletcher's Ice Island T-3 on Deep-Sea and Under-Ice Benthos of the Arctic Ocean". University of Southern California, Department of Biological Sciences, Los Angeles.
Gibson, J. A. E., Swadling, K. M. and Burton, H. R. (1997). Interannual variation in dominant phytoplankton species and biomass near Davis Station, East Antarctica. *Proceedings of the NIPR Symposium on Polar Biology* **10**, 78–90.
Gloersen, P., Campbell, W. J., Cavalieri, D., Comiso, J. C., Parkinson, C. L. and Zwally, H. J. (1993). Satellite passive microwave observations and analysis of arctic and Antarctic sea ice, 1978–1987. *Annals of Glaciology* **17**, 149–154.
Golikov, A. N. and Averintzev, V. G. (1977). Distribution patterns of benthic and ice biocoenoses in the high latitudes of the Polar Basin and their part in the biological structure of the world ocean. *In* "Polar Oceans" (M. J. Dunbar, ed.), pp. 331–364. Arctic Institute of North America, Canada.
Golikov, A. N. and Scarlato, O. A. (1973). Comparative characteristics of some ecosystems of the upper regions of the shelf in tropical, temperate and Arctic waters. *Helgoländer Wissenschaftliche Meeresuntersuchungen* **24**, 219–234.
Gonzalez, H. and Smetacek, V. (1994). The possible role of the cyclopoid copepod *Oithona* in retarding vertical flux of zooplankton material. *Marine Ecology Progress Series* **113**, 233–246.
Gradinger, R. (1999). Integrated abundance and biomass of sympagic meiofauna in Arctic and Antarctic pack ice. *Polar Biology* **22**, 169–177.
Gradinger, R. and Bluhm, B. (2004). *In-situ* observations on the distribution and behavior of amphipods and Arctic cod (*Boreogadus saida*) under the sea ice of the High Arctic Canada Basin. *Polar Biology* **27**, 595–603.
Gradinger, R., Spindler, M. and Henschel, D. (1991). Development of Arctic sea ice organisms under graded snow cover. *Polar Research* **10**, 295–307.

Gradinger, R., Spindler, M. and Weissenberger, J. (1992). On the structure and development of Arctic pack ice communities in Fram Strait: A multivariate approach. *Polar Biology* **12**, 727–733.
Gradinger, R., Friedrich, C. and Spindler, M. (1999). Abundance, biomass and composition of the sea ice biota of the Greenland Sea pack ice. *Deep-Sea Research II* **46**, 1457–1472.
Gradinger, R., Meiners, K., Plumley, G., Zhang, Q. and Bluhm, B. A. (2005). Abundance and composition of the sea-ice meiofauna in off-shore pack ice of the Beaufort Gyre in summer 2002 and 2003. *Polar Biology* **28**, 171–181.
Graeve, M., Kattner, G. and Piepenburg, D. (1997). Lipids in Arctic benthos: Does the fatty acid and alcohol composition reflect feeding and trophic interactions. *Polar Biology* **18**, 53–61.
Grainger, E. H. (1962). Zooplankton of the Foxe Basin in the Canadian Arctic. *Journal of the Fisheries Research Board of Canada* **19**, 377–400.
Grainger, E. H. and Hsiao, S. I. C. (1990). Trophic relationships of the sea ice meiofauna in Frobisher Bay, Arctic Canada. *Polar Biology* **10**, 283–292.
Grainger, E. H. and Mohammed, A. A. (1986). Copepods in arctic sea ice. *In* "Proceedings of the second International Conference on Copepoda" (G. Schiever, H. K. Schminke and C.-T. Shih, eds), pp. 303–310. National Museums of Canada, Ottawa.
Grainger, E. H. and Mohammed, A. A. (1990). High salinity tolerance in sea ice copepods. *Ophelia* **31**, 177–185.
Grainger, E. H. and Mohammed, A. A. (1991). Some diagnostic characters of copepodid stages of the cyclopoid copepod *Cyclopina schneideri* T. Scott and adults of arctic marine Cyclopinidae. *Canadian Journal of Zoology* **69**, 2365–2373.
Grainger, E. H., Mohammed, A. A. and Lovrity, J. E. (1985). The sea ice fauna of the Frobisher Bay, Arctic Canada. *Arctic* **38**, 23–30.
Green, J. M. and Steele, D. H. (1977). "Observations on Marine Life Beneath Sea Ice, Resolute Bay, N.W.T. Proceedings of the Circumpolar Conference on Northern Ecology, Ottawa, 1975". National Research Council Canada.
Griffiths, W. B. and Dillinger, R. E. (1981). Invertebrates. *In* "Environmental Assessment of the Alaskan Continental Shelf". Final reports of principal investigators, Vol. 8. Biological studies.
Grønvik, W. and Hopkins, C. C. E. (1984). Ecological investigations of the zooplankton community of Balsfjorden, Northern Norway: Generation cycle, seasonal vertical distribution, and seasonal variations in body weight and carbon and nitrogen content of the copepod *Metridia longa* (Lubbock). *Journal of Experimental Marine Biology and Ecology* **80**, 93–107.
Gruzov, E. N. (1977). Seasonal alterations in coastal communities in the Davis Sea. *In* "Adaptations within Antarctic Ecosystems" (G. A. Llano, ed.), pp. 263–278. Gulf Publishing Co., Houston.
Gruzov, Y. N., Propp, M. V. and Pushkin, A. F. (1967). Biological associations of coastal areas of the Davis Sea (based on the observations of divers). *Soviet Antarctic Expedition Information Bulletin* **6**, 523–533.
Gruzov, Y. N., Propp, M. V. and Pushkin, A. F. (1968). Hydrobiological diving work in the Antarctic. *Soviet Antarctic Expedition Information Bulletin* **6**, 405–408.
Gulliksen, B. (1984). Under-ice fauna from Svalbard waters. *Sarsia* **69**, 17–23.
Gulliksen, B. and Lønne, O. J. (1989). Distribution, abundance, and ecological importance of marine sympagic fauna in the Arctic. *Rapports et Procès-Verbaux des Réunions, International Council for the Exploration of the Sea* **188**, 133–138.

Gulliksen, B. and Lønne, O. J. (1991). Sea ice macrofauna in the Antarctic and the Arctic. *Journal of Marine Systems* **2**, 53–61.
Gulliksen, B., Palerud, R., Brattegard, T. and Sneli, J.-A. (1999). "Distribution of marine benthic macro-organisms at Svalbard (including Bear Island) and Jan Mayen", Research Report for DN 1999-4. Directorate for Nature Management.
Günther, S., George, K. H. and Gleitz, M. (1999). High sympagic metazoan abundance in platelet layers at Drescher Inlet, Weddell Sea, Antarctica. *Polar Biology* **22**, 82–89.
Gurjanova, E. (1936). Beiträge zur Amphipodenfauna des Karischen Meeres. *Zoologischer Anzeiger* **116**, 145–152.
Gurjanova, E. F. (1951). "Beach Hoppers of the Seas of the USSR and Adjacent Waters (Amphipoda: Gammaridea)". Fauna USSR, Academy of Science of the USSR, Moscow.
Hagen, W. (1988). Zur Bedeutung der Lipide im antarktischen Zooplankton (on the significance of lipids in Antarctic zooplankton). *Berichte zur Polarforschung* **49**, 1–129.
Hagen, W. and Schnack-Schiel, S. B. (1996). Seasonal lipid dynamics in dominant Antarctic copepods: Energy for overwintering or reproduction? *Deep-Sea Research I* **43**, 139–158.
Hagen, W., Kattner, G. and Graeve, M. (1993). *Calanoides acutus* and *Calanus propinquus*, Antarctic copepods with different lipid storage modes via wax esters or triacylglycerols. *Marine Ecology Progress Series* **97**, 135–142.
Hagen, W., Kattner, G., Terbrüggen, A. and VanVleet, E. S. (2001). Lipid metabolism of the Antarctic krill *Euphausia superba*. *Marine Biology* **139**, 95–104.
Hall, M. O. and Bell, S. S. (1993). Meiofauna on the seagrass *Thalassia testudinum*—population characteristics of harpacticoid copepods and associations with algal epiphytes. *Marine Biology* **116**, 137–146.
Hamner, W. M., Hamner, P. P., Strand, S. W. and Gilmer, R. W. (1983). Behaviour of Antarctic krill, *Euphausia superba*: Chemoreception, feeding, schooling, and molting. *Science* **220**, 433–435.
Hamner, W. M., Hamner, P. P., Obst, B. S. and Carleton, J. H. (1989). Field observations on the ontogeny of schooling of *Euphausia superba* furciliae and its relationship to ice in Antarctic waters. *Limnology and Oceanography* **34**, 451–456.
Harrington, S. A. and Thomas, P. G. (1987). Observations on spawning by *Euphausia crystallorophias* from waters adjacent to Enderby Land (East Antarctica) and speculations on the early ontogenetic ecology of neritic euphausiids. *Polar Biology* **7**, 93–95.
Hempel, I and Hempel, G. (1986). Field observations on the developmental ascent of larval *Euphausia superba* (Crustacea). *Polar Biology* **6**, 121–126.
Herman, Y. and Hopkins, D. (1980). Arctic Ocean climate in late Cenozoic time. *Science* **209**, 557–562.
Hicks, G. and Coull, B. (1983). The ecology of marine meiobenthic harpacticoid copepods. *Oceanography and Marine Biology: An Annual Review* **21**, 67–175.
Hirche, H.-J. (1997). Life cycle of the copepod *Calanus hyperboreus* in the Greenland Sea. *Marine Biology* **128**, 607–618.
Hirche, H.-J. and Kattner, G. (1993). Egg production and lipid content of *Calanus glacialis* in spring: Indication of a food-dependent and food-independent reproductive mode. *Marine Biology* **117**, 615–622.
Hirche, H.-J. and Niehoff, B. (1996). Reproduction of the Arctic copepod *Calanus hyperboreus* in the Greenland Sea—field and laboratory observations. *Polar Biology* **16**, 209–219.

Hofmann, E. E., Wiebe, P. H., Costa, D. P. and Torres, J. J. (2004). Integrated ecosystem studies of western Antarctica Peninsula continental shelf waters and related southern ocean regions. *Deep-Sea Research II* **51**, 1921–2344.

Holloway, G. and Sou, T. (2001). Is Arctic sea ice rapidly thinning? *Ice and Climate News* **1**. Available online at acsys.npolar.no/news/2001/No1_p2.htm.

Holt, E. W. and Tattersall, W. M. (1906). Preliminary notice of the Schizopoda collected by H.M.S. Discovery in the Antarctic Region. *Annals and Magazine of Natural History* **7**, 1–11.

Hop, H., Poltermann, M., Lønne, O. J., Falk-Petersen, S., Krosnes, R. and Budgell, W. P. (2000). Ice amphipod distribution relative to ice density and under-ice topography in the northern Barents Sea. *Polar Biology* **23**, 357–367.

Hop, H., Pearson, T., Hegseth, E. N., Kovacs, K. M., Wiencke, C., Kwasniewski, S., Eiane, K., Mehlum, F., Gulliksen, B., Wlodarska-Kowalczuk, M., Lydersen, C., Weslawski, J. M., Cochrane, S., Gabrielsen, G. W., Leakey, R. J. G., Lønne, O. J., Zajaczkowski, M., Falk-Petersen, S., Kendall, M., Wängberg, S.-Å., Bischof, K., Voronkov, A. Y., Kovaltchouk, N. A., Wiktor, J., Poltermann, M., die Prisco, G., Papucci, C. and Gerland, S. (2002). The marine ecosystem of Kongsfjorden, Svalbard. *Polar Research* **21**, 167–208.

Hopkins, T. L. (1985). Food web of an Antarctic midwater ecosystem. *Marine Biology* **89**, 197–212.

Hopkins, T. (1987). Midwater food web in McMurdo Sound, Ross Sea, Antarctica. *Marine Biology* **96**, 93–106.

Hopkins, C. C. E., Tande, K. S., Gronvik, S. and Sargent, J. R. (1984). Ecological investigations of the zooplankton community of Balsfjorden, northern Norway: An analysis of growth and overwintering tactics in relation to niche and environment in *Metridia longa* (Lubbock), *Calanus finmarchicus* (Gunnerus), *Thysanoessa inermis* (Krøyer) and *T. rashi* (M. Sars). *Journal of Experimental Marine Biology and Ecology* **82**, 77–99.

Horner, R. A. (1972). Ecological studies on arctic sea ice organisms. Institute of Marine Science, University of Alaska, Fairbanks. Report R 72-17.

Horner, R. A. (1985). "Sea Ice Biota". CRC Press, Florida.

Horner, R., Ackley, S. F., Dieckmann, G. S., Gulliksen, B., Hoshiai, T., Legendre, L., Melnikov, I. A., Reeburgh, W. S., Spindler, M. and Sullivan, C. W. (1992). Ecology of sea ice biota 1. Habitat, terminology and methodology. *Polar Biology* **12**, 417–427.

Hoshiai, T. and Tanimura, A. (1986). Sea ice meiofauna at Syowa Station, Antarctica. *Memoirs of the National Institute of Polar Research, Special Issue* **44**, 118–124.

Hoshiai, T., Tanimura, A. and Watanabe, K. (1987). Ice algae as food of an Antarctic ice-associated copepod, *Paralabidocera antarctica* (I. C. Thompson). *Proceedings of the NIPR Symposium on Polar Biology* **1**, 105–111.

Hoshiai, T., Tanimura, A., Fukuchi, M. and Watanabe, K. (1989). Feeding by the nototheniid fish, *Pagothenia borchgrevinki* on the ice-associated copepod, *Paralabidocera antarctica*. *Proceedings of the National Institute of Polar Research Symposium on Polar Biology* **2**, 61–64.

Ikeda, T. (1986). Preliminary observations on the development of the larvae of *Euphausia crystallorophias* Holt and Tattersall in the laboratory (extended abstract). *Memoirs of the National Institute of Polar Research Special Issue* **40**, 183–186.

Ikeda, T. and Dixon, P. (1982). Body shrinkage as a possible over-wintering mechanism of the Antarctic krill, *Euphausia superba* Dana. *Journal of Experimental Marine Biology and Ecology* **62**, 143–151.

Ikeda, T. and Kirkwood, R. (1989). Metabolism and body composition of two euphausiids (*Euphausia superba* and *E. crystallorophias*) collected from under the pack-ice off Enderby Land, Antarctica. *Marine Biology* **100**, 301–308.

Ito, T. and Fukuchi, M. (1978). *Harpacticus furcatus* Lang from the Antarctic Peninsula, with reference to the copepodid stages (Copepoda: Harpacticoida). *Antarctic Record* **61**, 40–64.

Johnson, G. L. (1990). Morphology and plate tectonics: The modern polar oceans. *In* "Geological History of the Polar Oceans: Arctic versus Antarctic" (U. Bleil and J. Thiede, eds), pp. 11–28. North Atlantic Treaty Organization, ASI Series C.

Johnson, M. W. and Olson, J. B. (1948). The life history and biology of a marine harpacticoid copepod, *Tisbe furcata* (Baird). *Biological Bulletin* **95**, 320–332.

Jones, R. H. and Flynn, K. J. (2005). Nutritional status and diet composition affect the value of diatoms as copepod prey. *Science* **307**, 1457–1459.

Ju, S.-J. and Harvey, H. R. (2004). Lipids as markers of nutritional condition and diet in the Antarctic krill *Euphausia superba* and *Euphausia crystallorophias* during austral winter. *Deep-Sea Research II* **51**, 2199–2214.

Kattner, G. and Hagen, W. (1995). Polar herbivorous copepods—different pathways in lipid biosynthesis. *ICES Journal of Marine Science* **52**, 329–335.

Kattner, G. and Hagen, W. (1998). Lipid metabolism of the Antarctic euphausiid *Euphausia crystallorophias* and its ecological implications. *Marine Ecology Progress Series* **170**, 203–213.

Kattner, G., Graeve, M. and Hagen, W. (1994). Ontogenetic and seasonal changes in lipid and fatty acid/alcohol compositions of the dominant Antarctic copepods *Calanus propinquus, Calanoides acutus* and *Rhincalanus gigas*. *Marine Biology* **118**, 637–644.

Kattner, G., Albers, C., Graeve, M. and Schnack-Schiel, S. B. (2003). Fatty acid and alcohol composition of the small polar copepods, *Oithona* and *Oncaea*: Indication on feeding modes. *Polar Biology* **26**, 666–671.

Kawaguchi, S. and Nicol, S. (eds), (2003). Proceedings of the International Workshop on Understanding Living Krill for Improved Management and Stock Assessment. *Marine and Freshwater Behaviour and Physiology* **36**, 1–307.

Kern, J. C. and Carey, A. G., Jr. (1983). The faunal assemblage inhabiting seasonal sea ice in the nearshore Arctic Ocean with emphasis on copepods. *Marine Ecology Progress Series* **10**, 159–167.

Kirkwood, J. M. (1993). Zooplankton community dynamics and diel vertical migration in Ellis Fjord, Vestfold Hills, Antarctica. PhD thesis, Monash University, Victoria, Australia.

Kirkwood, J. M. (1996). The developmental rate of *Euphausia crystallorophias* larvae in Ellis Fjord, Vestfold Hills, Antarctica. *Polar Biology* **16**, 527–530.

Kirkwood, J. M. and Burton, H. R. (1987). Three new zooplankton nets designed for under-ice sampling; with preliminary results of collections made from Ellis Fjord, Antarctica during 1985. *Proceedings of the NIPR Symposium on Polar Biology* **1**, 112–122.

Kittel, W. and Ligowski, R. (1980). Algae found in the food of *Euphausia crystallorophias* (Crustacea). *Polish Polar Research* **1**, 129–137.

Klekowski, R. Z., Opalinski, K. W. and Rakusa-Suszczewski, S. (1973). Respiration of Antarctic amphipoda *Paramoera walkeri* Stebbing during the winter season. *Polskie Archiwum Hydrobiologii* **20**, 301–308.

Klekowski, R. Z. and Weslawski, J. M. (1991). "Atlas of the Marine Fauna of Southern Spitsbergen", Vol. 2, "Invertebrates, Part I". Polish Academy of Science, Gdansk.

Kotwicki, L. (2002). Benthic Harpacticoida (Crustacea, Copepoda) from the Svalbard archipelago. *Polish Polar Research* **23**, 185–191.

Krembs, C., Mock, T. and Gradinger, R. (2001). A mesocosm study of physical–biological interactions in artificial sea ice: Effects of brine channel surface evolution and brine movement on algal biomass. *Polar Biology* **24**, 356–364.

Krembs, C., Tuschling, K. and Juterzenka, K. V. (2002). The topography of the ice-water interface—its influence on the colonization of sea ice by algae. *Polar Biology* **25**, 106–117.

Kurbjeweit, F. (1993). Reproduktion und Lebenszyklen dominanter Copepodenarten aus dem Weddellmeer, Antarktis (reproduction and life cycles of dominant copepod species from the Weddell Sea, Antarctica). *Berichte zur Polarforschung* **129**, 1–238.

Kurbjeweit, F., Gradinger, R. and Weissenberger, J. (1993). The life cycle of *Stephos longipes*—an example of cryopelagic coupling in the Weddell Sea (Antarctica). *Marine Ecology Progress Series* **98**, 255–262.

Lee, R. F. (1975). Lipids of arctic zooplankton. *Comparative Biochemistry and Physiology* **51**, 263–266.

Liess, M., Champeau, O., Riddle, M. J., Schulz, R. and Duquense, S. (2001). Combined effects of ultraviolet-B radiation and food shortage on the sensitivity of the Antarctic amphipod *Paramoera walkeri* to copper. *Environmental Toxicology and Chemistry* **20**, 2088–2092.

Loeb, V., Siegel, V., Holm-Hansen, O., Hewitt, R., Fraser, W., Trivelpiece, W. and Trivelpiece, S. (1997). Effects of sea ice extent and krill or salp dominance on the Antarctic food web. *Nature* **387**, 897–900.

Lønne, O. J. (1988). A diver-operated electric suction sampler for sympagic (= under-ice) invertebrates. *Polar Research* **6**, 135–136.

Lønne, O. J. and Gulliksen, B. (1991a). On the distribution of sympagic macro-fauna in the seasonally ice covered Barents Sea. *Polar Biology* **11**, 457–469.

Lønne, O. J. and Gulliksen, B. (1991b). Sympagic macro-fauna from multiyear sea-ice near Svalbard. *Polar Biology* **11**, 471–477.

Lubin, D. and Massom, R. (2005). "Polar Remote Sensing: Atmosphere and Polar Oceans". Springer Praxis Books, Berlin (Germany) and Chichester (UK).

MacGinitie, G. E. (1955). Distribution and ecology of the marine invertebrates of Point Barrow, Alaska. *Smithsonian Miscellaneous Collection* **128**, 1–201.

MacLellan, D. C. (1967). The annual cycle of certain calanoid species in West Greenland. *Canadian Journal of Zoology* **45**, 101–105.

Mangel, M. and Nicol, S. (eds), (2000). Proceedings of the Second International Krill Symposium, Santa Cruz, California, August 1999. *Canadian Journal of Fisheries and Aquatic Sciences Supplement* **3**, 1–202.

Marr, J. W. S. (1962). The natural history and geography of the Antarctic krill (*Euphausia superba* Dana). *Discovery Reports* **32**, 33–464.

Marschall, H.-P. (1988). The overwintering strategy of Antarctic krill under the pack-ice of the Weddell Sea. *Polar Biology* **9**, 129–135.

Matthews, J. B. L., Hestad, L. and Bakke, J. L. W. (1978). Ecological studies in Korsfjorden, Western Norway. The generations and stocks of *Calanus hyperboreus* and C. *finmarchicus* in 1971–1974. *Oceanologica Acta* **1**, 277–284.

Mauchline, J. (1998). The biology of calanoid copepods. *Advances in Marine Biology* **33**, 1–710.

Maykut, G. A. (1985). The ice environment. *In* "Sea Ice Biota" (R. A. Horner, ed.), pp. 21–82. CRC Press, Florida.

Mayzaud, P., Albessard, E. and Cuzin-Roudy, J. (1998). Changes in lipid composition of the Antarctic krill *Euphausia superba* in the Indian sector of the Antarctic Ocean: Influence of geographical location, sexual maturity stage and distribution among organs. *Marine Ecology Progress Series* **173**, 149–162.

McAllen, R. and Block, W. (1997). Aspects of the cryobiology of the intertidal harpacticoid copepod *Tigriopus brevicornis* (O. F. Müller). *Cryobiology* **35**, 309–317.

McConville, M. J., Mitchell, C. and Wetherbee, R. (1985). Patterns of carbon assimilation in a microalgal community from annual sea ice, East Antarctica. *Polar Biology* **4**, 135–141.

McGrath-Grossi, S., Kottmeier, S. T., Mo, R. L., Taylor, G. T. and Sullivan, C. W. (1987). Sea ice microbial communities. VI. Growth and primary production in bottom ice under graded snow cover. *Marine Ecology Progress Series* **35**, 153–164.

McLaren, I. A. (1969). Population and production ecology of zooplankton in Ogac Lake, a landlocked fjord on Baffin Island. *Journal of the Fisheries Research Board of Canada* **26**, 1485–1559.

Melnikov, I. A. (1989). Ecology of Arctic Ocean cryopelagic fauna. *In* "The Arctic Seas—Climatology, Oceanography, Geology and Biology" (Y. E. Hermann, ed.), pp. 235–256. Van Nostrand Reinhold, New York.

Melnikov, I. A. (1997). "The Arctic Sea Ice Ecosystem". Gordon and Breach Science Publishers, Amsterdam.

Melnikov, I. A. and Kulikov, A. S. (1980). The cryopelagic fauna of the central Arctic Basin. *In* "Biology of the Central Arctic Basin" (M. E. Vinogradov and I. A. Melnikov, eds), pp. 97–111. Nauka, Moscow.

Melnikov, I. A. and Spiridonov, V. A. (1996). Antarctic krill under perennial sea ice in the western Weddell Sea. *Antarctic Science* **8**, 323–329.

Melnikov, I. A., Zhitina, L. S. and Kolosova, H. G. (2001). The Arctic sea ice biological communities in recent environmental changes. *Memoirs of the National Institute of Polar Research* **54**, 409–416.

Melnikov, I. A., Kolosova, E. G., Welch, H. E. and Zhitina, L. S. (2002). Sea ice biological communities and nutrient dynamics in the Canada Basin of the Arctic Ocean. *Deep-Sea Research I* **49**, 1623–1649.

Menshenina, L. L. and Melnikov, I. A. (1995). Under-ice zooplankton of the western Weddell Sea. *Polar Biology* **8**, 126–138.

Metz, C. (1995). Seasonal variation in the distribution and abundance of *Oithona* and *Oncaea* species (Copepoda, Crustacea) in the southeastern Weddell Sea, Antarctica. *Polar Biology* **15**, 187–194.

Metz, C. (1996). Lebensstrategien dominanter antarktischer Oithonidae (Cyclopoida, Copepoda) und Oncaeidae (Poecilostomatoida, Copepoda) im Bellingshausenmeer. (Life strategies of dominant Antarctic Oithonidae (Cyclopoida, Copepoda) and Oncaeidae (Poecilostomatoida, Copepoda) in the Bellingshausen Sea). *Berichte zur Polarforschung* **207**, 1–109.

Metz, C. (1998). Feeding of *Oncaea curvata* (Poecilostomatoida, Copepoda). *Marine Ecology Progress Series* **169**, 229–235.

Metz, C. and Schnack-Schiel, S. B. (1995). Observations on carnivorous feeding in Antarctic calanoid copepods. *Marine Ecology Progress Series* **129**, 71–75.

Meyer, B. and Oettl, B. (2005). Effects of short-term starvation on composition and metabolism of larval Antarctic krill *Euphausia superba*. *Marine Ecology Progress Series* **292**, 263–270.

Meyer, B., Atkinson, A., Stübing, D., Oettl, B., Hagen, W. and Bathmann, U. V. (2002). Feeding and energy budgets of Antarctic krill *Euphausia superba* at the onset of winter—I. Furcilia III larvae. *Limnology and Oceanography* **47**, 943–952.

Michel, C., Nielsen, T. G., Nozais, C. and Gosselin, M. (2002). Significance of sedimentation and grazing by ice micro- and meiofauna for carbon cycling in annual sea ice (northern Baffin Bay). *Aquatic Microbial Ecology* **30**, 57–68.

Michels, J. and Schnack-Schiel, S. B. (2005). Feeding in dominant Antarctic copepods—does the morphology of the mandibular gnathobases relate to diet? *Marine Biology* **146**, 483–495.

Mohr, J. L. and Tibbs, J. (1963). Ecology of ice substrates. *In* "Proceedings of the Arctic Basin Symposium, October 1962", pp. 245–248. Arctic Institute of North America.

Montagna, P. A. and Carey, A. G., Jr. (1978). Distributional notes on Harpacticoida (Crustacea: Copepoda) collected from the Beaufort Sea (Arctic Ocean). *Astarte* **11**, 117–122.

Mueller, D. R., Vincent, W. F. and Jeffries, M. O. (2003). Break-up of the largest Arctic ice shelf and associated loss of an epishelf lake. *Geophysical Research Letters* **30**(20), 2031.

Mumm, N. (1991). Zur sommerlichen Verteilung des Mesozooplanktons im Nansen-Becken, Nordpolarmeer. *Berichte zur Polarforschung* **154**, 1–90.

Nansen, F. (1897). "Furthest North". Archibald Constable & Co., London.

Nicol, S. (1995). Krill, Antarctic. *In* "Encyclopedia of Environmental Biology" (W. Nierenberg, ed.), pp. 389–402. Academic Press, San Diego.

Nicol, S., Pauly, T., Bindoff, N. L., Wright, S., Thiele, D., Hosie, G. W., Strutton, P. G. and Woehler, E. (2000a). Ocean circulation off east Antarctica affects ecosystem structure and sea-ice extent. *Nature* **406**, 504–507.

Nicol, S., Kitchener, J., King, R., Hosie, G. W. and de la Mare, W. K. (2000b). Population structure and condition of Antarctic krill (*Euphausia superba*) off East Antarctica (80–150°E) during the austral summer of 1995/96. *Deep-Sea Research II* **47**, 2489–2518.

Nicol, S., Virtue, P., King, R., Davenport, S. R., McGaffin, A. F. and Nichols, P. (2004). Condition of *Euphausia crystallorophias* off East Antarctica in winter in comparison to other seasons. *Deep-Sea Research II* **51**, 2215–2224.

Niehoff, B., Schnack-Schiel, S., Cornils, A. and Brichta, M. (2002). Reproductive activity of two dominant Antarctic copepod species, *Metridia gerlachei* and *Ctenocalanus citer*, in late autumn in the eastern Bellingshausen Sea. *Polar Biology* **25**, 583–590.

Nordhausen, W. (1994). Winter abundance and distribution of *Euphausia superba, E. crystallorophias*, and *Thysanoessa macrura* in Gerlache Strait and Crystal Sound, Antarctica. *Marine Ecology Progress Series* **109**, 131–142.

Norrbin, M. F., Olsen, R.-E. and Tande, K. S. (1990). Seasonal variation in lipid class and fatty acid composition of two small copepods in Balsfjorden, northern Norway. *Marine Biology* **105**, 205–211.

Nozais, C., Gosselin, M., Michel, C. and Tita, G. (2001). Abundance, biomass, composition and grazing impact of the sea-ice meiofauna in the North Water, northern Baffin Bay. *Marine Ecology Progress Series* **217**, 235–250.

O'Brien, D. P. (1987). Direct observations of the behaviour of *Euphausia superba* and *Euphausia crystallorophias* (Crustacea: Euphausiacea) under pack ice during the Antarctic spring of 1985. *Journal of Crustacean Biology* **7**, 437–448.

Ólafsson, E., Ingólfsson, A. and Steinarsdóttir, M. B. (2001). Harpacticoid copepod communities of floating seaweed: Controlling factors and implications for dispersal. *Hydrobiologia* **453/454**, 189–200.

Opalinski, K. W. (1974). Standard, routine and active metabolism of the Antarctic amphipod—*Paramoera walkeri* Stebbing. *Polskie Archiwum Hydrobiologii* **21**, 423–429.

Paffenhöfer, G.-A. (2002). An assessment of the effects of diatoms on planktonic copepods. *Marine Ecology Progress Series* **227**, 305–310.

Pakhomov, E. and Perissinotto, R. (1996). Antarctic neritic krill *Euphausia crystallorophias*: Spatio-temporal distribution, growth and grazing rates. *Deep Sea Research I* **43**, 59–87.

Palmisano, A. C. and Sullivan, C. W. (1983). Sea ice microbial communities (SIMCO). 1. Distribution, abundance, and primary production of ice microalgae in McMurdo Sound, Antarctica in 1980. *Polar Biology* **2**, 171–177.

Parkinson, C. L. (2000). Variability of Arctic sea ice: The view from space, an 18-year record. *Arctic* **53**, 341–358.

Pasternak, A. F. and Schnack-Schiel, S. B. (2001). Feeding patterns of dominant Antarctic copepods: An interplay of diapause, selectivity, and availability of food. *Hydrobiologia* **453/454**, 25–36.

Pavshtiks, E. A. (1980). About some regularities in plankton life of the central Arctic Ocean. *In* "Biology of the Central Arctic Basin" (M. E. Vinogradov and I. E. Melnikov, eds), pp. 142–154. Nauka, Moscow.

Perissinotto, R., Gurney, L. and Pakhomov, E. A. (2000). Contribution of heterotrophic material to diet and energy budget of Antarctic krill, *Euphausia superba*. *Marine Biology* **136**, 129–135.

Pike, D. G. and Welch, H. E. (1990). Spatial and temporal distribution of sub-ice macrofauna in the Barrow Strait area, Northwest Territories. *Canadian Journal of Fisheries and Aquatic Sciences* **47**, 81–91.

Pohnert, G. (2005). Diatom/copepod interactions in plankton: The indirect chemical defense of unicellular algae. *ChemBioChem* **6**, 946–959.

Poltermann, M. (1997). Biologische und ökologische Untersuchungen zur kryopelagischen Amphipodenfauna des arktischen Meereises. *Berichte zur Polarforschung* **225**, 170.

Poltermann, M. (1998). Abundance, biomass and small-scale distribution of cryopelagic amphipods in the Franz Josef Land area (Arctic). *Polar Biology* **20**, 134–138.

Poltermann, M. (2000). Growth, production and productivity of the Arctic sympagic amphipod *Gammarus wilkitzkii*. *Marine Ecology Progress Series* **193**, 109–116.

Poltermann, M. (2001). Arctic sea ice as feeding ground for amphipods—food sources and strategies. *Polar Biology* **24**, 89–96.

Poltermann, M., Hop, H. and Falk-Petersen, S. (2000). Life under Arctic sea ice—reproduction strategies of two sympagic (ice-associated) amphipod species, *Gammarus wilkitzkii* and *Apherusa glacialis*. *Marine Biology* **136**, 913–920.

Proshutinsky, A. Y. and Johnson, M. A. (1997). Two circulation regimes of the wind-driven Arctic Ocean. *Journal of Geophysical Research* **102**, 12493–12514.

Quetin, L. B. and Ross, R. M. (2003). Episodic recruitment in Antarctic krill *Euphausia superba* in the Palmer LTER study region. *Marine Ecology Progress Series* **259**, 185–200.

Rakusa-Suszczewski, S. (1972). The biology of *Paramoera walkeri* Stebbing (Amphipoda) and the Antarctic sub-fast ice community. *Polskie Archiwum Hydrobiologii* **19**, 11–36.

Rakusa-Suszczewski, S. (1982). The biology and metabolism of *Orchomene plebs* (Hurley, 1965) (Amphipoda: Gammaridea) from McMurdo Sound, Ross Sea, Antarctica. *Polar Biology* **1**, 47–54.

Rakusa-Suszczewski, S. and Dominas, H. (1974). Chemical composition of the Antarctic amphipoda *Paramoera walkeri* Stebbing and chromatographic analysis of its lipids. *Polskie Archiwum Hydrobiologii* **21**, 261–268.

Rakusa-Suszczewski, S. and Klekowski, R. Z. (1973). Biology and respiration of the Antarctic amphipoda (*Paramoera walkeri* Stebbing) in the summer. *Polskie Archiwum Hydrobiologii* **20**, 475–488.

Ray, C. (1966). Stalking seals under Antarctic ice. *Geographic Magazine, London* **129**, 54–65.

Razouls, S., Razouls, C. and de Bovée, F. (2000). Biodiversity and biogeography of Antarctic copepods. *Antarctic Science* **12**, 343–362.

Reid, K. and Croxall, J. P. (2001). Environmental response of upper trophic-level predators reveals a system change in an Antarctic marine ecosystem. *Proceedings of the Royal Society of London B* **268**, 377–384.

Reimnitz, E., Clayton, J. J., Kempema, E. W., Payne, J. R. and Weber, W. S. (1993). Interaction of rising frazil with suspended particles: Tank experiments with application to nature. *Cold Regions Science Technology* **21**, 117–135.

Richardson, M. G. and Whitaker, T. M. (1979). An Antarctic fast-ice food chain: Observations on the interaction of the amphipod *Pontogeneia antarctica* Chevreux with ice-associated micro-algae. *British Antarctic Survey Bulletin* **47**, 107–115.

Rigor, I. G., Wallace, J. M. and Colony, R. L. (2002). On the response of sea ice to the Arctic Oscillation. *Journal of Climate* **15**, 2648–2668.

Ross, J. C. (1847). "A Voyage of Discovery and Research in the Southern and Antarctic Regions". John Murray, London.

Ross, R. M. and Quetin, L. B. (1982). *Euphausia superba*: Fecundity and physiological ecology of its eggs and larvae. *Antarctic Journal of the United States* **17**, 166–167.

Ross, R. M. and Quetin, L. B. (1986). How productive are Antarctic krill? *BioScience* **36**, 264–269.

Ross, R. M. and Quetin, L. B. (1989). Energetic cost to develop to the first feeding stage of *Euphausia superba* Dana and the effect of delays in food availability. *Journal of Experimental Marine Biology and Ecology* **133**, 103–127.

Ross, R. M., Quetin, L. B., Newberger, T. and Oakes, S. A. (2004). Growth and behaviour of larval krill (Euphausia superba) under the ice in late winter 2001 west of the Antarctic Peninsula. *Deep-Sea Research II* **51**, 2169–2184.

Rothrock, D. A., Yu, Y. and Maykut, G. A. (1999). Thinning of the Arctic ice cover. *Geophysical Research Letters* **26**, 3469–3472.

Runge, J. A. and Ingram, R. G. (1988). Underice grazing by planktonic, calanoid copepods in relation to a bloom of ice microalgae in southeastern Hudson Bay. *Limnology and Oceanography* **33**, 280–286.

Runge, J. A. and Ingram, R. G. (1991). Under-ice feeding and diel migration by the planktonic copepods *Calanus glacialis* and *Pseudocalanus minutus* in relation to the ice algal production cycle in southeastern Hudson Bay. *Marine Biology* **108**, 217–226.

Sagar, P. M. (1980). Life cycle and growth of the Antarctic gammarid amphipod *Paramoera walkeri* (Stebbing, 1906). *Journal of the Royal Society of New Zealand* **10**, 259–270.

Sakshaug, E. and Walsh, J. (2000). Marine biology: Biomass, productivity distributions and their variability in the Barents and Bering Seas. *In* "The Arctic: Environment, People and Policy" (M. Nuttall and T. V. Callaghan, eds), pp. 163–196. Harwood Academic Publications, Amsterdam.

Sainte-Marie, B. (1992). Foraging of scavenging deep-sea lysianassoid amphipods. *In* "Deep-Sea Food Chains and the Global Carbon Cycle" (G. T. Rowe and V. Pariente, eds), pp. 105–124. Kluwer Academic Publishers, Netherlands.

Sargent, J. R. and Falk-Petersen, S. (1988). The lipid biochemistry of calanoid copepods. *Hydrobiologia* **167/168**, 101–114.

Sars, G. O. (1900). Crustacea. *In* "The Norwegian North Pole Expedition 1893–1896, Scientific Research" (F. Nansen, ed.), Vol. 1, pp. 1–137. Bergen Museum, Bergen.

Schizas, N. V. and Shirley, T. C. (1996). Seasonal changes in structure of an Alaskan intertidal meiofaunal assemblage. *Marine Ecology Progress Series* **133**, 115–124.

Schmidt, K., Atkinson, A., Stübing, D., McClelland, J. W., Montoya, J. P. and Voss, M. (2003). Trophic relationships among Southern Ocean copepods and krill: Some uses and limitations of a stable isotope approach. *Limnology and Oceanography* **48**, 277–289.

Schnack-Schiel, S. B. (2001). Aspects of the study of the life cycles of Antarctic copepods. *Hydrobiologia* **453/454**, 9–24.

Schnack-Schiel, S. B. (2002). The macrobiology of sea ice. *In* "Sea Ice: An Introduction to Its Physics, Chemsitry, Biology and Ecology" (D. N. Thomas and G. S. Dieckmann, eds), pp. 211–239. Blackwell Publishing, Oxford.

Schnack-Schiel, S. B. and Hagen, W. (1994). Life cycle strategies and seasonal variations in distribution and population structure of four dominant calanoid copepod species in the eastern Weddell Sea, Antarctica. *Journal of Plankton Research* **16**, 1543–1566.

Schnack-Schiel, S. B. and Mizdalski, E. (1994). Seasonal variations in distribution and population structure of *Microcalanus pygmaeus* and *Ctenocalanus citer* (Copepoda: Calanoida) in the eastern Weddell Sea. *Marine Biology* **119**, 357–366.

Schnack-Schiel, S. B., Hagen, W. and Mizdalski, E. (1991). A seasonal comparison of *Calanoides acutus* and *Calanus propinquus* (Copepoda, Calanoida) in the eastern Weddell Sea, Antarctica. *Marine Ecology Progress Series* **70**, 17–27.

Schnack-Schiel, S. B., Thomas, D. N., Dieckmann, G. S., Eicken, H., Gradinger, R., Spindler, M., Weissenberger, J., Mizdalski, E. and Beyer, K. (1995). Life cycle strategy of the Antarctic calanoid copepod *Stephos longipes*. *Progress in Oceanography* **36**, 45–75.

Schnack-Schiel, S. B., Thomas, D. N., Dahms, H.-U., Haas, C. and Mizdalski, E. (1998). Copepods in Antarctic sea ice. *Antarctic Research Series* **73**, 173–182.

Schnack-Schiel, S. B., Dieckmann, G. S., Gradinger, R., Melnikov, I. A., Spindler, M. and Thomas, D. N. (2001a). Meiofauna in sea ice of the Weddell Sea (Antarctica). *Polar Biology* **24**, 724–728.

Schnack-Schiel, S. B., Thomas, D. N., Haas, C., Dieckmann, G. and Alheit, R. (2001b). The occurrence of copepods *Stephos longipes* (Calanoidea) and *Drescheriella glacialis* (Harpacticoida) in summer sea ice in the Weddell Sea, Antarctica. *Antarctic Science* **13**, 150–157.

Scott, C. L., Falk-Petersen, S., Sargent, J. R., Hop, H., Lønne, O. J. and Poltermann, M. (1999). Lipids and trophic interactions of ice fauna and pelagic zooplankton in the marginal ice zone of the Barents Sea. *Polar Biology* **21**, 65–70.

Scott, C. L., Kwasniewski, S., Falk-Petersen, S. and Sargent, J. R. (2000). Lipids and life strategies of *Calanus finmarchicus, Calanus glacialis* and *Calanus hyperboreus* in late autumn, Kongsfjorden, Svalbard. *Polar Biology* **23**, 510–516.

Scott, C. L., Kwasniewski, S., Falk-Petersen, S. and Sargent, J. (2002). Lipids and fatty acids in the copepod *Jaschnovia brevis* (Jaschnov) and in particulates from Arctic waters. *Polar Biology* **25**, 65–71.

Siegel, V. (1987). Age and growth of Antarctic Euphausiacea (Crustacea) under natural conditions. *Marine Biology* **96**, 483–495.

Siegel, V. (2000). Krill (Euphausiacea) life history and aspects of population dynamics. *Canadian Journal of Fisheries and Aquatic Sciences* **57**, 130–150.

Siegel, V. and Loeb, V. (1994). Length and age at maturity of Antarctic krill. *Antarctic Science* **6**, 479–482.

Siegel, V. and Loeb, V. (1995). Recruitment of the Antarctic krill *Euphausia superba* and possible causes for its variability. *Marine Ecology Progress Series* **123**, 45–56.

Siegel, V. and Harm, U. (1996). The composition, abundance, biomass and diversity of the epipelagic zooplankton communities of the southern Bellingshausen Sea (Antarctic) with special reference to krill and salps. *Archive of Fishery and Marine Research* **44**, 115–139.

Siegel, V., Loeb, V. and Groger, J. (1998). Krill (*Euphausia superba*) density, proportional and absolute recruitment and biomass in the Elephant Island region (Antarctic Peninsula) during the period 1977 to 1997. *Polar Biology* **19**, 393–398.

Siferd, T. D., Welch, H. E., Bergmann, M. A. and Curtis, M. F. (1997). Seasonal distribution of sympagic amphipods near Chesterfield Inlet, N.W.T., Canada. *Polar Biology* **18**, 16–22.

Skerratt, J. H., Nichols, P. D., McMeekin, T. A. and Burton, H. R. (1995). Seasonal and inter-annual changes in planktonic biomass and community structure in eastern Antarctica using signature lipids. *Marine Chemistry* **51**, 93–113.

Smetacek, V. and Nicol, S. (2005). Polar ocean ecosystems in a changing world. *Nature* **437**, 362–368.

Smith, S. L. (1990). Egg production and feeding by copepods prior to the spring bloom of phytoplankton in Fram Strait, Greenland Sea. *Marine Biology* **106**, 59–69.

Søreide, J. E., Hop, H., Falk-Petersen, S., Gulliksen, B. and Hansen, E. (2003). Macrozooplankton communities and environmental variables in the Barents Sea marginal ice zone in late winter and spring. *Marine Ecology Progress Series* **263**, 43–64.

Spindler, M. (1990). A comparison of Arctic and Antarctic sea ice and the effects of different properties on sea ice biota. *In* "Geological History of the Polar Oceans: Arctic versus Antarctic" (U. Bleil and J. Thiede, eds), pp. 173–186. Kluwer Academic Publishers, Netherlands.

Steele, D. H. (1972). Some aspects of the biology of *Gammarellus homari* (Crustacea, Amphipoda) in the Northwestern Atlantic. *Journal of the Fisheries Research Board of Canada* **29**, 1340–1343.

Steele, D. H. and Steele, V. J. (1974). The biology of *Gammarus* (Crustacea, Amphipoda) in the northwestern Atlantic. VIII. Geographic distribution of the northern species. *Canadian Journal of Zoology* **52**, 1115–1120.

Steele, D. H. and Steele, V. J. (1975). The biology of *Gammarus* (Crustacea, Amphipoda) in the northwestern Atlantic. IX. *Gammarus wilkitzkii* Birula, *Gammarus stoerensis* Reid, and *Gammarus mucronatus* Say. *Canadian Journal of Zoology* **53**, 1105–1109.

Steinarsdóttir, M. B., Ingólfsson, A. and Ólafsson, E. (2003). Seasonality of harpacticoids (Crustacea, Copepoda) in a tidal pool in sub-Arctic south-western Iceland. *Hydrobiologia* **503**, 211–221.

Stepnik, R. (1982). All-year populational studies of Euphausiacea (Crustacea) in the Admiralty Bay (King George Island, South Shetland Islands, Antarctic). *Polish Polar Research* **3**, 49–68.

Stevens, C. J., Deibel, D. and Parrish, C. C. (2004). Species-specific differences in lipid composition and omnivory indices in Arctic copepods collected in deep water during autumn (North Water Polynya). *Marine Biology* **144**, 905–915.

Stockton, W. L. (1982). Scavenging amphipods from under the Ross Ice Shelf, Antarctica. *Deep-Sea Research* **29A**, 819–835.

Stübing, D. and Hagen, W. (2003). On the use of lipid biomarkers in marine food web analyses: An experimental case study on the Antarctic krill, *Euphausia superba*. *Limnology and Oceanography* **48**, 1685–1700.

Swadling, K. M. (1998). Influence of seasonal ice formation on life cycle strategies of Antarctic copepods. PhD thesis, University of Tasmania, Hobart, Australia.

Swadling, K. M. (2001). Population structure of two Antarctic ice-associated copepods, *Drescheriella glacialis* and *Paralabidocera antarctica*, in winter sea ice. *Marine Biology* **139**, 597–603.

Swadling, K. M., Gibson, J. A. E., Ritz, D. A. and Nichols, P. D. (1997a). Horizontal patchiness in sympagic organisms of the Antarctic fast ice. *Antarctic Science* **9**, 399–406.

Swadling, K. M., Gibson, J. A. E., Ritz, D. A., Nichols, P. D. and Hughes, D. E. (1997b). Grazing of phytoplankton by copepods in eastern Antarctic waters. *Marine Biology* **128**, 39–48.

Swadling, K. M., McPhee, A. D. and McMinn, A. (2000a). Spatial distribution of copepods in fast ice of eastern Antarctica. *Polar Bioscience* **13**, 55–65.

Swadling, K. M., Nichols, P.D, Gibson, J. A. E. and Ritz, D. A. (2000b). Role of lipid in the life cycles of ice-dependent and ice-independent populations of the copepod *Paralabidocera antarctica*. *Marine Ecology Progress Series* **208**, 171–182.

Swadling, K. M., McKinnon, A. D., De'ath, G. and Gibson, J. A. E. (2004). Life cycle plasticity and differential growth and development in marine and lacustrine populations of an Antarctic copepod. *Limnology and Oceanography* **49**, 644–655.

Tande, K. S. and Henderson, R. J. (1988). Lipid-composition of copepodite stages and adult females of *Calanus glacialis* in arctic waters of the Barents Sea. *Polar Biology* **8**, 333–339.

Tanimura, A., Fukuchi, M. and Ohtsuka, H. (1984a). Occurrence and age composition of *Paralabidocera antarctica* (Calanoida, Copepoda) under the fast ice near Syowa Station, Antarctica. *NIPR Special Issue* **32**, 81–86.

Tanimura, A., Hoshiai, T. and Fukuchi, M. (1996). The life cycle strategy of the ice-associated copepod, *Paralabidocera antarctica* (Calanoida, Copepoda), at Syowa Station, Antarctica. *Antarctic Science* **8**, 257–266.

Tanimura, A., Hoshino, K., Nonaka, Y., Miyamoto, Y. and Hattori, H. (1997). Vertical distribution of *Oithona similis* and *Oncaea curvata* (Cyclopoida, Copepoda) under sea ice near Syowa Station in the Antarctic winter. *Proceedings of the NIPR Symposium on Polar Biology* **10**, 134–144.

Tanimura, A., Hoshiai, T. and Fukuchi, M. (2002). Change in habitat of the sympagic copepod *Paralabidocera antarctica* from fast ice to seawater. *Polar Biology* **25**, 667–671.

Tanimura, A., Minoda, T., Fukuchi, M., Hoshiai, T. and Ohtsuka, H. (1984b). Swarm of *Paralabidocera antarctica* (Calanoida, Copepoda) under sea ice near Syowa Station, Antarctica. *Nankyoku Shiryo (Antarctic Record)* **82**, 12–19.

Thistle, D. (2003). Harpacticoid copepod emergence at a shelf site in summer and winter: Implications for hydrodynamic and mating hypotheses. *Marine Ecology Progress Series* **248**, 177–185.

Thomas, D. N. and Dieckmann, G. S. (2003). "Sea Ice: An introduction to its Physics, Chemistry, Biology and Geology." Blackwell, Oxford.

Thomas, D. N., Lara, R. J., Haas, C., Schnack-Schiel, S. B., Dieckmann, G. S., Kattner, G., Nöthig, E.-M. and Mizdalski, E. (1998). Biological soup within decaying summer sea ice in the Amundsen Sea, Antarctica. *Antarctic Research Series* **73**, 161–171.

Torres, J. J., Aarset, A. V., Donnelly, J., Hopkins, T. L., Lancraft, T. M. and Ainley, D. G. (1994a). Metabolism of Antarctic micronektonic Crustacea as a function of depth of occurrence and season. *Marine Ecology Progress Series* **113**, 207–219.

Torres, J. J., Donnelly, J., Hopkins, T. L., Lancraft, T. M., Aarset, A. V. and Ainley, D. G. (1994b). Proximate composition and over-wintering strategies of Antarctic micronektonic Crustacea. *Marine Ecology Progress Series* **113**, 221–232.

Tourangeau, S. and Runge, J. A. (1991). Reproduction of *Calanus glacialis* under ice in spring in southeastern Hudson Bay, Canada. *Marine Biology* **108**, 227–233.

Tucker, M. J. and Burton, H. R. (1988). The inshore marine ecosystem off the Vestfold Hills, Antarctica. *Hydrobiologia* **165**, 129–139.

Uye, S., Aoto, I. and Onbé, T. (2002). Seasonal population dynamics and production of *Microsetella norvegica*, a widely distributed but little-studied marine planktonic harpacticoid copepod. *Journal of Plankton Research* **24**, 143–153.

Volkman, J. K., Jeffrey, S. W., Nichols, P. D., Rogers, G. I. and Garland, C. D. (1989). Fatty acid and lipid composition of 10 species of microalgae used in mariculture. *Journal of Experimental Marine Biology and Ecology* **128**, 219–240.

Vaughan, D. G. and Doake, C. S. M. (1996). Recent atmospheric warming and retreat of ice shelves on the Antarctic Peninsula. *Nature* **379**, 328–331.

Vinogradov, G. M. (1999). Deep-sea near-bottom swarms of pelagic amphipods *Themisto*: Observations from submersibles. *Sarsia* **84**, 465–467.

Wadhams, P. (2000). "Ice in the Ocean". Gordon and Breach Science Publishers, Amsterdam.

Waghorn, E. J. (1979). Two new species of Copepoda from White Island, Antarctica. *New Zealand Journal of Marine and Freshwater Research* **13**, 459–470.

Waghorn, E. J. and Knox, G. A. (1988). Summer tide-crack zooplankton at White Island, McMurdo Sound, Antarctica. *New Zealand Journal of Marine and Freshwater Research* **22**, 577–582.

Ward, P. and Shreeve, R. (1998). Egg hatching times of Antarctic copepods. *Polar Biology* **19**, 142–144.

Ward, P., Shreeve, R. S. and Cripps, G. C. (1996). *Rhincalanus gigas* and *Calanus simillimus*: Lipid storage patterns of two species of copepod in the seasonally ice-free zone of the Southern Ocean. *Journal of Plankton Research* **18**, 1439–1454.

Watanabe, K. (1988). Sub-ice microalgal strands in the Antarctic coastal fast ice near Syowa Station. *Japanese Journal of Phycology* **36**, 221–229.

Werner, I. (1997a). Ecological studies on the Arctic under-ice habitat—colonization and processes at the ice-water interface. Berichte aus dem Sonderforschungsbereich 313. *Christian-Albrechts-Universität zu Kiel* **70**, 167.

Werner, I. (1997b). Grazing of Arctic under-ice amphipods on sea-ice algae. *Marine Ecology Progress Series* **160**, 93–99.

Werner, I. (2000). Faecal pellet production by Arctic under-ice amphipods—transfer of organic matter through the ice/water interface. *Hydrobiologia* **426**, 89–96.

Werner, I. and Arbizu, P. M. (1999). The sub-ice fauna of the Laptev Sea and the adjacent Arctic Ocean in summer 1995. *Polar Biology* **21**, 71–79.

Werner, I., Auel, H. and Friedrich, C. (2002a). Carnivorous feeding and respiration of the Arctic under-ice amphipod *Gammarus wilkitzkii*. *Polar Biology* **25**, 523–530.

Werner, I., Meiners, K. and Schünemann, H. (2002b). Copepods in Arctic pack ice and the underlying water column: Living conditions and exchange processes. *In* "Ice in the Environment" (V. Squire and P. Langhorne, eds), "Proceedings of the 16th IAHR International Symposium on Ice, Dunedin, New Zealand, 2nd–6th December 2002", pp. 30–40. International Association of Hydraulic Engineering and Research, Dunedin, New Zealand.

Werner, I., Auel, H. and Kiki, R. (2004). Occurrence of *Anonyx sarsi* (Amphipoda: Lysianassoidea) below Arctic pack ice: An example of cryo-benthic coupling? *Polar Biology* **27**, 474–481.

Weslawski, J. M. (1994). *Gammarus* (Crustacea, Amphipoda) from Svalbard and Franz Josef Land. Distribution and density. *Sarsia* **79**, 145–150.

Weslawski, J. M. and Legezynska, J. (2002). Life cycles of some Arctic amphipods. *Polish Polar Research* **23**, 253–264.

Weslawski, J. M., Kwasniewski, S. and Wiktor, J. (1991). Winter in a Svalbard fiord ecosystem. *Arctic* **44**, 115–123.

Weslawski, J. M., Kwasniewski, S., Wiktor, J. and Zajaczkowski, M. (1993). Observations on the fast ice biota in the fjords of Spitsbergen. *Polish Polar Research* **14**, 331–342.

Weslawski, J. M., Pedersen, G., Petersen, S. F. and Porazinski, K. (2000). Entrapment of macroplankton in an Arctic fjord basin, Kongsfjorden, Svalbard. *Oceanologia* **42**, 57–69.

Whitaker, T. M. (1977). Sea ice habitats of Signy Island (South Orkneys) and their primary productivity. *In* "Adaptations within Antarctic Ecosystems: Proceedings of the Third SCAR Symposium on Antarctic Biology" (G. A. Llano, ed.), pp. 75–82. Gulf Publishing Co., Houston.

Williams, R., Kirkwood, J. M. and O'Sullivan, D. B. (1986). ADBEX I cruise zooplankton data. *Australian National Antarctic Research Expeditions, Research Notes* **31**, 1–108.

Wing, B. L. (1976). Ecology of *Parathemisto libellula* and *P. pacifica* (Amphipoda: Hyperiida) in Alaskan coastal waters. Northwest Fisheries Center, SeattleNWFC Processed Report.

TAXONOMIC INDEX

SUBJECT INDEX